Edited by Mark Wigglesworth and Terry Wood

Management of Chemical and Biological Samples for Screening Applications

Related Titles

Wu, G.

Assay Development

Fundamentals and Practices

2010

ISBN: 978-0-470-19115-6

Faller, B., Urban, L. (Eds.)

Hit and Lead Profiling

Identification and Optimization of Drug-like Molecules

2009

ISBN: 978-3-527-32331-9

Wolfbeis, O. S. (Ed.)

Fluorescence Methods and Applications

Spectroscopy, Imaging, and Probes

2008

ISBN: 978-1-57331-716-0

Haney, S. A. (Ed.)

High Content Screening

Science, Techniques and Applications

2008

ISBN: 978-0-470-03999-1

Edited by Mark Wigglesworth and Terry Wood

Management of Chemical and Biological Samples for Screening Applications

WILEY-VCH Verlag GmbH & Co. KGaA

The Editors

Dr. Mark Wigglesworth
GlaxoSmithKline
Medicines Research Centre
Gunnels Wood Road
Stevenage
Hertfordshire
SG1 2NY
United Kingdom

Dr. Terry Wood
TP & AAW Consultancy
8, Arundel Road
Cliftonville
Kent, CT9 2AW
United Kingdom

Library of Congress Card No.: applied for

British Library Cataloguing-in-Publication Data
A catalogue record for this book is available from the British Library.

Bibliographic information published by the Deutsche Nationalbibliothek
The Deutsche Nationalbibliothek lists this publication in the Deutsche Nationalbibliografie; detailed bibliographic data are available on the Internet at <http://dnb.d-nb.de>.

Print ISBN: 978-3-527-32822-2
ePDF ISBN: 978-3-527-64527-5
ePub ISBN: 978-3-527-64526-8
Mobi ISBN: 978-3-527-64528-2
oBook ISBN: 978-3-527-64525-1

Cover Design Grafik-Design Schulz, Fußgönheim
Typesetting Laserwords Private Limited, Chennai, India
Printing and Binding Markono Print Media Pte Ltd, Singapore

Dedication

We would first like to thank everyone that has contributed to this book and hope that it is a text which helps promote and project forward the science of Sample Management. We have both found our work within the pharmaceutical industry rewarding not least because it is an opportunity to make the medicines that help other people. To this end we pledge our editorial honoraria to Cancer Research UK, which is a beneficiary close to both our hearts. Finally we would like to dedicate this work to our families; we thank them for their patience and hope that we have made this world a better place for them.

The royalties from the sale of this book will be donated to Cancer Research UK.

Cancer Research UK is the world's leading cancer charity dedicated to saving lives through research. Our ground breaking work into the prevention, diagnosis and treatment of cancer has seen survival rates double in the last 40 years. But more than one in three of us will still get cancer at some point in our lives. Our research, entirely funded by the public, is critical to ensuring more people beat it.

The views and opinions within this book are those of the Authors, and are independent from the work of Cancer Research UK.

in aid of
CANCER RESEARCH UK

Contents

Preface

Within this book we present a modern, practice-orientated overview of concepts, technology, and strategies for the management of large entity collections, encompassing both chemical and biological samples. This book reports expert opinion in this area and documents the evolution, current best practice, and future goals of Sample Management in a form never previously achieved.

The field of Sample Management has evolved in the last 20 years from a necessary, if somewhat haphazard, occupation of a screening scientist into a highly controlled, scientific discipline, incorporating logistics and automation management. This evolved scientific discipline is a pivotal part of every pharmaceutical and biotechnological organization across the globe, yet holding these samples in huge warehouses has no intrinsic value in itself. The samples must be subjected to High-Throughput Screening, population-based clinical research, and/or many other techniques that form part of the drug discovery pipeline before that one single chemical structure is identified which will lead to a life-changing discovery (see Figure 1). These single samples, at the end of many years of research, are the ones of value, and nurturing them and ensuring that you are able to find them within your collection is the role of the Sample Manager.

This book will guide the reader through the complex paths of Sample Management, starting with a view of what it represents for both chemical and biological samples and the reasons why this discipline has had to be developed. We present views on sample quality and the importance of quality in both establishing collections and maintaining them once created. We present the rationale for the subdivision of collections for efficient screening, and provide an overview of automation, from large-scale storage devices through to bench-top liquid handling technology for compound dispense. We further examine the latest and most advanced technologies available and how these are being implemented within the industry. Rarely do organizations exist in isolation; hence we examine the logistics of sample storage and transportation, taking in the practical and legal elements. One of the biggest potential issues within Sample Management is the tracking of data and samples within an inventory, from sample receipt through dissolution, dispense, and utilization. An IT system that interacts with automation and tracks the movement of every sample is key to establishing reliable delivery of samples

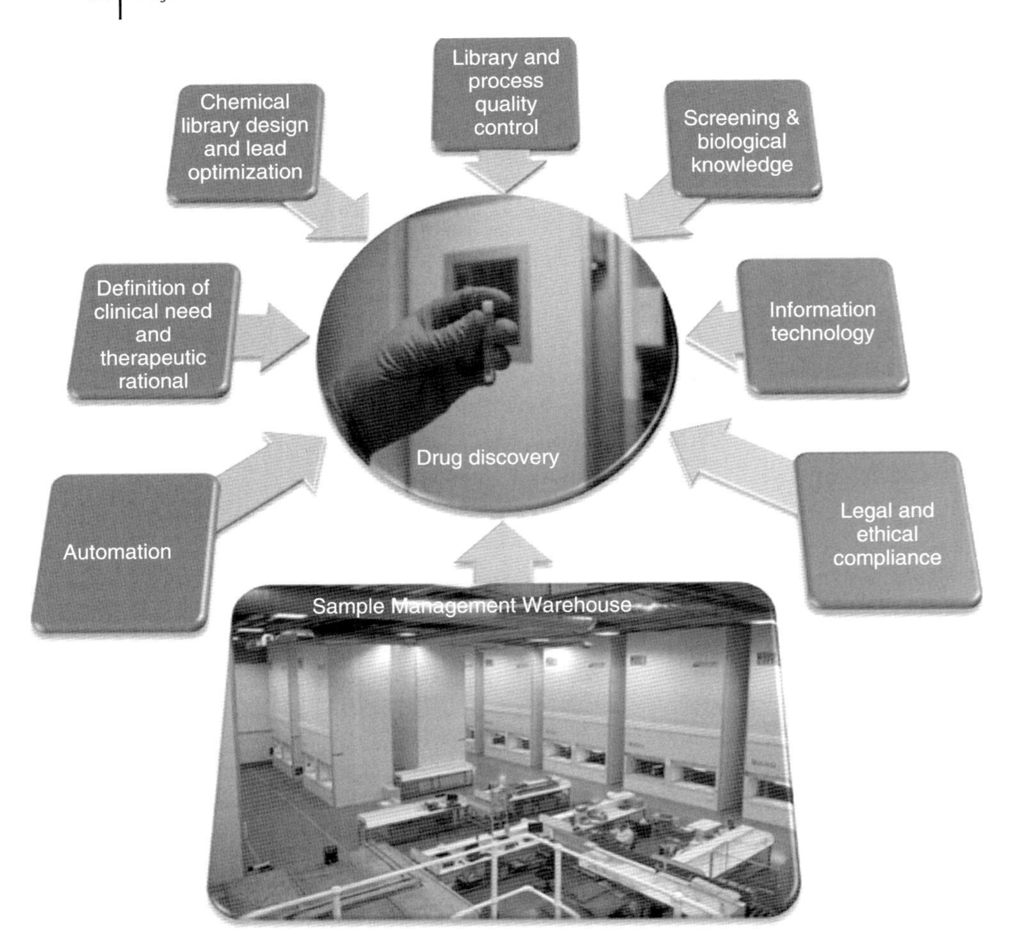

Figure 1 A schematic showing Sample Management as a contributor to drug discovery: only when many elements work cohesively together can the value in what we do be realized.

and data integrity. Hence, we present two chapters introducing bespoke database systems through to examples of the off-the-shelf systems that fulfill this need.

In order to survive in an increasingly complex business environment, many companies are turning to process efficiency techniques that were made popular by the automotive industry, such as LeanSigma. Sample Management has many similarities to a production activity, and hence we examine these new techniques and show how they can be applied to improve process efficiencies and deliver key insights. In a further drive for efficiency, the pharmaceutical industry has focused on reducing attrition. This in turn has focused on changing the chemical properties of small-molecule collections, leading to new thought on how collections should be generated.

For the management of biological samples, which present their own, unique set of issues, we examine the challenges of obtaining tissues of high and comparable quality, looking at automation currently in use in this field as well as examining the potential of future technologies to assist biobanking. Tissue biobanking as well as the management of biological materials in the form of cell lines used within biological assays is discussed. We finish with a projected view of Sample Management, where outsourcing opportunities have delivered benefits to both chemical and biological sample management organizations and how utilization of the skills within Sample Management facilitate operating large scale processes. We also offer opinion on what the Sample Management department of the future might look like and how alterations in the drug discovery process may affect the process of Sample Management.

Above all we hope that this book will be a useful tool for any Sample Management organization, large or small, and will challenge you to take a fresh look at your organization and what it is doing. Think not just about how you will fulfill the next request, but also how you, in your role as a Sample Manager, can continue to expedite the essential business of drug discovery in years to come.

United Kingdom, November 2011 *Mark Wigglesworth, Terry Wood*

List of Contributors

Michael Allen
PenCHORD (Peninsula
Collaboration for Health
Operational Research &
Development)
Peninsula college of
Medicine & Dentistry
Veysey Building
Salmon Pool Lane
Exeter, EX2 4SG
UK

María J. Artiga
Spanish National Cancer Centre
(CNIO)
Tumor Bank Unit
Molecular Pathology Programme
Melchor Fernández Almagro, 3
28045 Madrid
Spain

Colin Bath
AstraZeneca plc.
Alderley Park
Macclesfield
Cheshire, SK10 4TG
UK

Clive Battle
Titian Software Ltd.
2 Newhams Row
London, SE1 3UZ
UK

François Bertelli
Pfizer Global Research &
Development
Sandwich Laboratories
Ramsgate Road
Sandwich
Kent CT13 9NJ
UK

Stephen Besley
GlaxoSmithKline
Medicines Research Centre
Gunnels Wood Road
Stevenage
Hertfordshire, SG1 2NY
UK

Snehal Bhatt
GlaxoSmithKline
Sample Management
Technologies
Upper Providence R&D
1250 S. Collegeville Road
Collegeville
PA 19426
USA

Zoe Blaxill
GlaxoSmithKline
Medicines Research Centre
Gunnels Wood Road
Stevenage
Hertfordshire, SG1 2NY
UK

Sylviane Boucharens
Discovery Performance &
Strategy Ltd.
25 Shiel Drive
Larkhall ML9 2TJ
Scotland
UK

Brian Brooks
Unilever Centre for Molecular
Sciences Informatics
Department of Chemistry
University of Cambridge
Lensfield Road
Cambridge CB2 1EW
UK

Laura Cereceda
Spanish National Cancer Centre
(CNIO)
Tumor Bank Unit
Molecular Pathology Programme
Melchor Fernández Almagro, 3
28045 Madrid
Spain

Isabel Charles
AstraZeneca PLC
Alderley Park
Macclesfield
Cheshire SK10 4TG
UK

Philip B. Cox
Abbott Laboratories
Global Pharmaceutical Research
and Development
Advanced Technology
100 Abbott Park Road
Abbott Park, IL 60064
USA

Raymond H. Cypess
ATCC (American Type
Culture Collection)
10801 University Boulevard
Manassas, VA 20110-2209
USA

Holger Eickhoff
Scienion AG
Otto Hahn Str. 15
44227 Dortmund
Germany

Richard Ellson
Labcyte Inc.
1190 Borregas Avenue
Sunnyvale
CA 94089
USA

Paul A. Gosnell
GlaxoSmithKline
Upper Providence R&D
1250 S. Collegeville Road
Collegeville
PA 19426
USA

Michael Gray
Medicines Research Centre
Gunnels Wood Road
Stevenage
Hertfordshire, SG1 2NY
UK

Neil Hardy
GlaxoSmithKline
Medicines Research Centre
Gunnels Wood Road
Stevenage
Hertfordshire, SG1 2NY
UK

Sue Holland-Crimmin
GlaxoSmithKline
Upper Providence R&D
1250 S. Collegeville Road
Collegeville, PA 19426
USA

William P. Janzen
The University of North Carolina
at Chapel Hill
Eshelman School of Pharmacy
Center for Integrative Chemical
Biology and Drug Discovery
Division of Chemical Biology and
Medicinal Chemistry
2092 Genetic Medicine Building
120 Mason Farm Road
Campus Box 7363
Chapel Hill, NC 27599-7363
USA

Lisa B. Miranda
Biobusiness Consulting Inc.
Greater Boston Area
Massachusetts
USA

Manuel M. Morente
Spanish National Cancer Centre
(CNIO)
Tumor Bank Unit
Molecular Pathology Programme
Melchor Fernández Almagro, 3
28045 Madrid
Spain

Joe Olechno
Labcyte Inc.
1190 Borregas Avenue
Sunnyvale, CA 94089
USA

Ioana Popa-Burke
GlaxoSmithKline
5 Moore Drive, PO Box 13398
Research Triangle Park
NC 27709-3398
USA

Frank P. Simione
ATCC (American Type
Culture Collection)
10801 University Boulevard
Manassas, VA 20110-2209
USA

Eric Tang
AstraZeneca plc.
Alderley Park
Macclesfield
Cheshire, SK10 4TG
UK

Johann van Niekerk
Brooks Life Science Systems
Weststrasse 12
3672 Oberdiessbach
Switzerland

Anil Vasudevan
Abbott Laboratories
Global Pharmaceutical Research
and Development
Advanced Technology
100 Abbott Park Road
Abbott Park, IL 60064
USA

Amelia Wall Warner
Merck Research Laboratories
Merck & Co., Inc.
351 N Sumneytown Pike
UG4D-34
North Wales, PA 19454 USA

Gregory J. Wendel
The Siegel Consulting
Group, Inc.
111 Dutton Rd.
Sudbury, MA 01776
USA

Mark Wigglesworth
GlaxoSmithKline
Medicines Research Centre
Gunnels Wood Road
Stevenage
Hertfordshire, SG1 2NY
UK

Terry Wood
TP & AAW Consultancy
8, Arundel Road
Cliftonville
Kent, CT9 2AW
UK

Shoufeng Yang
University of Southampton
Faculty of Engineering and the
Environment
University Road
Southampton
SO17 1BJ
UK

Ji Yi Khoo
University of Southampton
Faculty of Engineering and the
Environment
University Road
Southampton
SO17 1BJ
UK

Andy Zaayenga
HighRes Biosolutions
Business Development
1730 W Circle Dr
Martinsville
NJ 08836-2147
USA

1
Introduction to Sample Management

William P. Janzen and Andy Zaayenga

At its simplest level sample management is just inventory – where can one find a given item and retrieve it? But in the context of modern discovery efforts, be they drug discovery, agricultural, protein therapeutic, biobanking, or the plethora of other disciplines that collect and manage samples, the problem is far more complex. Today, sample management may have to manage millions of samples in a library that spans several continents but will also have to contend with a worldwide customer base. To make the problem more difficult the content of the samples must also be managed, which may involve complex chemical structures and storage conditions that may vary from room temperature under inert atmosphere to storage in liquid nitrogen. At this level the storage of these samples must now involve complex informatics and automation systems. This volume will capture the best practices compiled from experts in the field of sample management and will hopefully serve as a guide to both novice sample managers who need to track a few thousand compounds in room-temperature vials to professionals in multinational organizations.

As long as there have been chemicals there has been a need for sample management. One could imagine that for a seventeenth century druggist this was simply an inventory of the herbal extracts and remedies he compounded into salves and potions and the location where they were stored. This could be done from memory in most cases and probably evolved to a written inventory when searching for needed components became too slow and cumbersome. Early sample management evolved in parallel with drug discovery. What we consider sample management today came into being as pharmaceutical companies began to amass chemical libraries and test these in disease-focused assays. As these companies synthesized compounds, they retained samples and began to amass collections of chemical compounds that numbered in the tens of thousands. At the same time, the testing of natural product extracts became common practice, significantly boosting the number of samples to be stored [1, 2]. As the number of samples exceeded 100 000 (at that time a seemingly immense number), automated systems were developed to store and catalog them. Initially, these were simple robotic units or adapted card file systems that would simply present entire drawers or boxes of samples to an operator. Chemical structures were often still paper copies and stored

Management of Chemical and Biological Samples for Screening Applications, First Edition.
Edited by Mark Wigglesworth and Terry Wood.

elsewhere, and the amount in the inventory was rarely accurate if tracked at all. Storage labware formats were standardized to accommodate the large volumes of samples moving through the system and to facilitate liquid handling and detection platform development [3]. Improvements in liquid handling and detection enabled increasingly higher labware densities, allowing tighter environmental control, and larger libraries.

Sample integrity became paramount with a focus on environmental conditions and consistent sample history both in the storage units and in the reformatting/analysis areas. Significant numbers of legacy compounds which had been subjected to variable temperatures, water, oxygen, and light were found to be compromised. Container seal adhesives and labware mold components could introduce interferents to the assay results. Compound managers realized that consistent sample quality was a key to valid scientific data. Programs were employed to provide cradle-to-grave care as well as purity monitoring to insure repeatable sample integrity.

With the advent of combinatorial chemistry, parallel synthesis made the creation of large compound sets numbering in the hundreds of thousands viable and raised the stakes for compound management. High-throughput screening (HTS) groups began requiring that compounds be presented in 96 well plates dissolved in Dimethyl sulfoxide (DMSO) and consumed these plates at an alarming rate. At about the same time, the electronic storage and representation of chemical structures became possible [4]. As the numbers of samples increased, chemists could no longer rely on visual inspection of structures, so tools were developed to analyze synthetic sets to determine their degree of similarity or difference [5]. Compound management groups, that had often become underfunded corporate backwaters, suddenly found themselves under the spotlight as the bottleneck in an exciting new process.

In answer to this challenge, funding was allocated to revamp chemical stores, and a plethora of bespoke systems of automation and data management appeared [1, 6–8]. The linkage between automated preparation systems and data systems was a slow process and the systems that were created varied widely in their architecture and success but shared a number of traits that embody today's samples management system:

- Sample registration
- Usage of enterprise-wide standardized labware
- Positive sample tracking, usually using barcodes
- Cradle-to-grave tracking of samples stored in both vials and plates
- Sample security with user access tracking and control
- Accurate quantity tracking of both mass and volume
- Storage of compounds in DMSO solutions
- Control of environmental conditions to minimize water uptake, oxygen and light degradation, and temperature fluctuation
- High speed automated storage and retrieval
- Reduced freeze/thaw cycles by efficient daughter plate production or by multiple aliquotting

- Regular purity monitoring
- Sample ordering systems
- Automated cherry picking and plate preparation systems
- Robust distribution methods and sample tracking outside of the storage system.

As HTS and ultra high throughput screening (uHTS) became ubiquitous tools in drug discovery, they also began to be used in other industries such as the discovery of agricultural agents (pesticides, animal health, etc.), catalysts, polymer discovery, fragrances, and flavorings. Similarly, the newly created science of sample management also found utility in many other areas. With the advent of the human genome project the need to store large numbers of biological samples became imperative [9]. Blood and tissue sample banks both for research and distribution grew to the point where they required similar techniques (Table 1.1). Today sample management is applied in industries as varied as hospital pharmacies, environmental repositories, and sperm banks. So let us now examine the techniques used in modern sample management.

Inventory: Probably the most critical factor in sample management remains inventory. But this has expanded well beyond the simple 'where is it' definition. Today's sample manager is more concerned with curation of the samples in their charge. As is discussed in Chapters 2–4, this includes the integrity of the samples on receipt, during storage, and even after delivery to end customers. To accomplish this, samples must be subjected to analytical tests for quality control (QC). In many cases this information will be provided by the supplier of the material. When that supplier is an internal group or a trusted partner this may be deemed

Table 1.1 Comparison of compound management and biobanking.

Compound management	**Biobanking**
Compounds are precious but for the most part replaceable	Specimens irreplaceable
Freeze/thaw cycles kept low, target 6–10	Freeze/thaw cycles very low or none
Large legacy libraries to be automated, which were added to en masse through library purchase or acquisitions/mergers	Small collections, few legacy specimens to be automated, samples added incrementally
Unregulated environment, compliance requirements low	Regulated environment, compliance requirements high
Low probability of cross organization exchange	High probability of cross organization exchange
Historically large budgets for R&D, low examination of return on investment (ROI), long-term funding available	Costs and resources may be subsidized, budgets and ROI examined closely, long term financing to cover length of studies difficult
Quality of legacy samples questionable	Quality of legacy samples questionable

sufficient and accepted but in other cases the purity of the material will have to be verified. For small molecule samples the most common method applied is Liquid Chromatography/Mass Spectrometry (LC/MS) analysis [10, 11].

In biobanking, the focus is on maintaining quality from collection to analysis. Here the primary problem is that one cannot sample the specimen regularly due to degradation during the aliquoting process. Also, the specimen volume is likely to be very small and therefore prized. Establishing the purity on receipt is a critical first step but is rarely sufficient. The quality of the samples in storage must be verified over time and, in many cases, after dissolution. The latter remains a largely unsolved problem as of the time of this writing. Dissolving a compound introduces a host of QC problems, particularly when the samples are transferred to plates. While it is possible to test the concentration and purity of samples dissolved in DMSO, it is not practical to test hundreds of thousands of samples on a regular basis using these techniques. In addition, it is nearly impossible to test small-molecule chemical samples in the environment used for HTS. Representative sampling of libraries has shown that a relatively high proportion (>20%) of the compounds in a sample set will be insoluble after a simple water dilution [12]. On the other hand, empirical data shows that many of these compounds will show activity in certain buffer or cellular testing systems implying that they are soluble under alternative conditions. The solution that many laboratories have adopted is to test subsets of the library and to test compounds that are determined to be active and are requested for further follow up. This approach is discussed in more detail in Chapters 2 and 15.

To make the problem even more complex, the samples may be subjected to various storage conditions and may be shipped to alternative sample management sites or end customers. The number of times the sample has been frozen and thawed and the storage temperatures may affect the stability of the sample set. There is not a clear body of literature on sample stability in DMSO [13, 14] and conflicting anecdotal evidence making the choice of storage conditions for DMSO samples difficult. The unusual physical properties of DMSO also complicate this matter [15]; DMSO will readily absorb water and oxygen from the atmosphere, which radically changes its freezing point and may affect the stability of compounds. The range of approaches in this area is widely varied. Some groups have established maximum freeze/thaw ranges and employ single-use plates in their process to minimize atmospheric exposure, while others have embraced room temperature storage and accepted the inevitability of water uptake by adding 10–20% water to their DMSO prior to sample dissolution [16–18]. This broad range of approaches and the fact that all have produced lead compounds makes establishing a true best practice impossible.

Tracking the location and history of samples is neither simple nor taken for granted. The use of barcodes is ubiquitous in sample management today. Barcodes are, in essence, very simple; they are simply a way of recording a serial number and rapidly and accurately entering that into a computer system. Barcode-based inventory systems, on the other hand, can be quite complex [19, 20]. They require the assignment of a tracking ID to every sample and a complex data model to register every manipulation of a sample from weighing through solubilization and any transfer from container to container. This system must always have not only the current volume of every sample but the historical record of every transfer from the lot submitted to the disposal of the last plate.

Data systems supporting sample management are discussed in Chapters 13 and 14. In addition to inventory, they will usually incorporate some mechanism for managing requests for samples. Customer ordering systems should always appear simple to end customers but may have quite complex internal management structures for the sample management professional. This can include 'pull' systems, where customer ordering software allows users to request samples and specify form (i.e., solid or liquid and concentration) and even location on a plate. Other aspects of a system may employ 'push' systems that automatically assign work to be performed on samples. For example, chemical samples that are synthesized as part of a specific medicinal chemistry program will usually have a prescribed group of assays that need to be conducted on each compound. When a chemist submits compounds, he or she may associate the molecules with a given program, and the IT system will automatically create work orders that create plates, tubes, and/or vials that are routed to the laboratories performing these tests.

The final aspect of sample management is automation systems, found in Chapters 10 and 12. While the management of samples does not require automation, it is virtually impossible to support the management of a large library without some degree of automation. Automated systems can range from simple liquid handling units that perform vial-to-plate transfers and plate-to-plate replication to large fully integrated systems that can perform all the aspects of sample preparation from sample dissolution to final microplate preparation. It should be noted that the one aspect of sample management that has never been efficiently automated is the weighing of samples. While significant resources have been devoted to this problem, automated solutions have been stymied by the highly varied nature of the chemical samples themselves. These samples can range from very dry, free flowing powders (which are easy to dispense) to tars that must be scraped or transferred by dipping a spatula, or proteins that are extremely hygroscopic and form light flakes that blow away easily. As a result almost all laboratories still employ a manual weighing process that is highly integrated with a data system and sample tracking to ensure accuracy – this is another reason that QC is so important. As with many HTS applications, this aspect of sample management has largely been solved. The systems have evolved from gymnasium-sized units that could store 1 million samples and process 10 000–20 000 samples per day to small unit stores that can be connected to provide the same storage capacity in a standard laboratory. Similarly, the problem of low-temperature storage has largely been solved. Systems that operate at temperatures down to vapor phase liquid nitrogen storage are now available and are discussed in the biobanking sections of this volume. Modern sample management systems enable the automated storage of virtually any sample from small-molecule chemicals to cells and tissues.

So, in conclusion, the field of sample management has grown both in importance and sophistication. The importance of this activity cannot be underestimated. The cost to replace a corporate chemical collection can be conservatively estimated at

$500 per sample. For a 1 million compound collection this might take four to five years and cost $500 000 000. Additionally, human and non-human biological specimens are irreplaceable. Even if a replacement specimen can be obtained, the biological state will have changed and the specimen will not be identical. When looked at in this light, it would be criminal to allow compounds or specimens to degrade or be lost.

References

1. Archer, J.R. (2004) History, evolution, and trends in compound management for high throughput screening. *Assay Drug Dev. Technol.*, **2** (6), 675–681.
2. Janzen, W.P. and Popa-Burke, I.G. (2009) Advances in improving the quality and flexibility of compound management. *J. Biomol. Screen.*, **14** (5), 444–451.
3. ANSI (2004) *New Microplate Standards Expected to Accelerate and Streamline Industry*, ANSI, New York, *http://www.ansi.org/news_publications/news_story.aspx?menuid=7&articleid=598* (accessed 2004). ANSI/SBS 1-2004: Footprint Dimensions ANSI/SBS, 2-2004: Height Dimensions ANSI/SBS, 3-2004: Bottom Outside Flange Dimensions ANSI/SBS 4-2004: Well.
4. Warr, W.A. (1991) Some observations on piecemeal electronic publishing solutions in the pharmaceutical industry. *J. Chem. Inf. Comput. Sci.*, **31** (2), 181–186.
5. Oprea, T.I. (2000) Property distribution of drug-related chemical databases. *J. Comput. Aided Mol. Des.*, **14** (3), 251–264.
6. Rutherford, M.L. and Stinger, T. (2001) Recent trends in laboratory automation in the pharmaceutical industry. *Curr. Opin. Drug Discov. Devel.*, **4** (3), 343–346.
7. Ray, B.J. (2001) Value your compound management team! *Drug Discov. Today.*, **6** (11), 563.
8. Janzen, W.P. (2002) *High Throughput Screening: Methods and Protocols*, Humana Press, Totowa, NJ.
9. Eiseman, E. and Haga, S. (1999) *Handbook of Human Tissue Sources: A National Resource of Human Tissue Samples*, Rand Corporation.
10. Ari, N., Westling, L., and Isbell, J. (2006) Cherry-picking in an orchard: unattended LC/MS analysis from an autosampler with >32,000 samples online. *J. Biomol. Screen.*, **11** (3), 318–322.
11. Letot, E., Koch, G., Falchetto, R., Bovermann, G., Oberer, L., and Roth, H.J. (2005) Quality control in combinatorial chemistry: determinations of amounts and comparison of the 'purity' of LC-MS-purified samples by NMR, LC-UV and CLND. *J. Comb. Chem.*, **7** (3), 364–371.
12. Popa-Burke, I.G., Issakova, O., Arroway, J.D., Bernasconi, P., Chen, M., Coudurier, L. *et al.* (2004) Streamlined system for purifying and quantifying a diverse library of compounds and the effect of compound concentration measurements on the accurate interpretation of biological assay results. *Anal. Chem.*, **76** (24), 7278–7287.
13. Kozikowski, B.A., Burt, T.M., Tirey, D.A., Williams, L.E., Kuzmak, B.R., Stanton, D.T. *et al.* (2003) The effect of freeze/thaw cycles on the stability of compounds in DMSO. *J. Biomol. Screen.*, **8** (2), 210–215.
14. Kozikowski, B.A., Burt, T.M., Tirey, D.A., Williams, L.E., Kuzmak, B.R., Stanton, D.T. *et al.* (2003) The effect of room-temperature storage on the stability of compounds in DMSO. *J. Biomol. Screen.*, **8** (2), 205–209.
15. Rasmussen, D.H. and Mackenzie, A.P. (1968) Phase diagram for the system water-dimethylsulphoxide. *Nature*, **220** (5174), 1315–1317.
16. Schopfer, U., Engeloch, C., Stanek, J., Girod, M., Schuffenhauer, A., Jacoby, E. *et al.* (2005) The Novartis compound

archive – from concept to reality. *Comb. Chem. High Throughput Screen.*, **8** (6), 513–519.

17. Jacoby, E., Schuffenhauer, A., Popov, M., Azzaoui, K., Havill, B., Schopfer, U. *et al.* (2005) Key aspects of the Novartis compound collection enhancement project for the compilation of a comprehensive chemogenomics drug discovery screening collection. *Curr. Top. Med. Chem.*, **5** (4), 397–411.
18. Engeloch, C., Schopfer, U., Muckenschnabel, I., Le Goff, F., Mees, H., Boesch, K. *et al.* (2008) Stability of screening compounds in wet DMSO. *J. Biomol. Screen.*, **13** (10), 999–1006.
19. Palmer, R.C. (1995) *The Bar Code Book: Reading, Printing, Specification, and Application of Bar Code and Other Machine Readable Symbols*, 3rd edn, Helmers Publishing Inc., Peterborough, NH.
20. Burke, H.E. (1990) *Automating Management Information Systems*, Van Nostrand Reinhold, New York.

2
Generating a High-Quality Compound Collection

Philip B. Cox and Anil Vasudevan

2.1
Defining Current Screening Collections

In part due to sustained effort aimed at adding drug-like compounds via internal/external synthesis but also complemented with mergers and acquisitions, most corporate collections comprise upwards of several hundred thousand compounds. Improvements in automation, miniaturization, and novel assay technologies have enabled ultra-high-throughput screening (uHTS) (>100 000 compounds per day) to become a routine strategy for hit identification. A recent analysis indicated that 104 compound candidates progressed into clinical studies from hits identified through uHTS prior to 2004, and 4 of these molecules are currently represented in approved drugs [1]. However, as the complexity of targets (or the confidence in them) evolves, uHTS is not always desirable due to cost effectiveness and other practical consideration, as discussed in Chapter 13. We provide our corroborative perspective on carefully selecting a screening set below.

The two commonly adopted approaches to match objective with screening effort are known as the focused and diversity methods of compound selection. Underlying both these methods is the 'similarity property principle' [2], which states that high-ranked structures are likely to have similar activity to that of the reference structure. Focused methods use this principle to try to find compounds that exhibit similar biological activity to a reference molecule or pharmacophore already known to be active. Diverse selection methods use the principle of choosing evenly across chemical space, hence maximizing the odds of finding diverse compounds. There are many methods used to assemble a diverse subset of molecules from a larger population for screening, a few of which are cluster-based selection [3], partition-based [4], or maximum dissimilarity [5]. Regardless of which screening philosophy is pursued, when the objective is to initiate a hit-to-lead effort, the most important outcome of any high throughput screening (HTS) campaign (or fast-follower approach) is the chemotype hit rate, distinct from the overall HTS hit rate [6].

In terms of a diverse screening approach, a productive follow-up strategy to subset screening is iterative follow-up on the initial screening results based on the cheminformatic clustering approaches described above. Karnachi and Brown [7]

Management of Chemical and Biological Samples for Screening Applications, First Edition.
Edited by Mark Wigglesworth and Terry Wood.

have described their results where a small representative subset of clusters were screened, and structural and descriptor properties around the clusters of HTS actives were used in follow-up screening, an approach which identified 97% of the structural classes while screening only approximately 25% of the uHTS collection. Blower and colleagues [8] performed a similar study where statistical models derived from the initial hits of subset screening were used to retrieve compounds for the next round of screening. Both these models are powerful examples of the value of intelligent subset screening followed by thorough cheminformatic follow-up of HTS actives.

When the targets are from traditionally well-studied protein families (e.g., kinases, G protein coupled receptor (GPCRs), and ion channels), pharmacophore information can be utilized to build a subset screening deck, replete with information, and options for hit-to-lead follow-up. The main advantage of this screening strategy, as observed at Abbott, is that hit rates from the target family libraries are often significantly higher than the hit rate from a random collection. The disadvantage, on the other hand, as medicinal chemists will often attest to, is that there is a much smaller chance of identifying structurally novel leads, which often becomes a major component of hit-to-lead efforts.

A very interesting recent study suggests that, within the kinome, the concept of privileged substructures for kinase inhibitors may be less applicable [9]. This study demonstrated that kinase family specificity depends less on the identity of the core scaffold and more on that of the specific substituent appended to the core. As such, the results suggest that, for a specific compound, the identity of the substituent may drive its activity pattern as much as or more than the identity of the core, even assuming rudimentary design criteria (flat aromatics with a hydrogen-bond donor/acceptor). This finding would seem to support the original work by Evans *et al.* [10], in that privileged substructures from gene family-specific analysis are probably more drug-like or receptor-privileged than broadly target family-privileged [11].

Organizations such as Abbott have utilized this scaffold-based approach to synthesize and successfully screen for active kinase inhibitors [12]. A recent publication on novel (6,5) and (6,6) fused heterocyclic moieties provides opportunities for generating novel screening libraries around these scaffolds for focused and diversity-based screening [13].

2.2
Design Criteria for Enriching a Compound Collection with Drug-Like Compounds

2.2.1
Physicochemical Tailoring of a Compound Collection

By inference, the concepts that underpin the design of drugs are also important when considering the design of compounds for a corporate screening collection. In this section we expand upon the important physicochemical properties that are also outlined in Chapter 13. These properties heavily influence the design of drugs and therefore chemical matter for corporate screening collections.

The optimization of hits-to-leads to clinical candidates and beyond relies on the ability of the medicinal chemist to adequately balance physicochemical, biochemical, pharmacokinetic (PK), and pharmacological properties of that compound. Optimizing physicochemical properties in order to generate chemical matter with good PK profiles and the desired levels of *in vitro* and *in vivo* potency is the ultimate objective of the medicinal chemist. While physicochemical properties have always been important in drug design, the emphasis on *in silico* design using physicochemical profiling has only been adopted and practiced (in part) by the medicinal chemist in recent years. Indeed, the concepts underpinning drug design are becoming more refined as the knowledge generated during drug discovery and development programs is fed back into new drug discovery projects. More and more, the medicinal chemist is becoming aware of the importance of balancing and optimizing multiple physicochemical parameters in order to achieve the desired lead-like and ultimately drug-like attributes. Moreover, this knowledge is being used to tailor corporate screening collections with the aim of designing and synthesizing molecules with higher odds of good overall PK (ultimately good oral bioavailability) and lower odds of promiscuity and, therefore, toxicity. A brief overview of some of the more important physicochemical properties that must be taken into consideration when designing compounds for a corporate screening collection is given below.

2.2.2
Lipophilicity Design Considerations

Since Lipinski's paper in 1997 [14] correlating high lipophilicity of a drug substance (as measured by the octanol/water partition coefficient LogP, or the calculated variant cLogP) with low absorption, permeability, and oral bioavailability, there have been many studies both vindicating and refining Lipinski's conclusions, as well as unearthing a multitude of liabilities associated with highly lipophilic compounds [15]. In particular, the comprehensive analysis by Leeson and Springthorpe re-affirmed the deleterious effects of high compound lipophilicity [16]. In this study, they correlated promiscuity with median cLogP of 2133 compounds (from the CEREP BioPrint database), where promiscuity was equated to the percentage of assays in which the compounds demonstrate greater than 30% inhibition at 10 μM (Figure 2.1).

Figure 2.1 clearly indicates the effects of increasing lipophilicity on promiscuity, and serves as a stark reminder of the perils associated with highly lipophilic compounds. It also leads to the conclusion that setting a lower threshold for cLogP is highly recommended when designing compounds for a screening collection, as ultimately both molecular weight (MW) and lipophilicity will likely increase during hit-to-lead and lead optimization. The results of the Leeson and Springthorpe study are consistent with Pfizer's [17] recent comprehensive study on preclinical *in vivo* animal toxicity data in which it was observed that compounds with a cLogP

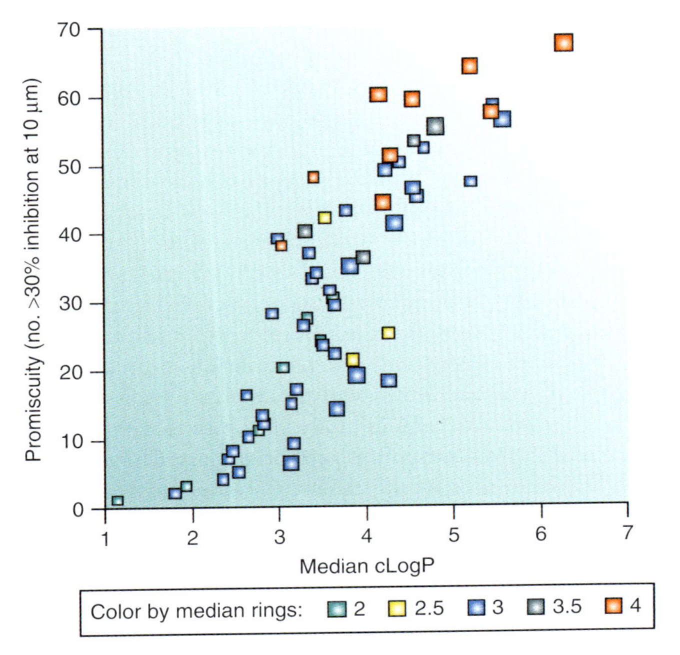

Figure 2.1 Plot of promiscuity versus median cLogP and molecular mass [16]. Points are sized by median molecular weight (281.3–582.9).

greater than three have an enhanced risk of toxicity (with the odds even higher for compounds which also have polar surface area less than 75). More recently, a similar analysis of pharmaceutical promiscuity was performed by Peters *et al.* from Roche, wherein promiscuity was again shown to track with increasing lipophilicity [18].

Highly lipophilic compounds have liabilities on a number of fronts, with many recent publications highlighting these issues. For example, highly lipophilic compounds (particularly bases) have an enhanced risk of cardiovascular side effects as a result of binding to human Ether-a-go-go-Related Gene (hERG) potassium channels and an increased likelihood for phospholipidosis (particularly amphiphilic (cationic lipophilic) compounds) [19].

Not surprisingly, there have been many studies aimed at further understanding the effect of lipophilicity on the absorption-distribution-metabolism-excretion (ADME) and PK properties of compounds. Johnson *et al.* from Pfizer recently published an interesting means of optimizing absorption and clearance using the 'Golden Triangle' visualization tool [20].

A comprehensive analysis of a data set of 47 K compounds shows that compounds with a higher probability of acceptable permeability and human microsomal clearance fall within a well-defined triangular area of a plot of MW versus LogD (either experimental, estimated, or calculated). The Golden Triangle area is defined by an apex at MW 450, with the base set at MW 200 between a LogD range of

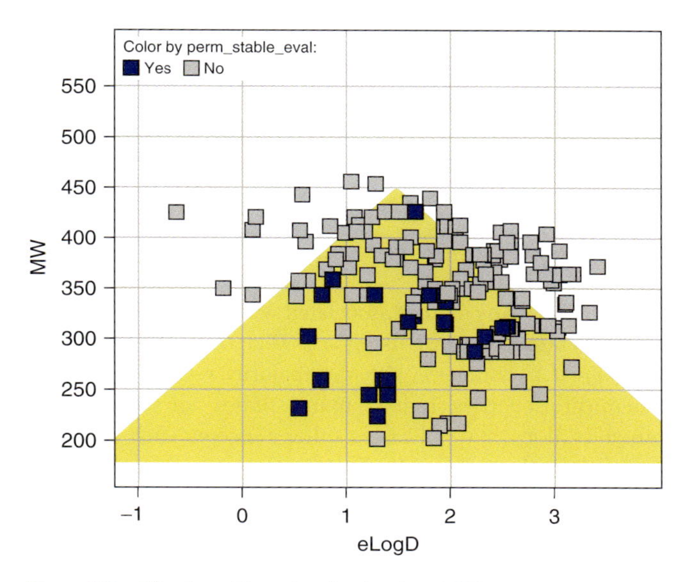

Figure 2.2 Pfizer's golden triangle visual tool [20].

−2 and 5 (Figure 2.2). In this analysis the authors point out that 25% of the compounds in the center of the Golden Triangle (LogD 1.5, MW 350) have acceptable Caco-2 permeability and microsomal stability versus just 3% for compounds outside the Golden Triangle (LogD 3 and 0 at MW 450). This type of tool is becoming important in the design process for medicinal chemists as it can be used to bias compound selection in library design and HTS triage.

A similar analysis by Waring [21] on the upper and lower limits of lipophilicity of drug candidates in the context of permeability indicated that both LogD and MW are key parameters that should be considered in combination to ensure higher levels of permeability.

Waring has also published a comprehensive review and assessment on the role of lipophilicity in drug discovery [22]. The effect of lipophilicity on many drug properties such as solubility, permeability, clearance (renal and hepatic), hERG channel blockade, promiscuity (bioprint data), *in vivo* rat toxicity, and phospholipidosis was reviewed with the conclusion that a narrow range of LogD (1–3) is required in order to design compounds with reduced odds of negatively affecting the above-mentioned drug properties and off-target activities. This is another stark indication that for compound collection design, in order to increase the odds of success, compound libraries should be designed to fall within a LogD range of 1–3. While ideal, in our experience, this is not practically feasible, especially when designing matrix arrays based on a diverse set of monomers. At Abbott, we have adopted a pragmatic approach where we design libraries to be minimally Ro4 (rule of 5 (Ro5) is defined in Chapter 13; Ro4 is equivalent but where values conform to a unit of 4) compliant in order to minimize overall risk, and yet cover reasonable diversity. For libraries focused on targets such as kinases, peptidic

GPCRs, and protein–protein interactions (PPIs) inhibitors, the physicochemical property requirements by necessity have to be relaxed in order to design libraries with a reasonable probability of modulating the target. For these 'less druggable' targets, we design libraries to be as close to Ro4 compliant as possible, with Ro5 compliance set as the upper limit.

2.2.3
Other Physicochemical Roadblocks

In recent years, there have been many studies associated with the negative impact of designing chemical matter outside of drug-like space. What constitutes drug-like space hinges on a number of parameters, with the majority related to a broad spectrum of physicochemical properties. Indeed, many studies not satisfied with physicochemical parameters alone have advanced other concepts such as ligand efficiency (LigE) [23], ligand lipid efficiency (LLE) [16], aromatic proportion (AP) [24] and the fraction of sp^3 hybridized carbons (Fsp^3) [25], in an attempt to enhance the scholarship around assessing potential drug-like chemical matter. Not surprisingly, oral drug space is very complex, with multiple parameters responsible for the overall pharmacodynamic and PK behavior of the drug substance in the body. Indeed, if it were the case that all compounds within Ro5 were orally bioavailable, then drug discovery would be rather more straightforward. The previous section highlighted the perils of highly lipophilic compounds, this section will focus on other physicochemical properties that, if not limited and heeded, may present significant roadblocks in terms of compound progressability and should be taken seriously when designing compounds for a screening collection.

A recent and very comprehensive assessment of the effect of aromatic ring count on compound 'developability' was published by Ritchie and MacDonald [26]. While it could be argued that the number of aromatic rings (NAR), *per se*, is not a true physicochemical property, the effect of increasing the NAR in a compound does have an impact on a wide range of physicochemical properties such as LogD, MW, polar surface area, and melting point (and therefore solubility). As the authors point out, the NAR is a simple guiding metric that does not require the use of a computational algorithm to calculate and allows the medicinal chemist to 'design by eye' when considering aromatic ring count. Not surprisingly, the GlaxoSmithKline (GSK) analysis shows, very clearly, a number of negative consequences of high aromatic ring count on a range of druggable factors such as solubility, cLopP, LogD, plasma protein binding, Cyp inhibition and activation, and hERG channel inhibition. Indeed, the study concludes that compounds with more than three aromatic rings have a lower probability of 'developability.' This study clearly shows that the mean NAR actually decreases as a function of project milestone in the GSK pipeline from candidate selection to clinical proof of concept (Table 2.1). It is no coincidence, therefore, that the average NAR for oral drugs is less than two. Care, therefore, should be exercised when designing compounds for a screening collection to restrict overall aromatic ring count, especially in the context of solubility.

Table 2.1 Mean aromatic ring count of candidates in the GSK pipeline [26].

	CS	FTIH	P1	P2	POC
Count[a]	50	68	35	53	96
Mean aromatic ring count	3.3	2.9	2.5	2.7	2.3

[a]Count is the number of compounds in the category.
CS, preclinical candidate selection;
FTIH, first time in human; P1, phase 1; P2, phase 2; and POC, proof-of-concept.

More recently, another comprehensive analysis from GSK [27] highlighted the effect of lipophilicity and aromatic ring count on aqueous solubility. This comprehensive analysis of 100 000 experimental kinetic solubility data points clearly demonstrates the importance of both LogD and NAR when considering overall intrinsic aqueous solubility. While lipophicity is inversely proportional to aqueous solubility, and is perceived to be the most common reason for poor solubility, the NAR has, arguably, as much of a pronounced effect on overall solubility. A pie plot of NAR versus binned LogD clearly shows a defined area of highest likelihood of good solubility (Figure 2.3), as increasing both parameters has a negative effect on solubility. From this, a simple formula was developed called the solubility forecast index (SFI) = LogD + NAR. In general if a compound has SFI $<$ 5 then it will have a reasonable chance of having good solubility. The authors make the point that, on average, drugs have a low SFI (around 3.3) and therefore are likely to have good aqueous solubility. At Abbott we have done a similar analysis and found that this cut-off for SFI (SFI < 5) is a fair metric in terms of defining a reasonable likelihood of solubility.

Another study on solubility adopting a similar policy, using the NAR or the degree of aromaticity (AP) as a main parameter in developing a solubility model, was recently reported by Lamanna *et al.* from Siena Biotech [24]. The recursive partitioning (RP) model for aqueous solubility built by the Siena group was developed as a quick filter for choosing compounds for their screening collection. They found that the most accurate (accuracy = 81%) and precise (precision = 75%) model utilized just two parameters, AP and MW, out of a total of five models that included combinations of seven different physicochemical descriptors (MW, number of rotatable bonds (RTB), number of hydrogen bond acceptors (HBA), number of hydrogen bond donors (HBD), polar surface area (PSA), ALogP, and AP). Again, this study demonstrates the importance of considering the overall degree of aromaticity in a compound, especially in the context of designing compounds with good aqueous solubility, and should certainly be taken into consideration when selecting or designing compounds for a screening collection.

Another interesting concept inversely related to AP and aromatic ring count was recently put forward by Lovering *et al.* from Wyeth [25], which considers the Fsp^3 as a function of total number of carbons in a molecule: thus, an Fsp^3 of 0.5 equates to 1 saturated carbon out of 2 total carbons. By analyzing the GVK

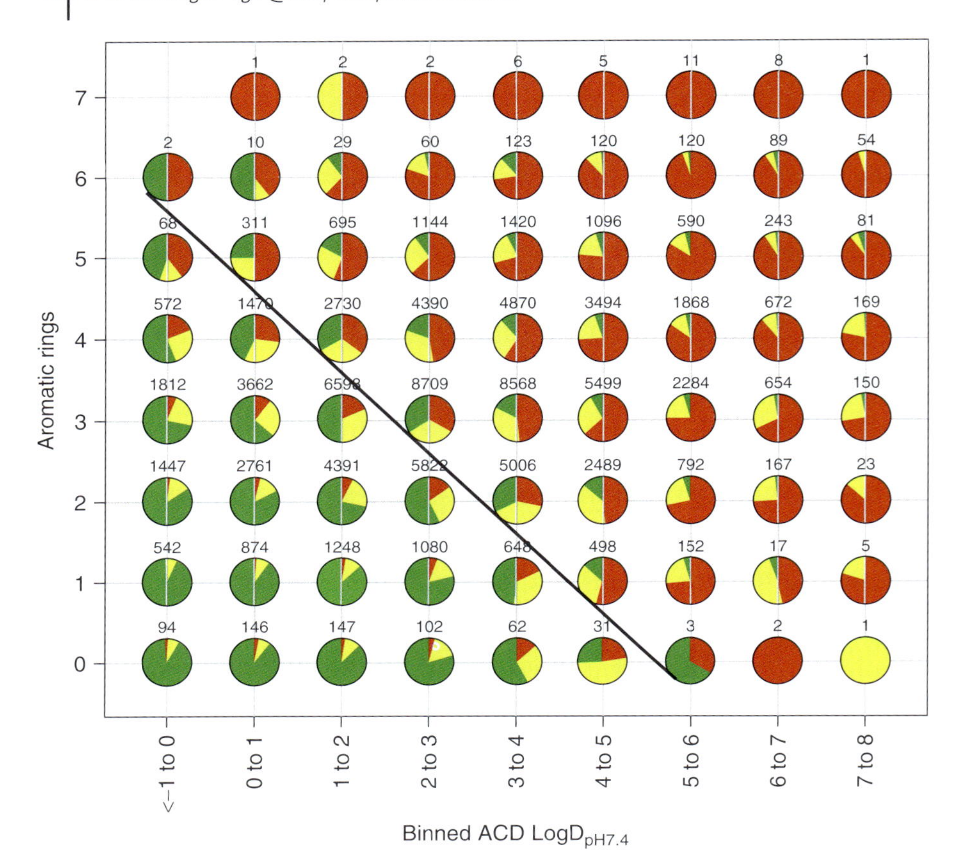

Figure 2.3 Pie matrix plot showing solubility as a function of LogD and number of aromatic rings [27]. Solubility category – high = green > 200 μM, medium = yellow 30–200 μM, low = red < 30 μM. Number above pie = compound count.

BIO database, the authors calculated the Fsp^3 for compounds across all stages of development from initial discovery to launched drugs. Given the GSK analysis, where the NAR decreases for each milestone transition, it is not surprising that the level of saturation (Fsp^3) increases. This inverse proportionality between NAR and Fsp^3 makes sense as the lower the Fsp^3, the less saturated a molecule is, and the more likely it is to have a higher number of unsaturated atoms and therefore aromatic rings. The authors also make the connection between low Fsp^3 and higher melting temperatures, and therefore low solubility, thus showing the positive effect of adding saturation to a molecule in terms of improving druggable physicochemical properties. The study also highlights the positive effect of chirality in a molecule in the context of improved physicochemical properties.

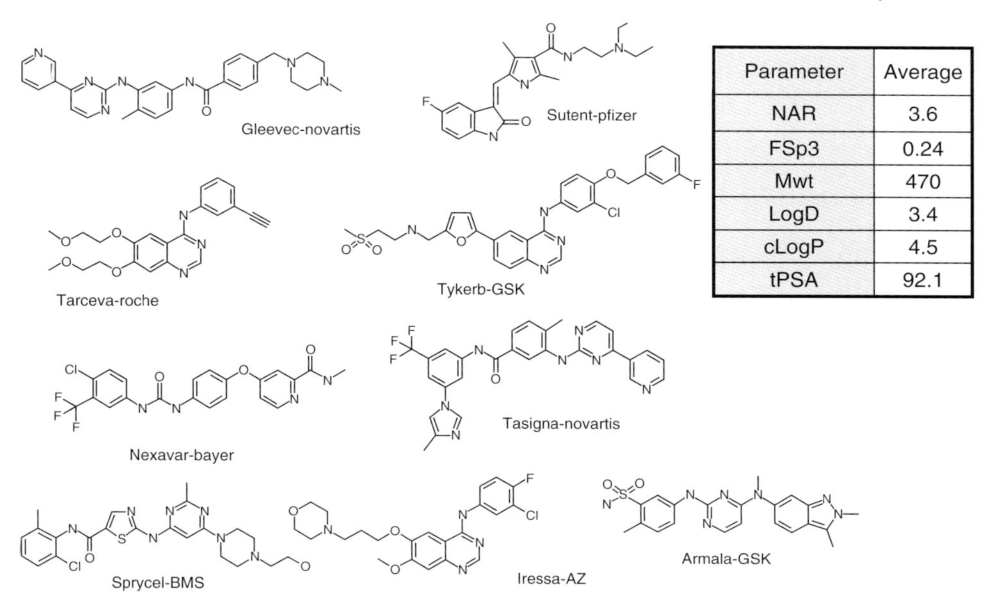

Parameter	Average
NAR	3.6
FSp3	0.24
Mwt	470
LogD	3.4
cLogP	4.5
tPSA	92.1

Figure 2.4 Average physicochemical properties of marketed kinase drugs.

A very recent perspective by Ishikawa and Hashimoto discusses the various ways in which disruption of molecular planarity/symmetry can be effective in increasing aqueous solubility, in some cases by up to 350-fold, despite increased hydrophobicity [28].

At Abbott we have conducted a comprehensive analysis of internal ADME data and correlated a number of properties to the NAR and Fsp^3. We have concluded that increasing the NAR has a negative effect on a variety of ADME properties (such as solubility, permeability, and plasma protein binding) and, not surprisingly, increasing Fsp^3, in general, has a beneficial effect on ADME properties. For compound collection design, limiting the NAR, whilst beneficial in terms of improving drug-like properties, may not be practical for some gene drug classes. If one considers kinase inhibitors in general, they have higher NAR and lower Fsp^3 compared to other oral drugs. In fact, across the board, kinase drugs have physicochemical properties that trend outside of optimal oral drug space (Figure 2.4).

Thus, for compound acquisition or library design around less druggable targets, a realization that modulating the target may in fact lead to chemical matter with sub-optimal physicochemical properties should be taken into consideration. Balancing ideal physicochemical properties with optimal on-target activity to produce chemical matter with higher odds of good drug metabolism and pharmacokinetics (DMPK) properties and reduced odds of promiscuity (and therefore toxicity) is the aim of any medicinal chemistry program. A well-conceived compound collection designed to preclude as many liabilities as possible and focused predominantly on a specific gene class should make the path forward in hit-to-lead and beyond less perilous. The next section deals with how risk, from a promiscuity standpoint, can be defined in terms of physicochemical property space and how we can use this knowledge in the context of future compound design.

2.2.4
Assessing Risk – from Rule of 5 to Rule of 3/75

The large majority of studies undertaken to understand drug-likeness from an assessment of either *in vitro* and or *in silico* physicochemical properties is aimed at one common goal – to ameliorate the risk of compound attrition. As mentioned earlier, if a simple rule of thumb, such as Ro5, was guaranteed to deliver development candidates, with perfect PK profiles and a low risk of *in vivo* toxicity, then drug discovery would be relatively straightforward. While we know the situation is rather more complex, the fact remains that the Ro5 and other studies have significantly enhanced our ability to design more drug-like chemical matter. Whether this translates into an overall reduction in compound attrition and a consequent increase in success for the pharmaceutical industry as a whole remains to be seen. Leeson and Springthorpe have clearly demonstrated in their study [16] that while our understanding of druggability has increased over the past decade, the overall quality, in terms of physicochemical properties, of compounds entering development has not necessarily improved. Irrespective of drug gene class, the overall MW and lipophilicity of development compounds has actually increased in recent years. While this may be attributed to a change in focus of the pharmaceutical industry to less druggable targets (such as peptidic GPCRs), the Leeson–Springthorpe study shows that for druggable aminergic GPCR targets the overall median cLogP and MW of compounds has increased significantly in recent years. The authors ascribe part of this worrying trend to the lack of attention by medicinal chemists to the physicochemical profiles of chemical matter put forward into development. Though there may be some truth to this assertion, it is clear that many drug discovery organizations are paying close attention to the physicochemical properties of compounds being advanced into development, and are carefully analyzing reasons why compounds fail in development. Pfizer's recent 'Rule of 3/75' paper [17] is an example of this, as the study disclosed trends from the analysis of over 200 pre-clinical animal toxicity studies. A clear correlation between lower odds of promiscuity (*in vitro*) and toxicity (*in vivo*) is evident for compounds with cLogP < 3 and topological polar surface area (tPSA) > 75 (Table 2.2).

Clearly, there is a vast amount of scholarship around defining the overall ideal structural and physicochemical characteristics of drug-like chemical matter. These ever-evolving characteristics should be taken seriously in the context of compound collection design too. Synthesizing or acquiring compounds without heeding and precluding both structural and physicochemical liabilities can lead to significant resource burden. Clearly, all risks must be taken into consideration, but ignorance in this context will only lead to further compound attrition and a continuation of the current productivity crisis in the pharmaceutical industry.

Table 2.2 Odds of toxicity as a function of cLogP and TPSA [17]. Odds of toxicity is the ratio of the number toxic to none toxic compounds. Number of associated compounds is shown in parentheses next to calculated odds.

Observed odds for toxicity versus cLogP/TPSA

Toxicity	Total-drug		Free-drug	
	TPSA > 75	TPSA < 75	TPSA > 75	TPSA < 75
cLogP < 3	0.39 (57)	1.08 (27)	0.38 (44)	0.5 (27)
cLogP > 3	0.41 (38)	2.4 (85)	0.81 (29)	2.59 (61)

2.2.5
Tools Enabling Desk Top *In Silico* Design

There are many tools available to the medicinal chemist that enable rapid calculation of *in silico* physicochemical properties – for example, *in silico* design tools compiled by the Swiss Institute of Bioinformatics (*http://www.isb-sib.ch/services/-software-tools.html*). This website contains a comprehensive list of commercial software with a brief overview of function and utility. Many drug companies have their own tailor-made platforms with a variety of custom features that make it straightforward for the medicinal chemist to calculate or access key physicochemical properties and also visualize the data in a format that will enable decision making. The workhorse of many design platforms, in the industry, is Pipeline Pilot. This multifaceted software enables the user to assemble tailored design protocols that can be formatted through a web port such that less computer-knowledgeable chemists can access key information very rapidly. In addition, for the more ambitious aspiring computational medicinal chemists, who lack programming expertise, Pipeline Pilot is set up in a way that makes it straightforward to put together, 'on the fly', custom design platforms without knowledge of programming language. Indeed the Pipeline Pilot simplistic plug and play platform has brought previously complex computational algorithms to the fingertips of medicinal chemists.

At Abbott, medicinal chemists have the option of using either customized Pipeline Pilot web ports that enable the calculation of a wide range of physicochemical properties, and related visualizations, or design their own protocols directly in Pipeline Pilot. This versatility enables a broad spectrum of medicinal chemists to effectively design chemical matter seamlessly.

By far the most widely used visualization tool in the industry is Spotfire™. In recent years, with the advent of knowledge-based drug discovery, the medicinal chemist has come to rely heavily on this very versatile visualization tool, as it can be used not only to visualize large data sets but also to allow the user to manipulate the data in a multitude of formats. Arguably, Spotfire™ has become one of the medicinal chemist's most versatile design tools. Not only has it been effectively utilized routinely in HTS triage, it is also now employed in all medicinal chemistry programs to track compound progress in hit-to-lead and lead optimization.

2.3 Concluding Remarks

The pharmaceutical industry has seen a steady decline in productivity over the past 15–20 years, despite significantly greater R&D investment. A detailed analysis of the reasons for this decline is beyond the scope of this chapter [29]. While a quality compound collection and effective screening strategies are important, it is critical that every experienced and knowledgeable medicinal chemist utilize all the tools and lessons learned from past successes and failures to design high-quality molecules with greater odds of survival through preclinical and clinical development.

References

1. Fox, S., Jones, S., Sopchak, L., Boggs, A., Nicely, H., Khoury, R., and Biros, M. (2006) *J. Biomol. Screen.*, **11**, 864.
2. Johnson, M. and Maggiora, G.M. (1990) *Concepts and Applications of Molecular Similarity*, John Wiley & Sons, Inc., New York.
3. Willett, P., Winterman, V., and Bawden, D. (1986) *J. Chem. Inf. Comput. Sci.*, **26**, 109–118.
4. Mason, J.S. and Pickett, S.D. (1997) *Perspect. Drug Discov. Des.*, **7/8**, 85–114.
5. Lajiness, M.S. (1997) *Perspect. Drug Discov. Des.*, **7/8**, 65–84.
6. Good, A.C., Cheney, D.L., Sitkoff, D.F., Tokarski, J.S., Stouch, T.R., Bassolino, D.A., Krystek, S.R., Li, Y., Mason, J.S., and Perkins, T.D. (2003) *J. Mol. Graphics Modell.*, **22**, 31–40.
7. Karnachi, P.S. and Brown, F.K. (2004) *J. Biomol. Screen.*, **9**, 678–686.
8. Blower, P.E., Cross, K.P., Eichler, G.S., Myatt, G.J., Weinstein, J.N., and Yang, C. (2006) *Comb. Chem. High Throughput Screen.*, **9**, 115–122.
9. Posy, S.L., Hermsmeier, M.A., Vaccaro, W., Ott, K.H., Todderud, G., Lippy, J.S., Trainor, G.L., Loughney, D.A., and Johnson, S.R. (2011) *J. Med. Chem.*, **54**, 54–66.
10. Evans, B.E., Rittle, K.E., Bock, M.G., Dipardo, R.M., Freidinger, R.M., Whiter, W.L., Lundell, G.F., Veber, D.F., Anderson, P.S., Chang, R.S., Lotti, V.J., Cerino, D.J., Chen, T.B., Kling, P.J., Kunkel, K.A., Springer, J.P., and Hirshfield, J. (1988) *J. Med. Chem.*, **31**, 2235–2246.
11. Schnur, D.M., Hermsmeier, M.A., and Tebben, A.J. (2006) *J. Med. Chem.*, **49**, 2000–2009.
12. Gracias, V., Ji, Z., Akritopoulou-Zanze., I., Abad-Zapatero, C., Huth, J.R., Song, D., Hajduk, P.J., Johnson, E.F., Glaser, K.B., Marcotte, P.A., Pease, L., Soni, N.B., Stewart, K.D., Davidsen, S.K., Michaelides, M., and Djuric, S.W. (2008) *Bioorg. Med. Chem. Lett.*, **18**, 2691–2695.
13. Pitt, W.R., Parry, D.M., Perry, B.G., and Groom, C.R. (2009) *J. Med. Chem.*, **14**, 2952–2963.
14. Lipinski, C.A., Lombardo, F., Dominy, B.W., and Feeney, P.J. (1997) *Adv. Drug Deliv. Res.*, **23**, 3.
15. (a) Veber, D.F., Johnson, S.R., Chang, H.Y., Smith, B.R., Ward, K.W., and Kopple, K.D. (2002) *J. Med. Chem.*, **45**, 2615–2623; (b) Wenlock, M.C., Austin, R.P., Barton, P., Davis, A.M., and Leeson, P.D. (2003) *J. Med. Chem.*, **46**, 1250.

16. Leeson, P.D. and Springthorpe, B. (2007) *Nat. Rev. Drug Discov.*, **6**, 881, and references therein.
17. Hughes, J.D., Blagg, J., Price, D.A., Bailey, S., DeCrescenzo, G.A., Devraj, R.V., Ellsworth, E., Fobian, Y.M., Gibbs, M.E., Gilles, R.W., Greene, N., Huang, E., Krieger-Burke, T., Loesel, J., Wager, T., Whiteley, L., and Zhang, Y. (2008) *Bioorg. Med. Chem. Lett.*, **18**, 4872.
18. Peters, J., Schnider, P., Mattei, P., and Kansy, M. (2009) *ChemMedChem*, **4**, 680.
19. (a) Ratcliffe, A.J. (2009) *Curr. Med. Chem.*, **16**, 2816; (b) Ploeman, J.-P., Kelder, J., Hafmans, T., van de Sandt, H., van Burgsteden, J.A., Saleminki, P.J., and van Esch, E. (2004) *Exp. Toxicol. Pathol.*, **55**, 347.
20. Johnston, T.W., Dress, K.R., and Edwards, M. (2009) *Bioorg. Med. Chem. Lett.*, **19**, 5560.
21. Waring, M.J. (2009) *Bioorg. Med. Chem. Lett.*, **19**, 2844.
22. Waring, M.J. (2010) *Expert Opin. Drug Discov.*, **5**, 235.
23. Hopkins, A.L., Groom, C.R., and Alex, A. (2004) *Drug Discov. Today*, **9**, 430.
24. Lamanna, C., Bellini, M., Padova, A., Westerberg, G., and Macari, L. (2008) *J. Med. Chem.*, **51**, 2891.
25. Lovering, F., Bikker, J., and Humblet, C. (2009) *J. Med. Chem.*, **52**, 6752.
26. Ritchie, J. and MacDonald, S.J. (2009) *Drug Discov. Today*, **14**, 1011.
27. Hill, A.P. and Young, R.J. (2010) *Drug Discov. Today*, **15**, 648–655.
28. Ishikawa, M. and Hashimoto, Y. (2011) *J. Med. Chem.*, **54**, 1539.
29. Paul, S.M., Mytelka, D.S., Dunwiddie, C.T., Persinger, C.C., Munos, B.H., Lindborg, S.R., and Schacht, A.L. (2010) *Nat. Rev. Drug Discov.*, **9**, 203–214.

3
Assessing Compound Quality

Ioana Popa-Burke, Stephen Besley, and Zoe Blaxill

3.1
Introduction

The term '*quality*' of a small-molecule compound[1)] has many different meanings. It can be related to the structure itself: druggable [1], optimal physico-chemical properties, lack of liability features, being in an intellectual property open space, and so on. Or, it can refer to its analytical chemistry properties: identity, purity, concentration, stability, and solubility. While both are equally important, here we will use the term '*compound quality*' as the analytical chemistry quality.

Drug discovery is a process which spans many phases and research groups, but it tends to follow a similar path: first target identification, next development of a biological assay to assess activity, and finally testing of (many or selected) compounds in that assay. Once initial positive results are obtained, chemistry resources are allocated to the project, and the medicinal chemistry cycle begins until a clinical trials candidate is declared. The quality of compounds tested plays a major role in this entire discovery process, but the impact of either poor or high quality will be very different depending on the stage of the discovery process, with the biggest impact being in the hit-to-lead discovery and the subsequent lead optimization stages, where compounds tend to undergo more testing (both biological and analytical). As illustrated in Figure 3.1, the number of compounds tested is highest in the earlier phases of the discovery process (especially in hit discovery), and it diminishes gradually as the medicinal chemistry cycles approach the candidate selection phase.

Even though fewer compounds are tested as programs advance, the design and synthesis become more complex, and the criteria for advancing to the next stage require multiple and more sophisticated assays, all of which dramatically increases the overall cost per compound. Earlier in the cycle, if a compound is missed in the hit discovery process, this may be the one compound that could make it to market. Therefore, screening a pure compound, at the right concentration to avoid

1) Small-molecule compound = a low molecular weight (typically up to 800 Da) organic compound, which is by definition not a polymer (Wikipedia).

Management of Chemical and Biological Samples for Screening Applications, First Edition.
Edited by Mark Wigglesworth and Terry Wood.

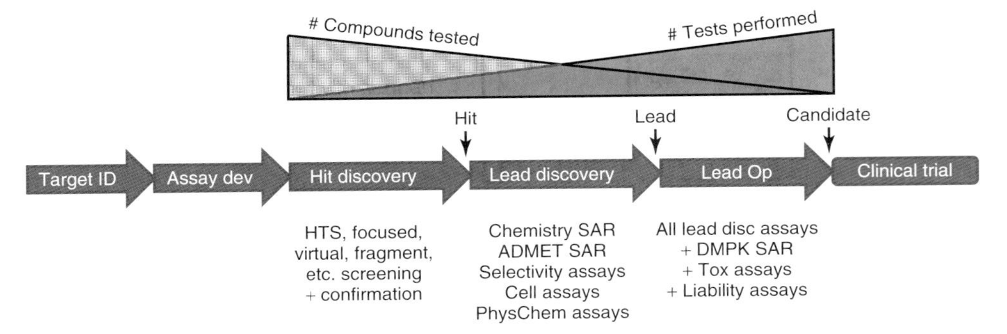

Figure 3.1 High-level schematic of a typical drug discovery process. Compound quality plays a critical role especially in the hit discovery, lead discovery, and lead optimization phases. As the cost per screened compound increases, so does the importance of screening 'the right compound, at the right concentration.'

costly false positives and negatives, is crucial at all discovery stages. However, ensuring that the solution being screened contains the right compound, of a high purity, at the intended concentration, is in itself a rather costly process. Clearly, a balance needs to exist between the cost of the analytical process and the cost and implications of screening a false positive/negative (cost in regards to both time and money). For each of the compound quality criteria – identity, purity, and concentration – we will discuss both 'how it is done' and criteria that can be used to determine an appropriate return on investment (ROI).

3.2
Process Quality and Analytical Quality in Compound Management

All three main properties generally associated with 'quality' of a compound – identity, purity, and concentration – are affected by how they are measured: what analytical technology is used, acceptance criteria for calibration curve, standards, and so on. Purity and concentration measurements also depend on the process used to make the solution being measured. For example, the actual purity value will depend on whether any impurity is more soluble in dimethyl sulfoxide (DMSO) than the compound itself, whether there is any carry-over in the liquid handling equipment used to make the solution, as well as the analytical method used to measure the purity. Similarly, the concentration of the DMSO stock will depend on the process used to make the solution (weighing the solid compound, adding an accurate amount of DMSO, and mixing), solubility of the compound and any impurity in DMSO, as well as the method used to measure the concentration.

The process quality is a reflection of the quality assurance (QA) procedures in place. QA embodies all the practices and operating protocols used to ensure that products of the process are within acceptable quality limits. Quality control (QC) evaluates the product of a process. In a compound quality context, QC is the analytical measurement of its identity, purity, and concentration, whereas QA refers to instrumentation checks performed routinely mainly on liquid handling equipment used in making the compound solutions.

3.2.1 Process Quality (QA)

Since both purity and concentration of a compound solution depend on the process used to make that solution, a rigorous QA program for the equipment used in the process is a must. Figure 3.2 depicts a common generic process used by compound management groups. The four main factors that influence the quality of a compound management process are: weighing, solubilization/mixing, liquid handling performance, and storage conditions.

Since the process (almost) always starts with a solid sample, weighing an aliquot of that sample is always the first step. Balances used are usually high-performance analytical balances, rated for 0.1 to 0.01 mg precision, which are calibrated on a regular basis for accuracy. It is important to remember, however, that the precision rating, even on high quality balances, does not reflect an actual compound management weighing process. For example, taking a 2 g tare vial on and off the scale to add the 1–5 mg of compound is certain to make the precision number worse than the manufacturer's rating. So while accuracy of the weight measurement can be ensured by a calibrated high performance balance, precision needs to be determined in the exact context of the process.

Once an aliquot of the solid has been placed in a vial destined for the solubilization process and the appropriate volume of DMSO has been added, a very important step to consider is the mixing of the solution to ensure full dissolution. Traditional mixing methods involve vortexing, sonication in a water bath, or solubilization through diffusion in time. More recently, adaptive focused acoustics (focused sonication) has been used successfully for mixing compound solutions. Two main

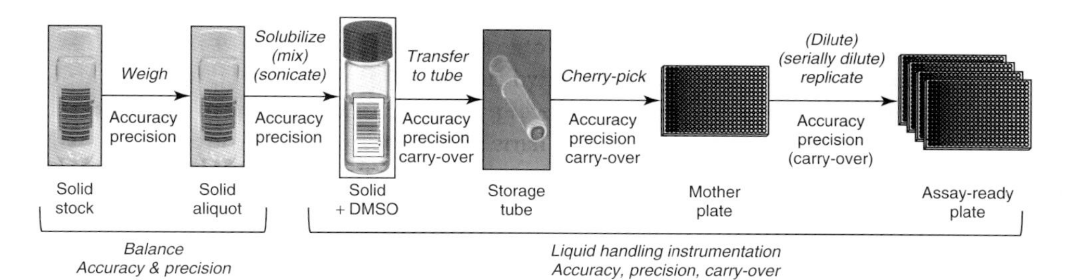

Figure 3.2 Generic compound management process. Balances used for solid weighing and all liquid handling equipment used for making and dispensing the solutions influence the quality of the sample used for biological testing.

types of high-throughput systems are available on the market: the SonicMan (Matrical, Spokane, WA) and C2000 (Covaris, Woburn, MA). The SonicMan uses disposable metal pins in a micro-titer plate format (96-, 384-, or 1536-well). The C2000's transducer is placed in a water bath and allows non-contact mixing of 4 mL vials or microtiter plates. Both instruments have proven that mixing is a crucial step in a compound management setting [2, 3].

As described by Oldenburg *et al.* [2] and Nixon *et al.* [3], mixing (in general, focused sonication in particular) will address kinetic solubility of a compound, but will not affect thermodynamic solubility much (if at all). What this means in practical terms is that if a solid will take hours to dissolve, sonication will speed that up to seconds/minutes (kinetic solubility). If a compound has crashed out of solution or it has limited DMSO solubility, sonication will have only a marginal effect on either bringing the solid back in solution, or making a more concentrated stock.

Performance of the liquid handling equipment used in making and dispensing any compound solution is also critical [4]. As illustrated in Figure 3.2, once a solid is weighed, all subsequent steps are performed with – and depend on – the various types of liquid handling equipment used. In most laboratories a software application will calculate the volume of DMSO needed to make an exact millimolar solution stock; however, the liquid handler instrument needs to dispense that volume accurately and precisely to each and every sample. So accuracy and precision [5–7] are the two main criteria used in assessing the performance of all liquid handlers. A third factor, sometimes overlooked because it applies to fixed (reusable) tip instruments only, is carry-over [8]. Since these types of instruments involve washing in between each step, but use the same tips for all dispenses, this can be a critical factor for optimal performance. Note: in liquid chromatography (LC) systems the autosampler is also a fixed-tip liquid handler, so the same factors will influence its performance – accuracy, precision, and carry-over.

For any robust liquid handling QA program, it is very important that the acceptance criteria (specification limits [7]) are appropriately set. Many laboratories consider a 10% precision and 10% accuracy of liquid dispense as adequate. However, it is not common that the question 'how does this translate into variability of the biological potency result?' is asked. It is very difficult to reverse-calculate the liquid handling specification limits for each of the multiple steps of the process based on acceptable variability in a biological potency. The most common procedure for setting these limits is through internal communications between screening groups, chemistry groups, project teams, and compound management.

Establishing specification limits for carry-over is somewhat different. For example, if screening is performed at a 10 μM maximum dose-response concentration, the requirement might be that 1 nM compound pipetted just before an inactive (10 μM) compound should not result in a false positive. The calculation is straightforward: carry-over should be less than $1\ \text{nM}/10\ \mu\text{M} = 10^{-4} = 0.01\%$. What is more difficult for carry-over QA is to use the right ('sticky') dye, in the appropriate linear range, to allow measurement of a 0.01% carry-over level. The most appropriate dye will be different when looking at carry-over of DMSO solutions versus buffer solutions. Tartrazine, Rhodamine green, Fluorescein, and Alexa 488 have all been used in reported studies. Each of these dyes has its own advantages and disadvantages in terms of linear dynamic range and 'stickyness.' Of these commonly used dyes, tartrazine is least soluble (hence 'sticky') in DMSO, while Rhodamine green has the worst buffer solubility. In our laboratories, tartrazine is the dye of choice for carry-over measurements on instruments where the system liquid is DMSO.

Over the past 10 years, there have been several literature reports on storage conditions and their influence on stability of compounds in solution. While storage in neat DMSO, at −4 or −20 °C, in dehumidified air seems to be the most commonly used systems, there are several other methods which are being successfully used. As previously pointed out [9], the most important thing is keeping the process consistent, which, in turn, will ensure that biological results will agree from run to run.

3.2.2 Analytical Quality (Sample QC)

It is widely agreed that solid samples are more stable in storage than their solution counterparts; however, all screening activities require solutions. According to a recent survey [10], 97% of pharmaceuticals and biotech discovery organizations store their compounds as 100% DMSO solutions, the only reported alternatives being 10% water/DMSO solutions [11], or dried [12]. DMSO has good solubilizing properties, but it is very hygroscopic. Many (early) discovery compounds have poor water solubility. As DMSO absorbs water, solubility of stock solutions also starts becoming a factor. Even though both stability and solubility would be much less of an issue, there are high costs associated with making up a fresh solution from solid every time a request is made for screening. Therefore, in all published studies to date, compound quality is assessed on the DMSO solutions in long- or short-term storage.

The same principles described above for liquid handlers apply to analytical instrumentation. Accuracy, precision, and carry-over are also the main criteria affecting the response/result. However, as Lane *et al.* [13] point out, analytical approaches also need to be specific, sensitive, and rapid. Specificity refers to the ability of analyzing the target compound, where the response is not being affected by impurities or reaction/method artifacts – this can be achieved mainly through chromatographic separation. Chromatographic separation, in turn, influences the speed and throughput of analysis, making it much slower than screening – minutes for a single sample versus minutes for a multi-well plate of 96–1536 samples. Sensitivity is another important factor to consider, with the various types of common detectors differing significantly from one to another.

As shown in Table 3.1, there are several different ways to assess the analytical quality of a compound. All three main properties measured – identity, purity, and concentration – depend on the analytical technology (detector) used, as well as the

Table 3.1 Ways of assessing analytical Quality/Identity, Purity/Stability, Concentration.

	Separation	Detection
Identity	(HPLC, UPLC)	MS, NMR, EA
Purity/Stability	HPLC, UPLC, SFC	UV, DAD, ELSD, DAD/ELSD, CLND, NMR
Concentration/Solubility	(HPLC)	CLND, ELSD, UV

Abbreviations: CLND, chemiluminescent nitrogen detector; DAD, diode-array detector; EA, elemental analysis; ELSD, evaporative light scattering detector; HPLC, high performance liquid chromatography; MS, mass spectrometry; NMR, nuclear magnetic resonance spectroscopy; SFC, supercritical fluid chromatography; UPLC, ultra-high performance liquid chromatography; UV, ultraviolet-visible spectroscopy.

performance of that instrument. Below we discuss types of detectors and separation instruments, and their respective performance in the context of assessing quality of a small molecule compound.

3.3 Identity

In today's drug discovery laboratories any small molecule synthesized (or isolated, in the case of natural products) has its structure identified by the chemist who made it. Purchased compounds from chemistry suppliers or other contract research organizations (CROs) also generally come with some form of a certificate of analysis (usually liquid chromatography–mass spectrometry (LC–MS) or nuclear magnetic resonance spectroscopy (NMR) indicating identity and purity). As previously mentioned, it has become an increasingly common practice across drug discovery organizations to employ an additional analytical QC for the DMSO solutions in the compound management (CM) repository [14–19].

Most discovery compounds are produced in small amounts (generally up to 100 mg), enough to allow basic testing of biological activity. One of the advantages of only synthesizing these small amounts is that it renders the process amenable to prep-LC purification, which is a fast and convenient method for small scale purifications [20]. Prep-LC purification systems have an MS detector, which means that a new compound has, from the very beginning, a confirmation of its identity.

Only a few years ago, it was common practice, after the last purification step, for chemists to take an NMR spectrum of the compound made. This is because NMR is by far the ultimate tool for structure elucidation. With the recent advancements in high-throughput synthetic chemistry, however, there are many practical drawbacks to using NMR for analyzing every compound synthesized: the instruments are very expensive, data analysis is complex and tedious, throughput is in the order of minutes per sample, sensitivity is low, a pure sample is required (otherwise interpretation is almost impossible), and the sample needs to be dissolved in a deuterated solvent. Some solutions do exist to address some of these drawbacks. For example, software is now available to help with assignment of peaks; high magnetic field, shimming techniques, improved probes, and so on can be used to help increase sensitivity [21], but all of these come at a cost. Even with all these advances, NMR (and LC–NMR) is not yet at a point where it can be widely applied to assessing compound identity on the solution in a CM repository.

Another technique which was used extensively just a few years ago is elemental analysis (EA). For organic compounds, the most common form of EA is C, H, and N analysis (and sometimes halogens or sulfur), which is accomplished through complete combustion of the sample in an excess of oxygen. The gases formed (carbon dioxide, water, N_2, and other oxides of nitrogen) are trapped, and the weight of these is used to calculate the composition of the unknown sample. Although this is a very precise technique, these days it is seldom used because it is not very sensitive (milligrams of material are required) and it is a destructive analysis (no sample can be recovered post-analysis).

A much simpler analysis tool to assess identity is MS [22], especially when coupled to a separation technique such as high-performance liquid chromatography (HPLC) or ultra-high-performance liquid chromatography (UPLC). MS instruments have the advantage of being high-throughput. They are more sensitive than other techniques, are amenable to automated data analysis, and can handle impure samples. They are also easy to operate and maintain. An MS analysis gives a spectrum containing an ionized form of the molecular weight (MW) of the compound. Each ion has a particular ratio of mass to charge, or m/z value. For most ions this ratio is one, so that m/z gives the mass of the molecule. Plotting the relative intensities of signals produced at the various m/z values gives the mass spectrum.

The intensity of the molecular ion will not give any indication of either the amount of compound, or purity of the sample. Indeed, sometimes the molecular ion is not even observed, but a correct fragmentation pattern for that ion is seen. Automated/integrated software to identify the molecular ion of a compound within a mass spectrum is very common, making this technique the tool of choice for a quick identity check.

Ions may be produced by a variety of methods. Whereas older techniques for the production of ions relied, typically, on a beam of energetic electrons to bombard the sample, these methods tended to result in high levels of fragmentation which, although allowing a greater amount of structural information to be deduced, often made it difficult to identify the molecular ion. For the analysis of pharmaceutical compounds, where often all that is required is a quick confirmation of MW, a softer ionization technique is generally used, one which results in much lower levels of fragmentation and thus a more easily identifiable molecular ion. This choice is largely between electrospray ionization (ESI) and atmospheric pressure chemical ionization (APCI), both of which are atmospheric pressure ionization techniques and both of which are easily interfaced with existing LC platforms (Figure 3.3).

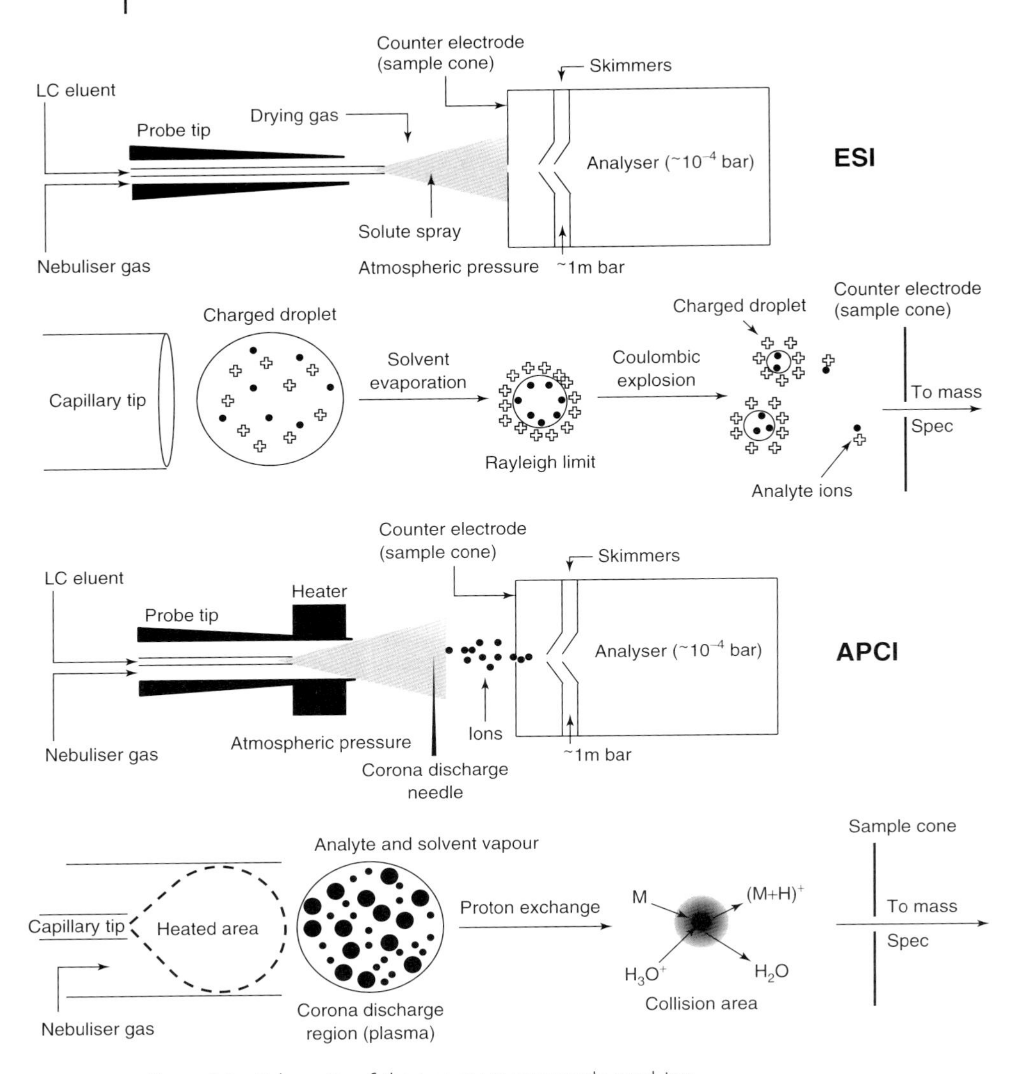

Figure 3.3 Schematic of the two most commonly used ionization techniques, ESI and APCI.

ESI is the favored ionization technique for small amounts of large and/or labile molecules such as peptides, proteins, organometallics, and polymers. In ESI a strong electrical charge is applied to the eluent as it emerges from a nebulizer, producing a spray of charged droplets. A heated sheath gas helps with solvent evaporation and reduces the size of the droplets until a sufficient charge density (the Rayleigh limit) makes the ejection of sample ions from the surface of the droplets possible (a 'Coulombic explosion' or ion evaporation). Ions can be produced singly or multiply charged, and the mass analyzer sorts them by m/z ratio. High-MW compounds are typically measured as ions with multiple charges. This allows mass spectra to be obtained for high-MW molecules. Both positive- and negative-ion spectra can be obtained, and ESI is commonly used in routine QC as diverse compounds are being analyzed. For positive-ion mode, 0.1% formic acid is usually added into the analyte solution to enhance protonation and increase sensitivity [22].

With APCI, protonated or deprotonated molecular ions are generally produced from the sample via a proton transfer (positive ions) or proton abstraction (negative ions) mechanism. The sample is vaporized in a heated nebulizer, again in the presence of a sheath gas, before flowing into a plasma consisting of solvent ions formed within the atmospheric source by a corona discharge. Proton transfer then takes place between the solvent ions and the sample.

In addition to the different types of ionization available, different types of mass detectors are also available. The most commonly used for pharmaceutical applications is the quadrupole mass filter, which consists of four circular rods arranged parallel to each other such that the ion beam from the source is directed axially between them (Figure 3.4). A DC current is applied to two opposite rods and an alternating RF field to the remaining two. Ions produced in the source are focused and pass between the rods. The motion of each ion is dependent on the electric fields applied to the rods such that only ions of a particular m/z will pass along the length of the rods to the detector. The mass range is scanned by changing the DC and RF components but keeping the DC/RF ratio constant, thus building up the complete mass spectrum.

Since ESI and APCI are soft ionization techniques and reveal little structural information, a triple quadrupole mass spectrometer is often used to induce fragmentation and thus provide more information for structural characterization. These are essentially two mass spectrometers in series, with a collision quadrupole in between. A number of experiments can be performed, for example, ions of a specific m/z can be filtered using the first quadrupole, these can be fragmented in the second quadrupole, and the third is set to scan the entire m/z range to detect the fragment produced.

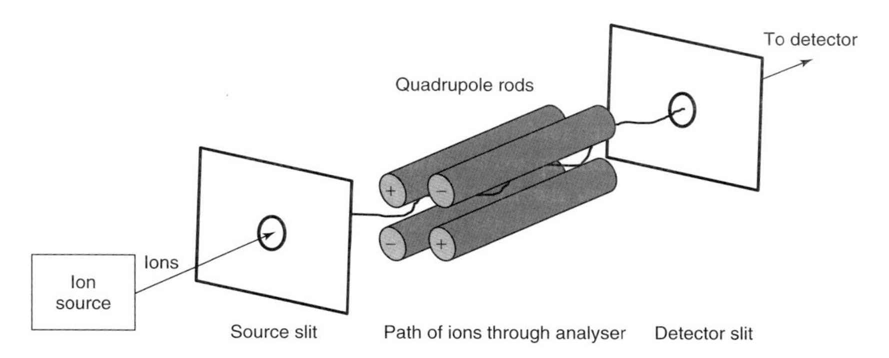

Figure 3.4 Basic schematic of a quadrupole analyzer.

Time of flight (TOF) is an alternative method of mass measurement that has been used in the past for analysis of combinatorial chemistry libraries [14]. Because selective ion monitoring (SIM) is not an efficient application of the TOF-MS [22], it has not been reported recently in any applications for identity or purity determinations in drug discovery.

Integrating the two techniques of LC and MS allows a combined purity and identity check to be performed in the same analysis. Although some employ a flow-injection analysis (no chromatographic separation) for purity check [15], the vast majority of analyses now carried out are performed by 'hyphenated' (i.e. integrated) LC–MS systems. Recent advances in column chemistry and manufacturing methods have allowed the production of smaller particle sizes, resulting in improved chromatography. The development of new systems (UPLC) able to utilize these advances means that improved separations/resolutions are now available at higher flow rates than would otherwise be possible, meaning that the time for individual analyses can be reduced to 2 min or less, with improved resolution over the longer method.

Previous studies [15], as well as data from our laboratories, indicate that, for many reputable, major suppliers of chemical libraries, there is a greater than 90% agreement between the compound identity and relative purity value determined by the supplier and the in-house confirmation. These studies were straightforward to accomplish because suppliers usually provide LC–MS data with the purchased compounds. The analytical confirmation of the CM samples is also an LC–MS analysis, so the data for both is readily available and the comparison is 'apples-to-apples.' Data presented at conferences indicate that the same is true of in-house samples: there is an approximately 90% agreement between the identity and relative purity data determined by the organic chemist and the confirmation study performed on the CM solution [23].

3.4
Purity/Stability

Purity and stability are almost always measured in the same way, purity being defined by the measurement at time zero (or at a specific time point) and stability by the measurement at a subsequent time point relative to the initial measurement. When looking at purity, there are two main factors to consider: (i) how is purity measured and how well was the target compound/peak separated from any impurities present and (ii) what should the cut-off criteria for initial purity be?

3.4.1
Measuring Purity

As Lemoff and Yan [24] point out, purity is always reported as a number (most commonly as a percentage of the target compound weight in the total sample weight), which does not have any meaning unless one knows exactly the details of how it was measured. The most common method involves reverse-phase chromatographic separation and an MS detector used to identify the exact peak corresponding to the target compound, as well as the actual detector (or detectors) used to determine 'weight' [15–25].

Chromatographic separation is the main factor limiting throughput of any analytical analysis. However, without separation, detectors would not be able to distinguish between mixtures, so it is a critical component of any method. While some indication of identity is possible (identification of target molecular ion, for example), purity analysis in particular could not be performed without separation. Table 3.2 below summarizes some of the applications, advantages and disadvantages of the main reported separation techniques used in a CM setting.

The most common detectors used for purity determination are the ultraviolet-visible spectroscopy/diode-array detector (UV/DAD) and the evaporative light-scattering detector (ELSD), or a combination thereof. The area under the curve (AUC) is automatically measured by the instrument's software for each peak in the chromatogram. The purity number is then expressed as a percentage of the AUC for the target compound versus the sum of the AUCs for all detected peaks

Table 3.2 Main separation techniques used to assess compound quality.

Chromatograph	Applications	Advantages	Disadvantages
HPLC	Open Access, routine QC, method development, manufacturing, ADME screening, bioanalysis, clinical, food safety, analysis of complex mixtures	Sensitivity, selectivity, robustness, lower cost	Low throughput due to longer run times
UPLC	Open Access, routine QC, method development, manufacturing, ADME screening, bioanalysis, clinical, food safety, metabo(l/n)omics, analysis of very complex mixtures	Increased resolution, speed and sensitivity. Smaller amounts of solvent required.	High pressure, robustness, cost
SFC[33]	Analysis and purification of low to moderate molecular weight, thermally labile molecules. Analysis of chiral compounds	No requirement for organic solvents. Fast high resolution separations. Easy separation of mobile phase from separated compounds.	Strongly polar and ionic molecules are not dissolved by supercritical gases

in the chromatogram. This will result in a relative purity number which assumes that all compounds give an equivalent response in the particular detector used. Note: generally, for isomers, the AUCs will be added up and the sum will be used to calculate purity [26].

Because there is no high-throughput detector (NMR is not included in this category) that will give an equivalent response to all compounds [13], the only way absolute purity can be measured is by obtaining the weight of the total sample and then measuring the weight of the target compound after chromatographic separation using an analytical technique to determine absolute concentration (see below). This is not practical for many reasons, including the small amounts of compound available and the throughput and limitations of absolute concentration detectors (NMR and chemiluminescent nitrogen detector (CLND)). All of the detectors used for relative purity measurement (Table 3.3) will therefore give only an estimate of the purity, due to the fact that the target compound and any impurities will not have the same exact response.

UV detectors are very commonly used to assess relative purity, either using a single wavelength, or a range of wavelengths from a DAD. A clear advantage of using a UV detector is that it is non-destructive, sensitive, simple, robust, and relatively inexpensive, and almost any LC system has one. For single-wavelength UV, the response factors (and hence the AUC in a chromatogram) of the target compound and any impurities present will depend on the extinction coefficients of their respective chromophore(s). Extinction coefficients tend to diverge more at higher wavelengths, so a wavelength between 200 and 220 nm is commonly used [14, 16, 18, 20, 26]. It has been shown [25] in a direct comparison that purity assessment using UV_{254} overestimates the relative purity values compared to UV_{214}, since 214 nm is more likely to be closer to the common chromophore for many organic compounds. Another approach used [17] involves averaging of relative purities determined at two different wavelengths (214 and 254 nm), but it is not clear how much better this approach is compared to using a single wavelength in the 200–220 nm range. In terms of UV detection, a DAD [27–29] is the best choice to estimate purity. While the reported wavelength range used varies somewhat (e.g., 220–400 [16], 260–320 [24], 240–320 [27], and 210–350 nm [28]), the overall approach ensures that most chromophores are covered within the range.

A close second to the UV detector in terms of its widespread availability and ease of use is the ELSD shown in Figure 3.5. While some laboratories use it as the primary detector for relative purity assessment [30, 31], others use it as a back-up detector for compounds with a weak UV chromophore at the wavelength or range of wavelengths used [15, 28]. Several reports have compared the performance of the ELSD as an universal (single calibrant) quantitative detector with other detectors [13, 23, 24]. The general consensus is that ELSD overestimates the purity measurement, most likely due to the fact that compounds with low MWs (a category into which many synthetic intermediates, starting materials, and decomposition products fit) are more volatile, which results in a lower response than the one for the target compound. An alternative approach [24] takes an average of the relative purities

Table 3.3 Detectors used in a drug discovery environment to determine purity of a compound.

Purity detector	Advantages	Disadvantages	Reference
Single wavelength UV	Simple, robust, non-destructive, integrated with all chromatographs	Compounds need to have a UV chromophore. Wavelengths of 200 – 220 nm are commonly used, and these tend to underestimate purity. Higher wavelengths (254 nm) overestimate purity	[14, 16, 18, 20, 25, 26]
DAD	Same as above, plus most accurate relative purity determination by UV	Still a relative purity value, actual purity will depend on extinction coefficients of compound of interest and all impurities	[15, 27, 28, 29]
ELSD	Simple, robust, integrated with HPLC and UPLC	Overestimates relative purity, many low molecular weight (<300 Da) compounds tend to give a weak response.	[30, 31]
Combination ELSD/DAD	Study indicates most accurate relative purity value	Two detectors being used simultaneously and data averaged. Assumes that ELSD always overestimates purity and DAD underestimates it.	[24]
CLND	Universal quantitative detector, relatively inexpensive	Compound needs to contain nitrogen, value not accurate for N-N bonds, instrument robustness, commercial detector only integrated with HPLC, can not use acetonitrile as mobile phase	[14]
NMR	Ultimate universal quantitative detector	Expensive, needs specialized operators, data interpretation not fully automated, low throughput, low sensitivity, deuterated solvent.	[13, 21]

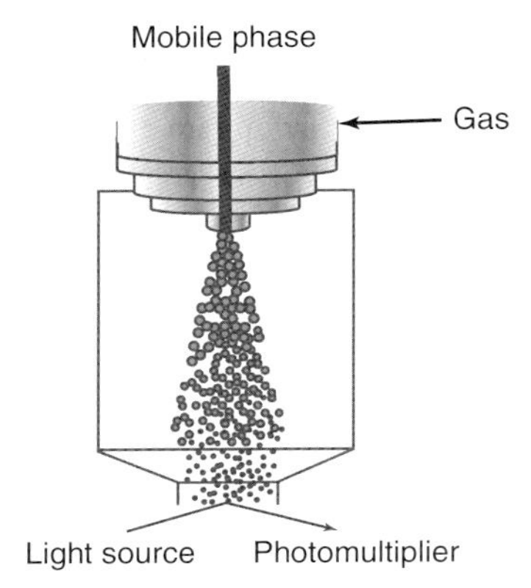

Figure 3.5 Basic schematic of an ELSD.

determined by DAD and ELSD. The premise for this is that DAD underestimates purity for the set of compounds tested, whereas ELSD overestimates purity, so an average of the two will come closest to absolute purity.

One more detector in the charged-aerosols space worth mentioning is the Corona charged-aerosol detector (CAD) [32]. Since this is an aerosol detector, it has the same limitations as the ELSD (very poor sensitivity for volatile (many low-MW) compounds), and response varies with mobile phase composition. However, it is a relatively simple and robust instrument and has the advantage of being able to detect certain compounds that are not seen by either UV or ELSD, but it does not have the widespread use of either one of the latter two detectors.

The CLND is mainly used for measurements of concentration/solubility rather than purity [13–15, 18, 19, 28]. It involves combustion of the sample in the presence of oxygen (Figure 3.6). The nitrogen in an analyte becomes nitric oxide. Nitric oxide is in turn reacted with ozone to form excited nitrogen dioxide, and the detector measures the decay of excitation. During the combustion process, significant amounts of water are formed. To allow analysis of the gases only, the water needs to be separated, a process which leads to high complexity of the instrument itself and results in high maintenance overheads for these detectors.

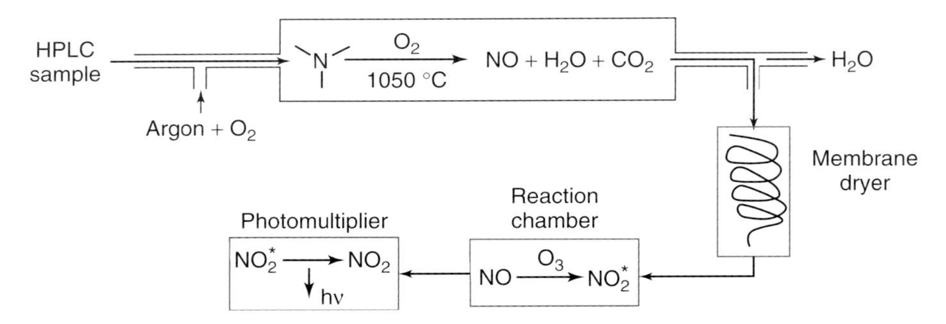

Figure 3.6 Basic schematic of a CLND.

The CLND provides a linear response to the absolute amount of nitrogen in a sample. The only exception to this is due to nitrogen–nitrogen double bonds (such as imidazoles, tetrazoles, etc.), which upon combustion yield in part N_2, which is not measured by the detector. Because the response is proportional to the number of moles of nitrogen in the analyte, the MW and the number of nitrogens needs to be known. The advantage of using a CLND detector is that, as a single calibrant detector, it provides an absolute concentration or purity, as opposed to the relative purity measurements offered by the UV or ELSD detectors.

The CLND is used as a hyphenated LC detector; however, the technique is limited by the fact that only non-nitrogen-containing solvents can be used, meaning that acetonitrile cannot be used (methanol/water is the most common solvent system). It has also not been successfully integrated with a UPLC system, so the throughput for this detector is significantly lower than in the case of UV and ELSD. In many applications, a flow-injection mode is used for the HPLC–CLND system to allow higher throughput. Even though several studies have evaluated the CLND for its potential as a universal quantitative detector (see Ref. [13] and references therein) and have found it to be superior to both UV and ELSD (and comparable to NMR) in terms of its accuracy, for the reasons outlined above, this detector is not widely used for either purity or concentration estimation.

3.4.2
Determining the Most Appropriate Purity Cut-Off for Solutions

What is (and is there) a 'right' cut-off? This is difficult to answer as again it is a cost/benefit balancing act. Initial purity of a sample is critical when considering the cost of screening/following-up on a false positive. It is also important as it affects the stability of the solution in storage (unpublished data from our laboratories clearly indicate that the less pure a compound is at time zero, the more likely it is to degrade faster in storage). Clearly, one would want to aim for as pure a compound as possible, but one also needs to be pragmatic and ensure purity is fit for purpose. For a screening collection the QC cut off is generally 60–80% (most commercial suppliers will only sell compounds of >90% relative purity), but most compounds are 95% or better (Figure 3.7). Raising the purity cut-off for compounds in a screening collection could potentially exclude hard-to-synthesize compounds and limit diversity. Therefore, for a screening collection, the most important thing to consider is the purity of an identified screen active to eliminate false positives.

For later drug discovery stages, where the number of compounds is lower, the diversity is more limited (only selected scaffolds are advanced) and many more screens are performed on any one compound, it is far more critical to have a high purity cut-off. It has a positive effect on stability of the sample in solution, it allows identification of an undesired synthetic impurity, and it is the best overall deterrent for false positive results with costly consequences. Data presented at conferences by several pharmaceutical and biotech companies indicate that most in-house chemists aim for a 95% cut off (and many compounds are actually better than this) in post-HTS phases.

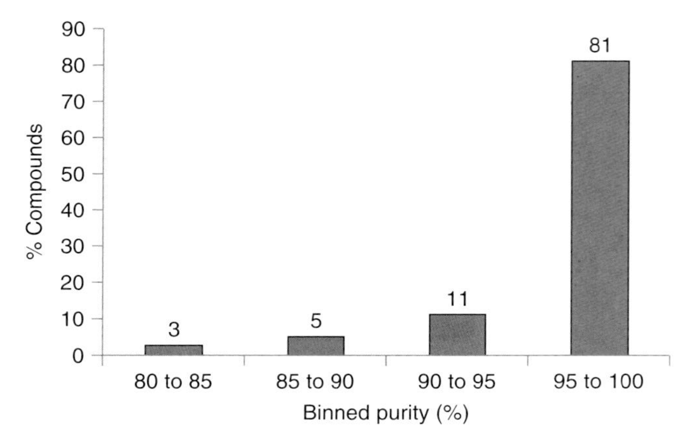

Figure 3.7 Purity distribution for >150 K samples measured at the time of addition to the GSK collection.

3.4.3
Stability of Solutions

Deciding how long a stored solution can be used for testing is another question that needs to be answered by each individual drug discovery organization, and it is intrinsically linked to the initial purity cut-off decision. The question of what is acceptable is an almost philosophical question. For example, is it OK if >95% of compounds are still >80% pure after *x* years? Or should >90% compounds be within 10% of initial purity? Or can a collection be stored for *x* years if the probability of >50% degradation is <5%? To add to this complexity, not all solutions are made at the same time, not all initial purity is the same, stability is usually predicted based on shorter-term studies and a limited number of compounds and is assumed to be the same across all chemical diversity. Even with all these complexities, it has been proven many times over that long-term storage is not only possible, it clearly provides high-quality solutions for testing at a relatively low cost (for a review of published studies, see Ref. [9]).

The lifetime of a DMSO solution depends on the storage conditions: temperature, number of freeze/thaws, and water absorption. There are very significant costs associated with both setting up a CM store to minimize any of the factors above and replenishing a solution collection from solid material. So the shelf life of solution sample collections has to be a balance between storage of compounds as long as possible and removing the compounds once they have reached the end of the shelf life. At GSK [28], we have shown that the rate of degradation decreases significantly after the first two years at −20 °C, so we store and maintain master 10 mM DMSO stocks for >7 years at −20 °C, while replenishing room-temperature 1 mM plate-based stocks for HTS supply every three months, as shown in Figure 3.8 below.

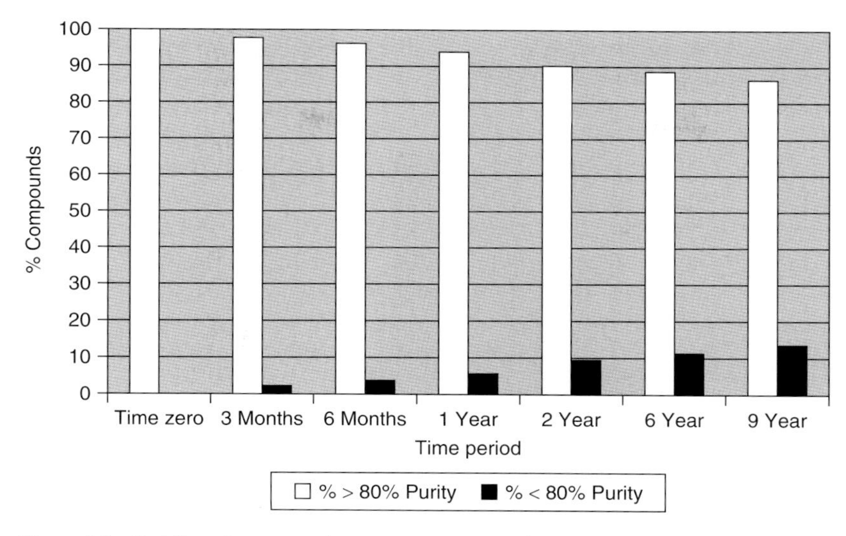

Figure 3.8 Stability of compounds over nine years in the GSK master liquid store.

In terms of analytical methods used to measure stability, there are two subtly different approaches: (i) measuring purity on the same solution at time zero and again at a subsequent time point and comparing the two values or (ii) measuring stability as a relative value of purity at time zero. Most published studies use the first approach, but the second approach [33] is certainly a very interesting proposition, which can be easily achieved using only an UV/DAD detector. The advantage of the first approach is that it gives an actual purity number at the time point of interest, independently of any measurement at time zero. It is the most appropriate approach to take when following up on hits from a campaign or if any false positive issue is suspected. The second approach may be the simplest to implement in large studies interrogating the stability of collections over time.

Overall, in the absence of a robust universal quantitative detector, there is no right or wrong way of measuring purity and/or stability or an absolute way of deciding what is acceptable in one's particular laboratory. It is a balance between the potential risk of obtaining false positives and the cost of screening only very pure compounds.

3.5
Concentration/Solubility

There is a real need in drug discovery for high-throughput quantification methods. In a CM setting, it is important to know the concentration of solutions stored, while in a screening laboratory an accurate potency determination cannot be made without knowing the actual concentration of a compound in solution. Without an actual measurement, it is always assumed that compounds are 100% soluble in

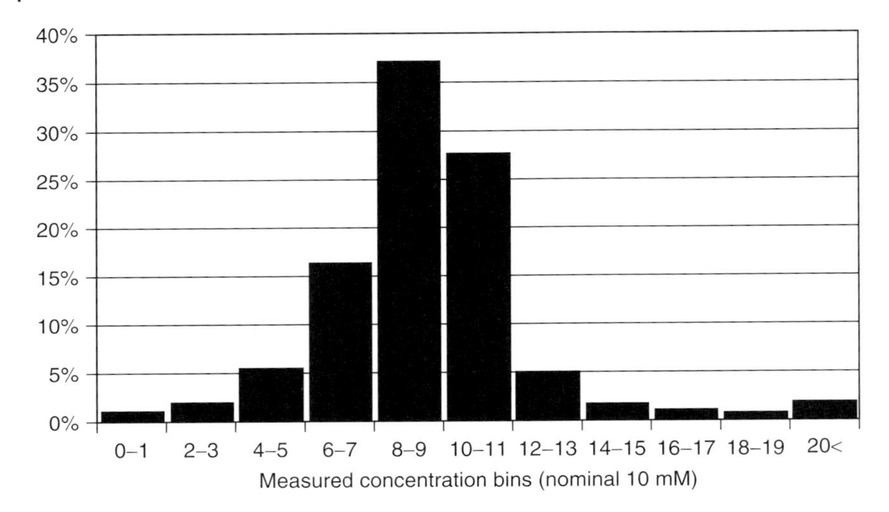

Figure 3.9 Concentration distribution of discovery compounds tested over a one-year time period at GSK.

DMSO, which is not always the case. Figure 3.9 shows an example of concentration distributions across samples processed in our laboratories over a one year period. The nominal concentration for all these samples is 10 mM.

To take it a step further, all biological screening is performed in aqueous solutions containing 1% or less DMSO, where solubility is far poorer than that in DMSO. It has been shown [15] that ~80% of a diverse set of compounds have a concentration of <20% in a buffer solution compared to the same compound set in DMSO.

To date, a true universal quantitative detector suitable for the high-throughput analysis of CM solutions does not exist. While the CLND comes closest to that requirement, it has many practical limitations: poor robustness due to the need to remove all water traces, it cannot be used with nitrogen-containing LC mobile phases, it has limited throughput because it has not been integrated with UPLC. To add to the practical limitations, the CLND can only detect nitrogen-containing analytes and underestimates the signal for compounds containing adjacent nitrogens. However, it is the most accurate high-throughput detector currently on the market, and it is being used by several groups to measure concentration of compounds in both DMSO [14, 19, 23] and aqueous [15] solutions. While the quest for the best universal detector continues, recent reported solutions include: limited use of the HPLC–CLND detector for late-stage discovery compounds only, use of the CLND detector without LC separation (flow-injection mode – concentration value obtained is subsequently adjusted for relative purity measured using a different detector [28]) to allow higher throughput, and limited use of the CLND detector for aqueous solubility measurements [15] and building prediction models [34].

Other potential approaches include: relative quantitation of a compound solution at a particular time-point or dilution compared to a time zero or a more concentrated stock solution (unpublished data from our laboratories), and a 'self-calibrated LC/MS/UV method' [33]. The former does not constitute a true concentration measurement, but it can be a very appropriate surrogate in certain applications and it allows concentration measurements at much lower concentrations than those measurable by CLND. The second approach is an actual concentration determination; however, this requires two analyses: first to establish the extinction coefficient of the compound, to then allow calculation of the concentration on a subsequent HPLC–UV analysis.

3.6 Conclusions

Process quality and analytical quality are both important. Process quality affects all stages of drug discovery equally (all samples prepared, stored, and handled in a centralized CM system, whether HTS, lead discovery, or lead optimization compounds), so even though expensive, the investment is worth it. An ROI analysis is needed to determine the size and scope of any capital investment in long-term storage, sophisticated liquid handling, and so on. There are many solutions available in the market place that will suit various needs. For example, for a small compound library, it may be perfectly acceptable to use a manual storage rack/upright freezer system. Or, if performing mainly high-volume, lower throughput assays, manual pipetting of solutions might give a high enough quality of dispensing without the need for automated low-volume dispensing.

Analytical quality has a different impact depending on the stage of an investigation, since it is a lot more costly to deal with a false positive or negative later in the discovery cycle. An ROI analysis is needed here also to determine the appropriate instrumentation (one might not need a state-of-the-art QTOF, but a single quad might suffice at a hit discovery stage), the throughput needed (an UPLC would probably not be necessary for lead optimization, where the number of samples analyzed is relatively low, so a 10–15 min LC run would be acceptable), and the type of analysis (identity and purity only, relative or absolute purity, concentration in DMSO, and concentration in buffer). It should also be noted that, on both the process quality and the analytical quality fronts, there are many reputable CROs which will handle and dispense compounds and which will measure purity and concentration of solutions, so capital investment is not always a must.

Acknowledgments

We gratefully thank Brenda Ray for her review of the manuscript and support of the project.

References

1. Lipinski, C. and Hopkins, A. (2004) Navigating chemical space for biology and medicine. *Nature*, **432**, 855–861.
2. Oldenburg, K., Pooler, D., Scudder, K., Lipinski, C., and Kelly, M. (2005) High throughput sonication: evaluation for compound solubilization. *Comb. Chem. High Throughput Screen.*, **8**, 499–512.
3. Nixon, E., Holland-Crimmin, S., Lupotsky, B., Chan, J., Curtis, J., Dobbs, K., and Blaxill, Z. (2009) Applications of adaptive focused acoustics to compound management. *J. Biomol. Screen.*, **14**, 460–467.
4. Hentz, N. (2008) The importance of liquid handling quality assurance through the drug discovery process. *Drug Discov. World Spring*, 27–31.
5. Taylor, P.B., Ashman, S., Baddeley, S., Batram, S.L., Battle, C.D., Bond, B.C., Clements, Y.M., Gaul, N.J., McAllister, E., Mostacero, J.A., Ramon, F., Wilson, J.M., Hertzberg, R.P., Pope, A.J., and Macarron, R. (2002) A standard operating procedure for assessing liquid handler performance in high-throughput screening. *J. Biomol. Screen.*, **7**, 554–569.
6. Quintero, C., Rosenstein, C., Hughes, B., Middleton, R., and Kariv, I. (2007) Quality control procedures for dose-response curve generation using nanoliter dispense technologies. *J. Biomol. Screen.*, **12**, 891–899.
7. Popa-Burke, I., Lupotsky, B., Boyer, J., Gannon, W., Hughes, R., Kadwill, P., Lyerly, D., Nichols, J., Nixon, E., Rimmer, D., Saiz-Nicolas, I., Sanfiz-Pinto, B., and Holland, S. (2009) Establishing quality assurance criteria for serial dilution operations on liquid-handling equipment. *J. Biomol. Screen*, **14**, 1017–1030.
8. Weibel, R., Iten, M., Konig, I., Beckbissinger, R., Benthien, T., Halg, W., Ingenhoven, N., Lehnert, A., Oeltjen, L., and Zaborosch, C. (2010) Development of standard test procedures for quantifying carry over procedures from fixed pipetting tips in liquid-handling systems. *J. Assoc. Lab. Autom.*, **15**, 369–389.
9. Janzen, W.P. and Popa-Burke, I.G. (2009) Advanced in improving the quality and flexibility of compound management. *J. Biomol. Screen.*, **14**, 444–451.
10. Comley, J. (2004) *Compound Management Trends Report*, HTStec.
11. Schopfer, U., Engeloch, C., Stanek, J., Girod, M., Schuffenhauer, A., Jacoby, E., and Acklin, P. (2005) The Novartis compound archive – from concept to reality. *Comb. Chem. High Throughput Screen.*, **5**, 513–519.
12. Waybright, T.J., Britt, J.R., and McCloud, T.G. (2009) Overcoming problems of compound storage in DMSO: solvent and process alternatives. *J. Biomol. Screen.*, **14**, 708–715.
13. Lane, S., Boughtflower, B., Mutton, I., Paterson, C., Farrant, D., Taylor, N., Blaxill, Z., Carmody, C., and Borman, P. (2005) Toward single-calibrant quantification in HPLC. A comparison of three detection strategies: evaporative light scattering, chemiluminescent nitrogen, and proton NMR. *Anal. Chem.*, **77**, 4354–4365.
14. Yurek, D.A., Branch, D.L., and Kuo, M.S. (2002) Development of a system to evaluate compound identity, purity, and concentration in a single experiment and its applications in quality assessment of combinatorial libraries and screening hits. *J. Comb. Chem.*, **4**, 138–148.
15. Popa-Burke, I.G., Issakova, O., Arroway, J.A., Bernasconi, P., Chen, M., Coudurier, L., Galasinski, S., Jadhav, A.P., Janzen, W.P., Lagasca, D., Liu, D., Lewis, R.S., Mohney, R.P., Sepetov, N., Sparkman, D.A., and Hodge, C.N. (2004) Streamlined system for purifying and quantifying a diverse library of compounds and the effect of compound concentration measurements on the accurate interpretation of biological assay results. *Anal. Chem.*, **76**, 7278–7287.
16. Choi, B.K., Ayer, M.B., Siciliano, S., Martin, J., Schwartz, R., and Springer, J.P. (2005) Interpretation of high-throughput liquid chromatography mass spectrometry data for

quality control analysis and analytical method development. *Comb. Chem. High Throughput Screen.*, **8**, 467–476.

17. Isbell, J.J., Zhou, Y., Guintu, C., Rynd, M., Jiang, S., Petrov, D., Micklash, K., Mainquist, J., Ek, J., Chang, J., Weselak, M., Backes, B.J., Brailsford, A., and Shave, D. (2005) Purifying the masses: integrating prepurification quality control, high-throughput LC/MS purification, and compound plating to feed high-throughput screening. *J Comb. Chem.*, **7**, 210–217.
18. Letot, E., Koch, G., Falchetto, R., Bovermann, G., Oberer, L., and Roth, H.J. (2005) Quality control in combinatorial chemistry: determinations of amounts and comparison of the 'purity' of LC-MS-purified samples by NMR, LC-UV and CLND. *J Comb. Chem.*, **7**, 364–371.
19. Lane, S.J., Eggleston, D.S., Brinded, K.A., Hollerton, J.C., Taylor, N.C., and Readshaw, S.A. (2006) Defining and maintaining a high quality screening collection: the GSK experience. *Drug Discov. Today*, **5/6**, 267–272.
20. Schaffrath, M., von Roedern, E., Hamley, P., and Stilz, H.U. (2005) High-throughput purification of single compounds and libraries. *J. Comb. Chem.*, **7**, 546–553.
21. Holzgrabe, U. (2010) Quantitative NMR spectroscopy in pharmaceutical applications. *Prog. NMR Spectrosc.*, **57**, 229–240.
22. Niessen, W.M.A. (2003) Progress in liquid chromatography – mass spectrometry instrumentation and its impact on high-throughput screening. *J. Chromatogr. A*, **1000**, 413–436.
23. Lewis, K. Paper presented at CoSMos 2006, San Diego, CA.
24. Lemoff, A. and Yan, B. (2008) Dual detection approach to a more accurate measure of relative purity in high-throughput characterization of compound collections. *J. Comb. Chem.*, **10**, 746–751.
25. Yan, B., Fang, L., Irving, M., Zhang, S., Boldi, A.M., Woolard, F., Johnson, C.R., Kshirsagar, T., Figliozzi, G.M., Krueger, C.A., and Collins, N. (2003) Quality control in combinatorial chemistry: determination of the quantity, purity, and quantitative purity of compounds in combinatorial libraries. *J. Comb. Chem.*, **5**, 547–559.
26. Matson, S.L., Chatterjee, M., Stock, D.A., Leet, J.E., Dumas, E.A., Ferrante, C.D., Monahan, W.E., Cook, L.S., Watson, J., Cloutier, N.J., Ferrante, M.A., Houston, J.G., and Banks, M.N. (2009) Best practices in compound management for preserving compound integrity and accurately providing samples for assays. *J. Biomol. Screen.*, **14** (5), 476–484.
27. Zitha-Bovens, E., Maas, P., Wife, D., Tijhuis, J., Hu, Q.-N., Kleinoder, T., and Gasteiger, J. (2009) COMDECOM: predicting the lifetime of screening compounds in DMSO. *J. Biomol. Screen.*, **14** (5), 557–565.
28. Blaxill, Z., Holland-Crimmin, S., and Lifely, R. (2009) Stability through the ages: the GSK experience. *J. Biomol. Screen.*, **14** (5), 547–556.
29. Bowes, S., Sun, D., Kaffashan, A., Zeng, C., Chuaqui, C., Hronowski, X., Buko, A., Zhang, X., and Josiah, S. (2006) Quality assessment and analysis of Biogen Idec compound library. *J. Biomol. Screen.*, **11** (7), 828–835.
30. MacArthur, R., Leister, W., Veith, H., Shinn, P., Southall, N., Austin, C.P., Inglese, J., and Auld, D.S. (2009) Monitoring compound integrity with cytochrome P450 assays and qHTS. *J. Biomol. Screen.*, **14** (5), 538–546.
31. Waybright, T.J., Britt, J.R., and McCloud, T.G. (2009) Overcoming problems of compound storage in DMSO: solvent and process alternatives. *J. Biomol. Screen.*, **14** (6), 708–715.
32. Gorecki, T., Lynen, F., Szucs, R., and Sandra, P. (2006) Universal response in liquid chromatography using charged aerosol detection. *Anal. Chem.*, **78**, 3186–3192.
33. Qi, M., Zhou, H., Ma, X., Zhang, B., Jefferies, C., and Yan, B. (2008) Feasibility of a self-calibrated LC/MS/UV method to determine the absolute amount of compounds in their storage

and screening lifecyle. *J. Comb. Chem*, **10**, 162–165.

34. Hill, A.P. and Young, R.J. (2010) Getting physical in drug discovery: a contemporary perspective on solubility and hydrophobicity. *Drug Discov. Today*, **15** (15/16), 648–655.

Further Reading

White, C. and Burnett, J. (2005) Integration of supercritical fluid chromatography into drug discovery as a routine support tool. *J. Chromatogr. A*, **1074**, 175–185.

4
Delivering and Maintaining Quality within Compound Management

Isabel Charles

4.1
Introduction

By the 1980s, pharmaceutical companies had established methods for storage and recording of their proprietary collections of solid and liquid samples that would enable faster access to compounds for biological screening. A centralized model, based on simple rack and shelf stores with accompanying paper records, was gaining preference over the bench-based sample storage regimes still found in many academic research laboratories. The accelerated growth of company collections in the 1990s, triggered by requirements to feed high-throughput screening (HTS), then confirmed the ascendancy of centralized custodianship of compounds, with trends toward standardization of vessel sizes, control of storage temperature and humidity, and electronic management of data. Weighing and delivery of dry formats, originally the major activity in compound dispensing, was soon exceeded by the liquid handling activities for HTS and secondary screening. A combination of requirements for storage and dispensing, together with the need to regulate compound registrations and sample requests, led many pharmas to invest in dedicated compound management (CM) facilities and staff. For more than 20 years such capabilities have sought to improve, and gain a leading edge in maintaining, the quality of the compounds in their care.

An ability to search electronic registration databases provided company scientists with much greater visibility of the compounds adding to inventories and collections, not only from the traditional in-house syntheses and external purchases but also from consolidation of samples during company mergers. The expanding choice of compounds was accompanied by an expectation that they would be available for wider screening, possibly at multiple geographic locations, and would remain accessible (and, by implication, retain their physical integrity) in the centralized collection over longer periods of time. With hindsight, this expectation has had a major impact in shaping CM strategies [1–5]. Thus, assessment of different storage and dispensing environments, monitoring of purities and concentrations, and improvements in handling technologies and techniques have all contributed hugely to our understanding and optimization of compound longevity.

Management of Chemical and Biological Samples for Screening Applications, First Edition.
Edited by Mark Wigglesworth and Terry Wood.

The integrity of samples derived from in-house chemistry and compound acquisition is now a primary focus for any CM group, and in particular those groups that share responsibility, alongside computational and medicinal chemistry groups, for determining membership of their company's solubilized collection for routine screening. Starting from a large bank of compounds, several million in some pharmas, critical decisions must be made in regard to which samples are eligible for long-term liquid storage and, consequentially, become generally available for the hit discovery process. A default approach is to create solutions of all compounds and to monitor them over time, replacing or removing samples from the solubilized collection if they fall short of required standards of purity and concentration. This carries a risk, however, of overlooking other attributes, and allowing eligibility by physical criteria alone to dominate the purposeful inclusion of samples in a screening set.

4.2 What is Quality from a Compound Management Perspective?

It is widely recognized that biological screening interests will rarely extend across an entire historical bank of compounds. Indeed, many large company collections are now divided into multiple subsets with varying orders of preference in relation to different screens, and screening the whole solubilized collection may be perceived as a last resort when the hit rate is negligible. Most collections of both liquids and solids also contain compounds that are rarely, if ever, requested. This invites an extended definition of quality, from merely physical parameters such as purity to a more complex assessment of intrinsic value. In our current business climate, with pressures mounting upon pharmaceutical companies to find new drugs more quickly, success will require not only confidence in the physical integrity of the compounds, but also a credible capability for drug-hunting in unexplored regions of chemical and biological space, and the generation of lead compound series that can rise to the challenges of the late stage drug pipeline. These indicators of quality from a medicinal chemistry perspective are no less important than physical integrity, and, in scrutinizing the current production lines feeding compound screening, we need to challenge some old assumptions about the independence of sample management and medicinal chemistry objectives.

For example, simply restricting screening to those compounds that are the best survivors in solution may impact the population of whole clusters of lead-like compounds, constraining collection diversity and limiting the spectrum of biological activities that can be probed. Although the elements of lead-like compounds are outside the scope of this chapter [6], taking a balanced approach that considers significance within the collection profile, as well as individual physical characteristics, should be part of our definition of high-quality compounds (Figure 4.1). A CM strategy that molds itself to the widest possible range of 'desirable attributes' may require new and inventive ways to nurture more eccentric candidates through the early stages of the pipeline, but the experience already amassed by CM groups

	Positives for sample handling		Positives for medicinal chemistry
O	Intrinsically stable to light.	✡	"Lead-like" properties e.g. Lipinski rule-of-five.
O	Highly soluble in DMSO.	✡	Populating unexplored chemical space.
O	No precipitation in ageing solutions or on dilution for assays.	✡	Similar compounds screened to develop structure-activity relationship hypotheses.
O	Stable in solution at ambient temperature.	✡	No toxicity indicators.
O	Not adsorbed from solution into labware vessel materials.	✡	3D properties e.g. pharmacophores, rotatable bonds, polar surface area.
O	Not affected by polymers leaching from labware materials.	✡	Screening complementary fragments.

Figure 4.1 A balanced approach to quality.

in optimizing and maintaining physical quality can be brought to bear upon this challenge.

4.3
Storage and Delivery of Samples in Solution

Achievements by CM groups in delivering and maintaining high-quality solutions are already extensively documented in earlier publications, with an excellent cross-section of pharma experience being provided by a special edition of the Journal of Biomolecular Screening in 2009 [7]. These experiences range widely across liquid storage formats both in tubes and in microtiter plates with different types of seals. Varying storage times are explored under different environmental conditions (including temperature, humidity, and number of freeze/thaw cycles), and the use of dimethyl sulfoxide (DMSO) as solvent is examined through differing approaches to the control of its water content [8–10].

Common drivers underlying these and other studies [1–5, 11–14] that have influenced the approaches taken by many CM groups, including AstraZeneca, are: maximizing the length of time that compounds are stored (and hence controlling the costs of consumables and resources), reducing the number of low-purity samples dispensed to screening, and optimizing the efficiency of dispensing and handling of solutions. As Janzen and Popa-Burke have articulated [15], the perfect compounds from a sample handling perspective would be those that are pure and unaffected by solubilization, dilution, storage, or dispensing. Under such ideal circumstances, modern automation and handling techniques would then commute the physical integrity into an accurate quantitative sample delivery [16–18].

In the real world of managing solutions, there is still much to learn about factors influencing compound solubility in DMSO [19]. This solvent was originally preferred for its ability to solvate an extensive range of compound types and has, by virtue of its success, become the default choice around which CM automation has developed. However, in comparison to our understanding of aqueous solubility [20] and creation of tools to predict it [21, 22], the forecasting of solubility in DMSO is still in its infancy. Cautious extrapolation from one compound to another has been attempted with some limited success for interrelated structures [23–25], but in general there are no calculations demonstrating reliable predictivity from structural parameters alone.

DMSO is notorious for being an extremely hygroscopic solvent, and, depending on the humidity of the environment, absorption of atmospheric moisture can increase its water content leading to a corresponding increase in solution volume. Alongside the propensity of some solvated compounds to precipitate once moisture is absorbed, this creates significant problems in maintaining steady sample concentrations in DMSO solutions, and CM strategies thus focus strongly on limiting the fluctuations from these causes. Microtiter storage plates are particularly vulnerable to moisture uptake if they have their plate seals frequently removed and replaced, for example, to allow repeated dispenses by acoustic droplet ejection [26, 27]. For this reason, storage time in plates is often limited to months, rather than the years anticipated for tube storage. The problem can be mitigated by short-term storage in daughter plates that are used for regular onward dispensing, allowing the mother plates to be desealed less frequently and thus delaying their moisture uptake.

Precipitates also commonly lead to blocked pipetting tips in contact liquid dispensing systems, resulting in inaccurate solution transfers and sometimes requiring system downtime to restore correct operation. Although more aggressive dissolution techniques such as sonication are often employed to increase the rate of solubilization for susceptible compounds, the combined risks of low concentration and disruption of sample throughputs may lead these compounds to be excluded from a solubilized collection. Furthermore, certain compounds display accelerated degradation in DMSO solution once moisture has been absorbed [10, 28], and the resultant purity losses may also be a signal for their withdrawal from screening. While these responses are entirely understandable, they highlight the impact of sample handling issues on the admission of compounds to screens, and pose the question of whether important chemical attributes for the collection profile should be recognized and weighted within this process of elimination. It is not unknown, for example, for some biological screening customers to accept assay-ready plates with visible precipitation in the wells, on the basis that the next steps of their assay entail addition of reagents causing the material to redissolve.

Assay-ready plates enjoy advantages over the plates and tubes that are subjected to longer term storage. Typically, sub-microliter transfers are performed from 'working' plates (which may be mother or daughter plates created from the main liquid store) into assay-ready output plates by acoustic droplet ejection in a controlled temperature and humidity environment, after which the output plates are quickly sealed and dispatched to the customer. Initial sample integrity in the prepared plates is therefore governed primarily by integrity in the main liquid store and intermediate working plates. Apart from rare opportunities to perform a snapshot analysis of the entire collection [2], the monitoring of sample purity and concentration in a large store tends to rely upon extrapolation of analytical results from a representative subset of the collection. Exceptions to this approach are granted to compounds entering the lead identification and optimization stages which demand better confidence. For example, in the AstraZeneca CM group at Alderley Park all such compounds are subjected to 'just-in-time' analyses by liquid chromatography–mass spectrometry (LCMS) coupled to an ultraviolet (UV) spectrophotometer detector and a charged aerosol detector (CAD) [29].

For newly-created intermediate and assay-ready plates it is generally assumed that sample purities mirror the results seen from the main liquid store. Delays introduced by additional sampling from these plates for analyses would not be acceptable to customers. The subsequent handling of plates within different companies then follows their preferences on temperature and humidity control for general liquid storage [1–5, 11–14]. One notable point of agreement is on thermal plate sealing rather than the use of plate lids.

Once assay-ready plates are dispatched to customers, the subsequent control of environmental conditions and handling protocols is critical. When immediate use by local customers is anticipated, plates will be delivered at ambient temperature, whereas shipped plates are most likely to be frozen in dry ice. It is standard practice at AstraZeneca to provide handling guidelines to all new customers, including advice on thawing and centrifuging under reduced humidity and desealing the plates at the latest possible time before use. Also, evidence of compounds leaching from certain types of plate materials into the sample solutions led to a recommendation to use the plates within a maximum of three weeks [29].

4.4 Intercepting Low Purity

Low-purity solutions may be derived from impure solids, or sometimes from a degradation occurring upon dissolution in DMSO, although identifying and separating these two types of event usually requires the investment of extra resource. Preliminary pointers for solution-triggered degradation, though, are provided by general monitoring of collection quality (typically by LCMS–UV). For example, a computational clustering of structures for those samples showing degradation over time will generate some indications of any structure-related liabilities. More directly, analyses of compound libraries have already shown that certain structural motifs clearly increase the propensity for degradation [30]. This information can be used in a variety of ways: firstly, to track down similar compounds in the whole collection that may be susceptible to DMSO solution instability; secondly, to guide the preferences of future library design where choices between scaffolds or substitution patterns are, in other respects, equivalent; and thirdly, where a structural element relating to instability is still a strongly desired component of the compound, we have advance indication of the need for special handling – perhaps solubilization on a 'just-in-time' basis, or the use of an alternative solvent. However, this third option requires a real commitment to include compounds that are considered high-value and worthy of additional CM resources.

When the solid itself is found to be impure, additional effort to purify the most valued members of the collection can be invoked. In 2008–2009 AstraZeneca undertook such a venture to resuscitate several thousand previously impure samples by chromatographic purification. The cost-benefit of revitalizing these important compounds, that would otherwise have been lost from screening, could be calculated in terms of alternative costs to resynthesize an equivalent number of valuable compounds, either in-house or externally.

The nature of certain compound impurities throws up opportunities to take preventative action. For example, trifluoroacetic acid (TFA) has been observed as a contaminant in compounds when the final purification stages of synthesis have been carried out by unwary preparative chromatographers (e.g., in some outsourced compound synthesis). TFA is a regular additive to analytical reversed-phase chromatography solvent systems to improve peak separation, but as a product contaminant it is well known to catalyze degradation once the sample has been dissolved [31, 32]. Substituting TFA with a more volatile trace acid in the preparative chromatography system will help avoid the unwanted contamination (Figure 4.2).

Another problem seen in some compound syntheses has been the presence of residual silica from preparative normal-phase thin-layer chromatography. Again, early detection enables alternative purification methods to be adopted. Such early warnings are readily provided by dedicated quality assurance/quality control staff working alongside CM. Typical analytical systems employ LCMS–UV detection and some type of non-UV recognition, such as charged aerosol or chemiluminescent nitrogen detection (CAD or CLND) [10, 29].

It has been shown that simply tightening the purity requirement for compounds entering the screening collection will improve the average lifetime of the solutions [28], and this must be seen as encouragement for synthetic chemistry to provide the highest possible physical quality before dissolution. It may even offer hope that improving the purity of desirable compounds that are notoriously unstable in solution may be a way to increase their survival rate. The consequences for expanding collection diversity could be significant.

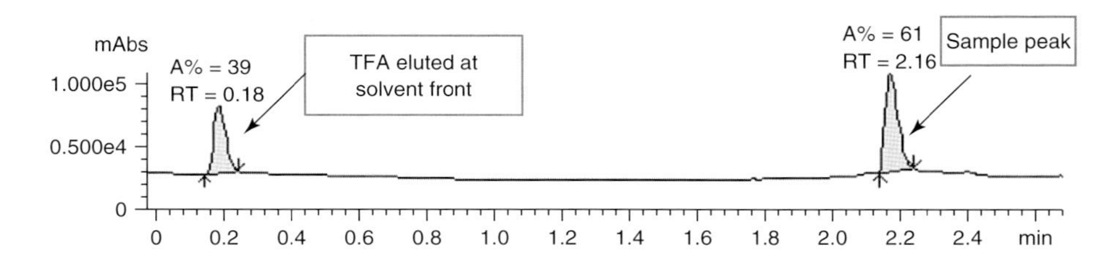

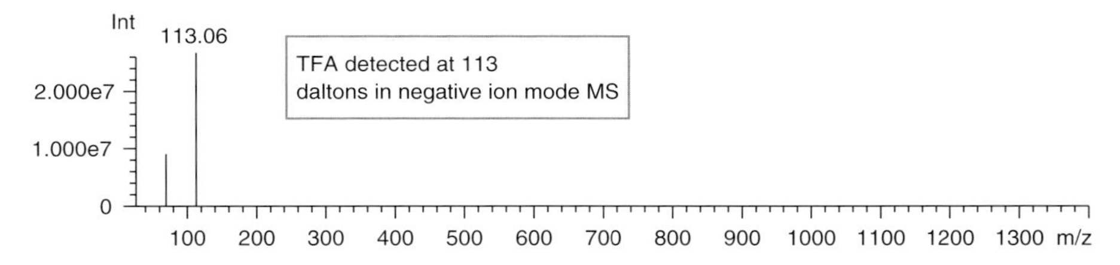

Figure 4.2 CAD chromatogram showing the presence of TFA in a sample, and the mass spectrum for this peak.

4.5
Storage and Delivery of Solids

Although the delivery format for biological screening leans heavily toward solutions, the numbers of stored solids in larger companies have continued to rise into the millions. Basic housekeeping such as expunging compounds defined within the company as 'ugly,' reducing the numbers of duplicate samples, or dissolving very low quantities of solids to store as solutions instead, will only hold back the tide temporarily. Rationalization of a modern solids store may only come to the fore when volume of growth threatens to overload the store capacity or efficiency. One solution to the threat is simply to create a larger archive space, either internally or by outsourcing some or all of the collection, as undertaken by Pfizer [5]. A logical next step, for those bold enough to consider it, would be to designate certain solids as expendable once they attain a certain lack of demand. For example, the number of solid requests on project-specific compounds often plummets once the lead optimization phase is complete. Apart from requests on material needed for *in vivo* formulations, or as analytical reference standards, residual demand mainly comes from the replenishment of fresh liquids in the general screening set. The quantities needed for such replenishments are small and could be satisfied by limiting solid to a single vial per compound, allowing the solids collection to be managed more interactively, like a library rather than a museum. The inevitable storage of solids in vials or bottles can also be challenged. Armed with a clear identification of compounds that are no longer required in multi-milligram quantities, CM can consider other storage options instead. For example, the sample could be transferred in solution to alternative vessels (e.g., microtubes or plate wells) and dried down to a film or lyophilized solid [33]. Very small aliquots of solid may thereby be stored and individually dispensed. They could also be re-dissolved in a solvent chosen for delivery, that need not be DMSO. Alternatively, a sample that is very soluble and stable in solution could be stored entirely as a liquid rather than a solid, at a high concentration that will be diluted at the point of dispense. The option to dry down and redissolve in a different solvent is also available.

These examples merely illustrate alternative processes to the historical default of weigh–dissolve–store–dispense. Efficiencies gained from molding a delivery strategy around one process flow should not obscure, or limit, our exploration along other avenues to meet customer expectations of sample delivery on demand, in the right format, with high confidence in physical integrity.

4.6
Automation Quality Control and Reliability

The use of a standard operating procedure (SOP) with every instrument and sample handling system is a mainstay of CM good practice that helps to minimize the variables introduced at manual stages within processes. Fortunately, automated storage and handling of samples is well advanced for many high-throughput CM activities, and they now require minimal human interaction. For example, the placement and retrieval of vessels within an automated solids or liquids store, the dispensing of liquids, and the updating of data for amounts recorded in input and output vessels are, in high-throughput environments, typically controlled and executed by a combination of robotics and software. Confidence in the automation is thus absolutely paramount. Whilst the robustness of dispensary hardware is reflected in manufacturer success and satisfied users over time [34], its operational stability relies upon a well-managed schedule of routine maintenance, visual and automated checks, and precision and accuracy testing.

Barcoding of every vessel is a prerequisite for efficient automated handling and problem resolution. For example, checking the barcodes of vessels on the way into and out of an automated store, and comparison with expected parameters, means that any issues in placing and retrieving samples should be highlighted immediately. Incidents such as dropped vessels or misalignments into their seating positions often cause automation downtime. Such occurrences can be usefully monitored by control charts [35] that will help the users to recognize systematic patterns and special-cause variations, using the principles of Statistical Process Control (SPC). A surge in vessel misalignments, for example, coinciding with a new batch of tube racks, may be linked to deviations in the rack dimensions. A problem found to occur on a regular basis may be linked to other activities that occur with the same frequency. The average lifetimes of hardware parts can, through such monitoring, become fairly predictable, so that breakdowns can be preempted by timely replacements. In summary, control charts can be an effective way to address multiple deviations that impact upon operational continuity (Figure 4.3).

The attention of staff to signs of spillage, dust accumulation, loose seals or stoppers, and so on, are of course inherent to good CM practice, and visual checks

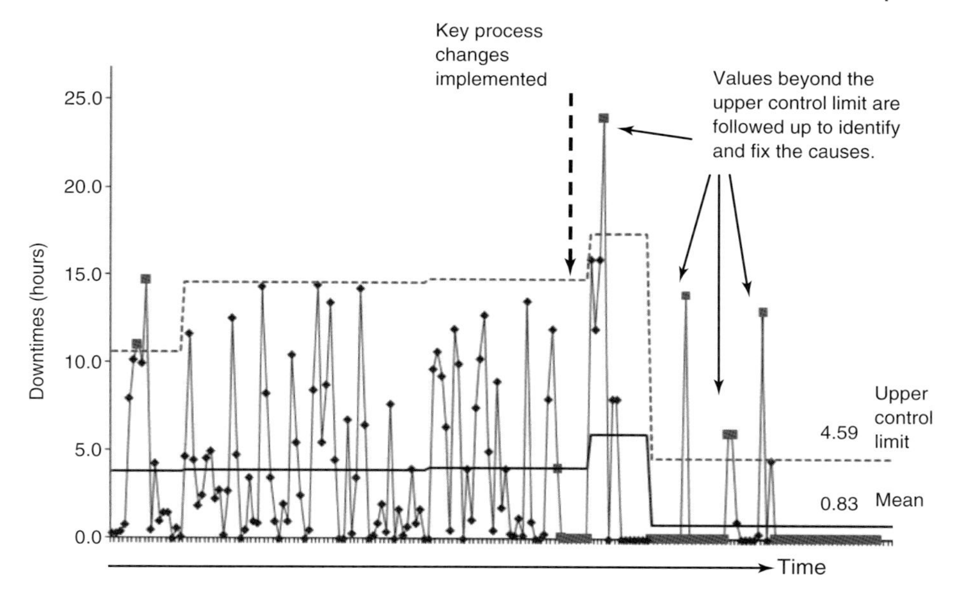

Figure 4.3 Example of a control chart addressing multiple interruptions to a process flow. Cumulative downtimes are recorded at regular intervals, for example, weekly, fortnightly, or monthly. Patterns and outliers are identified and their causes investigated. The impact of specific changes can be recognized.

will underpin quality assurance in the laboratory. Regular preventative maintenance on hardware is also essential, and may require planned downtime on one or more occasions per year, scheduled outside the peaks of customer demand to minimize the impact. Software changes require rigorous testing in a simulation environment before their release [36]. Purges on the operational data may also be needed: vessel transactions (relating to the instrument movements involved in dispensing, rather than the sample inventory) create a large excess of information over and above the data required to record the sample dispense history. Removing superfluous data on a scheduled basis helps to optimize the performance of the transactional database and the ordering system software.

For contact dispensing liquid handling systems that are expected to be continuously available for use, manual checks along with testing for precision and accuracy are routine. A key indicator of instrument robustness is repeatability within the normal range of volumes dispensed. To take a simple example, the reliability of a system transferring between 50 and 1000 μL by automated pipetting can be assessed using gravimetric analysis. The delivered amounts from a series of test dispenses are weighed, and a coefficient of variance (CV) calculated from the collected data. Accuracy can be measured at the same time, using the density of DMSO to calculate the weight predicted for the dispensed amount and comparing it with the actual weight. To avoid manual weighing, these operations are usually incorporated into the instrument software protocols, with appropriate hardware additions. For example, integrating a balance and a receptacle for repeated test dispenses allows data to be easily captured for a precision and accuracy read-out. The criteria for acceptable performance of the instrument must be set in relation to the process where it will be used [16, 17].

Recently, the ability to audit liquid content in tubes has been enhanced by the introduction of a camera-based technology that can be integrated into high-throughput settings [37]. Tubes used for liquid storage and dispensing may have had their content data revised on multiple occasions, and the instrument offers an opportunity to reconfirm the residual volumes. It also provides an alert to precipitation within solutions. Take-up and reliability of this technology will be followed with interest.

Quality control of acoustic dispensing equipment embraces a wider range of variables. As well as the accuracy of the dispense volume, the targeting of the droplet into the output plate is critical. Deviations off-center in the receiving well can mean that subsequent mixing of the droplet with added assay reagents is not efficient. Standard quality assurance routines therefore include checking the alignment and firing power for droplet ejection. The precision of repeated dispenses is readily measured using a fluorescent dye solution [27, 38]. Accuracy checks rely on the comparison of a time-of-flight measurement with standard parameters provided by the manufacturer.

4.7 High-Quality Data Management

It would be remiss to discuss the quality of a sample collection without mentioning the management of data. Not only compound transactions but also technical data can be managed by more efficient information systems. For example, some CM groups pool and share their data from routine precision and accuracy tests so that instrument models can be assessed on performance over time with a much wider variety of users and conditions. However, it is the management of sample data that has seen the biggest revolution. From thousands of solid weighings per year, through hundreds of thousands of plate wells with microliter liquid dispenses, to modern nanoliter dispensing into millions of wells in assay-ready plates, the number of samples requested for screening since the 1980s has rocketed. A capability to cross-reference samples in dispensed formats back to their original storage vessels, along with all their associated quality assurance data, is a major asset. It provides vital evidence in deciphering assay anomalies arising from different sources of samples, and helps to distinguish these events from problems with the assay reagents or from unexpected contamination by labware materials [39].

Opportunities to link information on samples occupying multiple locations at multiple sites can truly bring together global dispensing as a single cooperative.

It is relatively easy, for example, to use analytical capabilities at two sites in tandem so that low purity or concentration results on a sample at one location will automatically trigger a check on the same sample at other locations. AstraZeneca's two centralized compound stores in the United Kingdom and Sweden already collaborate in this way.

The use of modern informatics also allows comprehensive integration of compound data, including their selection preference ratings for screening based on structure and lead-like properties, as well as their physical quality and quantity. It is now feasible to identify those compounds that are considered high-value from a medicinal chemistry vantage point, but that clearly require different management techniques for their routine delivery to screening. While HTS has more relaxed criteria on 'acceptable' sample integrity (than, for example, a screen measuring concentration responses on individual compounds), the possibility of finding a cluster of similar structures with real hits will be aided by samples of high physical quality populating every possible cluster. The common strategy of removing samples from screening when they fail purity and concentration criteria may actually inhibit representation of all clusters, or deplete those that need more members to provide confidence in deriving a structure-activity relationship hypothesis. Population of a screening set with all possible clusters from the compound bank, containing representative compounds certified for purity and concentration, would indeed be a bold commitment.

4.8 Conclusion

The definition of high-quality compound delivery means substantially more than dispensing a correct compound at the expected purity and concentration. The approach of most pharmas has been to optimize the percentage of compounds from their collection of solids and liquids that can be readily dispensed on demand to meet the physical requirements of the screen. This has successfully powered the HTS engine and, with more individual customization, the lead identification and optimization stages of the drug pipeline. It may continue to do so for many years.

However, the balance of compounds entering screening should be kept under continuous review [40] and not be obscured by efforts to cultivate a screening set of high physical quality aimed primarily at dispensing more conveniently and quickly. It is questionable whether the smooth-running delivery machine could quickly adapt, for example, to dispensing large numbers of compounds that have DMSO solubility and stability challenges. Other formats for storage and dispensing, such as lyophilized solids or alternative solvents, should be explored further to gain an advantage in the more demanding areas. It is gratifying to reflect on past and current successes in handling and dispensing samples, but we should be mindful of challenges that could lie ahead, and respect those obstinate samples that could yet be our leads of the future.

Acknowledgments

Information provided by Ian Sinclair (Compound Management Group, AstraZeneca, Alderley Park) is most gratefully acknowledged.

References

1. Spencer, P.A. (2004) The challenges of managing a compound collection. *Eur. Pharm. Rev.*, **9** (2), 51–57.
2. Lane, S.J., Eggleston, D.S., Brinded, K.A., Hollerton, J.C., Taylor, N.L., and Readshaw, S.A. (2006) Defining and maintaining a high-quality screening collection: the GSK experience. *Drug Discov. Today*, **11** (5/6), 267–272.
3. Houston, J.G., Banks, M.N., Binnie, A., Brenner, S., O'Connell, J., and Petrillo, E.W. (2008) Case study: impact of technology investment on lead discovery at Bristol-Myers Squibb, 1998–2006. *Drug Discov. Today*, **13** (1/2), 44–51.
4. Schopfer, U., Engeloch, C., Stanek, J., Girod, M., Schuffenhauer, A., Jacoby, E., and Acklin, P. (2005) The Novartis compound archive – from concept to reality. *Comb. Chem. High Throughput Screen.*, **8** (6), 513–519.
5. Burr, I., Winchester, T., Keighley, W., and Sewing, A. (2009) Compound management beyond efficiency. *J. Biomol. Screen.*, **14** (5), 485–491.
6. Oprea, T.I. and Hann, M.M. (2004) Pursuing the leadlikeness concept in pharmaceutical research. *Curr. Opin. Chem. Biol.*, **8** (3), 255–263.
7. Rigby, C. and Holland-Crimmin, S. (ed) (2009) *J. Biomol. Screen.*, **14** (5).
8. Matson, S.L., Chatterjee, M., Stock, D.A., Leet, J.E., Dumas, E.A., Ferrante, C.D., Monahan, W.E., Cook, L.S., Watson, J., Cloutier, N.J., Ferrante, M.A., Houston, J.G., and Banks, M.N. (2009) Best practices in compound management for preserving compound integrity and accurately providing samples for assays. *J. Biomol. Screen.*, **14** (5), 476–484.
9. Pfeifer, M.J. and Scheel, G. (2009) Long-term storage of compound solutions for high-throughput screening by using a novel 1536-well microplate. *J. Biomol. Screen.*, **14** (5), 492–498.
10. Blaxhill, Z., Holland-Crimmin, S., and Lifely, R. (2009) Stability through the ages: the GSK experience. *J. Biomol. Screen.*, **14** (5), 547–556.
11. Kozikowski, B.A., Burt, T.M., Tirey, D.A., Williams, L.E., Kuzmak, B.R., Stanton, D.T., Morand, K.L., and Nelson, S.L. (2003) The effect of room temperature storage on the stability of compounds in DMSO. *J. Biomol. Screen.*, **8** (2), 205–209.
12. Kozikowski, B.A., Burt, T.M., Tirey, D.A., Williams, L.E., Kuzmak, B.R., Stanton, D.T., Morand, K.L., and Nelson, S.L. (2003) The effect of freeze/thaw cycles on the stability of compounds in DMSO. *J. Biomol. Screen.*, **8** (2), 210–215.
13. Cheng, X., Hochlowski, J., Tang, H., Hepp, D., Beckner, C., Kantor, S., and Schmitt, R. (2003) Studies on repository compound stability in DMSO under various conditions. *J. Biomol. Screen.*, **8** (3), 292–304.
14. Morand, K.L. and Cheng, X. (2004) Organic compound stability in large, diverse pharmaceutical screening collections, in *Analysis and Purification Methods in Combinatorial Chemistry*, vol. 63, (ed. B. Yan), John Wiley & Sons, Inc., **163**, pp. 323–350.
15. Janzen, W.P. and Popa-Burke, I.G. (2009) Advances in improving the quality and flexibility of compound management. *J. Biomol. Screen.*, **14** (5), 444–451.
16. Popa-Burke, I., Lupotsky, B., Boyer, J., Gannon, W., Hughes, R., Kadwill, P., Lyerly, D., Nichols, J., Nixon, E., Rimmer, D., Saiz-Nicholas, I., Sanfiz-Pinto, B., and Holland, S. (2009) Establishing quality assurance criteria for serial dilution operations on

liquid-handling equipment. *J. Biomol. Screen.*, **14** (8), 1017–1030.

17. Quintero, C., Rosenstein, C., Hughes, B., Middleton, R., and Kariv, I. (2007) Quality control procedures for dose-response curve generation using nanoliter dispense technologies. *J. Biomol. Screen.*, **12** (6), 891–899.
18. Eriksson, H., Brengdahl, J., Sandström, P., Rohman, M., and Becker, B. (2009) Validation of low-volume 1536-well assay-ready compound plates. *J. Biomol. Screen.*, **14** (5), 468–475.
19. Lipinski, C. (2006) Presentation at LRIG Conference, February 2, 2006, New Jersey. *http://lab-robotics.org/Presentations/Lipinski%200602/Lipinski_LRIG%20Feb%202,%202006.pdf* (accessed 20 December 2010).
20. Bhattachar, S.N., Deschenes, L.A., and Wesley, J.A. (2006) Solubility: it's not just for physical chemists. *Drug Discov. Today*, **11** (21/22), 1012–1018.
21. Delaney, J.S. (2005) Predicting aqueous solubility from structure. *Drug Discov. Today*, **10** (4), 298–295.
22. Katritzky, A.R., Kuanar, M., Slalov, S., Hall, C.D., Karelson, M., Kahn, I., and Dobchev, D.A. (2010) Quantitative correlation of physical and chemical properties with chemical structure: utility for prediction. *Chem. Rev.*, **110** (10), 5714–5789.
23. Balakin, K.V., Ivanenkov, Y.A., Skorenko, A.V., Nikolsky, Y.V., Savchuk, N.P., and Ivashchenko, A.A. (2004) In silico estimation of DMSO solubility of organic compounds for bioscreening. *J. Biomol. Screen.*, **9** (1), 22–31.
24. Balakin, K.V., Savchuk, N.P., and Tetco, I.V. (2006) In silico approaches to prediction of aqueous and DMSO solubility of drug-like compounds: trends, problems and solutions. *Curr. Med. Chem.*, **13** (2), 223–241.
25. Zitha-Bovens, Z., Maas, P., Wife, D., Tijhuis, J., Hu, Q., Kleinöder, T., and Gasteiger, J. (2009) COMDECOM: predicting the lifetime of screening compounds in DMSO solution. *J. Biomol. Screen.*, **14** (5), 557–565.
26. Zaragoza-Sundqvist, M., Eriksson, H., Rohman, M., and Greasley, P.J. (2009) High-quality cost-effective compound management support for HTS. *J. Biomol. Screen.*, **14** (5), 509–514.
27. Heron, E., Ellson, R., and Olechno, J. (2006) Acoustic Droplet Ejection in Drug Discovery. Drug Plus International (April/May 22–25).
28. Holland-Crimmin, S. (2010) Presentation at Cambridge Healthtech Institute Compound Management Forum, December 7–8, 2010, Providence, Rhode Isand.
29. Sinclair, I. and Charles, I. (2009) Applications of the charged aerosol detector in compound management. *J. Biomol. Screen.*, **14** (5), 531–537.
30. Sinclair, I. (2010) Presentation at Pharma IQ (IQPC) Compound Management and Integrity Conference, May 18–19, 2010, London.
31. Isbell, J. (2008) Changing requirements of purification as drug discovery programs evolve from hit discovery. *J. Comb. Chem.*, **10** (2), 150–157.
32. Hochlowski, J., Cheng, X., Sauer, D., and Djuric, D. (2003) Studies of the relative stability of TFA adducts vs non-TFA analogues for combinatorial chemistry library members in DMSO in a repository compound collection. *J. Comb. Chem.*, **5** (4), 345–349.
33. Bowes, S., Sun, D., Kaffashan, A., Zeng, C., Chuaqui, C., Hronowski, X., Buko, A., Zhang, X., and Josiah, S. (2006) Quality assessment and analysis of Biogen Idec compound library. *J. Biomol. Screen.*, **11** (7), 828–835.
34. Comley, J. (2005) Compound management in pursuit of sample integrity. *Drug Discov. World*, **6** (2), 59–78.
35. Tague, N.R. (2005) Seven basic quality tools, in *The Quality Toolbox (2nd ed)*, Quality Press, p. 15.
36. Walsh, C. and Ratcliffe, S. (2011) Automated compound management systems: maximizing return on investment through effective life-cycle planning. *J. Lab. Autom.*, **16** (5), 387–392.
37. (2010) Measuring sample volume and detecting precipitates. *Lab Manager Magazine* (May edition), p. 74.
38. Harris, D. and Mutz, M. (2006) Debunking the myth: validation of fluorescein for testing the precision of nanoliter

dispensing. *J. Assoc. Lab. Autom.*, doi: 10.1016/j.jala.2006.05.006

39. Watson, J., Greenough, E.B., Leet, J.E., Ford, M.J., Drexler, D.M., Belcastro, J.V., Herbst, J.J., Chatterjee, M., and Banks, M. (2009) Extraction, identification, and functional characterization of a bioactive substance from automated compound-handling plastic tips. *J. Biomol. Screen.*, **14** (5), 566–572.

40. Snowden, M. and Green, D.V.S. (2008) The impact of diversity-based, high-throughput screening on drug discovery: 'Chance favours the prepared mind'. *Curr. Opin. Drug Discov. Dev.*, **11** (4), 553–558.

5
Obtaining and Maintaining High-Quality Tissue Samples: Scientific and Technical Considerations to Promote Evidence-Based Biobanking Practice (EBBP)

Lisa B. Miranda

5.1
Introduction

5.1.1
Current Issues and Impediments to Benchmark Level Biospecimen Research

High-quality, well-characterized annotated biological samples, or 'biospecimens' are, without doubt, crucial to support a vast range of biomedical, bench, and clinical research platforms. Such platforms aim to foster and further advance key areas in personalized, molecular, and translational medicine. Key areas include, but may not be limited to, conduct of clinical trials, drug discovery and development, as well as biomarker identification and validation-based research. In recent years, it has been speculated that the poor quality of many biospecimen collections and related data may be impeding advances. A popular approach to addressing this issue has been the charge for cultivation of 'benchmark' level and/or evidence-based collections. Evidence-based collections are predicted to fulfill requisite refinements to accelerate progress of such research platforms.

However, several prominent issues, both in design and practice, currently impede the capability to cultivate benchmark level biospecimen collections. These issues plague academic, industrial, and government sectors alike. It should be noted that not only do these issues impede efficacious scientific advances and collaboration, they limit true understanding of the relevance and value of the specimen and biosample collections. From a fiscal perspective, such impediments contribute to potentially unsustainable sample management and biomedical research practices prone to tremendous economic risk in regards to return on investment.

One underlying issue is that no known global consensus has been agreed as to what parameters constitute a 'high-quality' biospecimen or 'well-characterized' annotation. Exacerbating this problem is the lack of publicly defined benchmark criteria for composition of specimen collection sets, discrete biospecimen requirements, associated clinical data points or common data elements. There are also

Management of Chemical and Biological Samples for Screening Applications, First Edition.
Edited by Mark Wigglesworth and Terry Wood.

no definitive or evidence-based quality parameters for evaluation of biospecimens and data. Initial recommendations might aim to remedy this via guidance as to standard sample sets of biospecimens and data collections. This can be referred to as a *minimal biospecimen collection and data set*. Following this guidance, customized criteria would also be helpful, particularly if categorized per organ type, disease under investigation, and research area of interest. Such guidance would then function as an essential element of benchmark level biobanking and sample management protocols.

Related to the ambiguity of what defines quality biospecimens or annotation, several key practical impediments remain prevalent in research protocol design and implementation. Such impediments dramatically limit the impact of biomedical research. The first is the range of variance in design of research protocols. Such variance in design formats results in greater risk of disparity in quality in sample collection and sample management. Second, compounding this issue is the fact that many research protocols are poorly documented, making results difficult to frame in absentia. Third, pre- and post-acquisition analytic variability has not historically been prospectively factored into design of protocols or related annotation, augmenting the difficulty of interpreting artifact in downstream analysis. Finally, lack of standardized reporting of research protocol-related methods and materials in the literature has also added to confusion in framing downstream quality assessment. All of the issues noted above contribute to gross limitation of functional and scientific proficiency to obtain an 'apples to apples' comparison of specimens or research results within, or across, collections. Such limitation in understanding dramatically restricts the quality and relevance of research and related findings, particularly in biomarker-based research.

5.1.2
The Role of the Research Protocol in Preserving Biospecimen Quality

'The biospecimen research protocol is the cornerstone of any research utilizing precious biological samples and serves as a crucial tool to support both large and small biobanking programs. Determining the quality of biological specimens continues to be a significant challenge for all research programs as investigators struggle to delineate biological change from specimen artifact. Both financial and specimen management are interrelated components of any archival strategy, and require careful management to ensure the quality and accessibility of renewable and, more importantly, nonrenewable biological resources. The management of biorepository protocols is one of the largest and most important considerations when calculating costs in biobanking programs today, putting sustainable quality operations of biospecimen resources at risk. While often overlooked, proactive management and design of biobanking protocols can dramatically improve and address critical issues, helping most programs achieve the quality required for serving their community' [1]. Therefore, the biospecimen protocol plays both a supporting and a leading role in preserving biospecimen quality, which begins with technical guidance as to best practice.

5.1.3
Rationale for Best Practice Integration into Sample Management Procedures and Protocols

An early practical approach toward alleviation of quality issues in biospecimen collection and sample management has been the creation of four prominent best practice-based guides: 'National Institutes of Health/National Cancer Institute's Office of Biorepositories and Biospecimen Research (NIH/NCI OBBR) Best Practices for Biospecimen Resources' (v. 2010); 'International Society of Biological and Environmental Repositories (ISBER) Best Practices for Repositories: Collection, Retrieval, and Distribution of Biological Materials for Research.' (v. 2008); 'Organisation for Economic Co-operation and Development (OECD) Best Practice Guidelines for Biological Resource Centers' (v. 2007); and the 'World Health Organization International Agency for Research on Cancer (IARC) Common Minimum Standards and Protocols for Biological Resource Centres Dedicated to Cancer Research' (v. 2007). Each best-practice document offers ranging yet complementary technical guidance related to bioresource-based laboratory operations and sample management practices.

The intention is that implementation of foundation level best practice should aim to support improvements in and/or establishment of requisite infrastructure necessary for benchmark level practice. It is estimated that many bioresources and some sample management laboratories may not currently operate at levels suitable to support baseline yet alone benchmark quality collections. Ideal research practices should aim to provide for the ability to standardize current methodologies via formal documentation and thoughtful design of sample management protocols. Real-time adoption of best practice has been relatively slow due to limited resources and related logistical constraints. Across the board, the majority of bioresources suffer from inadequate funding, lack of benchmark level physical and business infrastructure, limited operational free time, staffing, and management as well as inadequate practical training as how to implement best practices for ideal operations. Delay in implementation of best practice, combined with key impediments noted previously, have resulted in the inability to ascertain the level of scientific data that historically drives definition of many biological sample protocols and procedures. One solution may be to alleviate this issue via early integration of evidence-based data.

Over the last half decade, it has become evident that issues and concepts related to evidence-based biobanking practice (EBBP) may be abstract even to those who work in the field of sample management. Additionally, it appears unclear how to commence construction of sample management protocols utilizing data extrapolated from current literature. Biobanking sample management is positioned to lead the industry by acting as a bridge between scientific and clinical practices, driving improvements through lessons learned and observations drawn from biospecimen science driven EBBP. Clarification of such concepts is therefore significant and timely. In this chapter the author aims to provide preliminary insight as to how evidence-based protocol design may be approached via integration of technical and scientific considerations that may affect sample integrity.

5.2
The Path toward Integration of Evidence-based Biobanking Practice

5.2.1
Conceptual Foundations of Evidence-based Biobanking Practice

The trans-boundary application of evidence-based practice to biobanking and sample management is a relatively new concept. However, the idea of EBBP is not entirely novel in the sense that it mimics evidence-based clinical practice. Evidence-based clinical practice is inter-related and has long been utilized as part of bench-to-bedside care. However, the EBBP concept is still abstract and could benefit from formal definition. Therefore, if *evidence-based practice* is defined as 'the timely process of implementing validated scientific discoveries into daily techniques and procedures to support definition of relevant research gold standards, also referred to as benchmarks, with the aim of enabling scientific discoveries that foster refinements in patient care and overall treatment outcomes,' EBBP could then be viewed as an adjunct interdisciplinary practice that involves collection, interpretation, and integration of research-derived evidence from biospecimen analyses, otherwise known as *biospecimen science* [2].

The capability of EBBP to refine bench-to-bedside clinical practice is an important concept to grasp. EBBP aims to improve clinical outcomes by alleviating quality-related disparities in the research process, supporting sample management environment, and process chain. A resulting theory is that only via 'fit-for-purpose' protocol and process chain design may quality be improved significantly enough to enable increased relevance of downstream analysis and related research results. If this theory is valid, adjunct clinical decision-making capacities could also be extended and/or improved. The most immediate bench-to-bedside example that springs to mind is the practice of using biomarker research findings as a surrogate marker. In this scenario, diagnostic biomarkers are being utilized in practice to rule out error that may exist currently with traditional diagnostic and/or prognostic tests to guide a patient's course of treatment.

In personalized medicine-based trials, translational scientists are now coordinating clinical practice with oncologists to advance bench-to-bedside care. Traditionally, radiographic, histological, and serological findings have been utilized as tools to evaluate presence, recurrence, and metastasis of disease in the oncology clinic. With traditional methods, there is a likelihood that disease could exist but may not be observable. At the cellular level, biomarkers are able to function as surrogate markers for disease to demonstrate evidence of disease where it may be missed. If the clinician is able to more conclusively detect and/or rule out disease, he or she could potentially offer an intervention sooner or adjust treatment accordingly to prevent metastasis and future recurrence. This could result in longer survival times, opportunities for decreased rate of mortality and morbidity, as well as improved quality of life for the patient under treatment. This is just one example in which evidence-based biospecimen research in conjunction with clinical care has the potential to impact incidence and prevalence of disease. Efforts are underway in the biobanking community to actuate and track this early capability via formal definition of 'BioResource Economic Impact Factors' (BREIF) [3]. The BREIF, still in development, is being explored as an adjunct 'BioResource Impact Factor' (BRIF) [4] index aimed at formal qualification and quantification of factors that affect impact of biobanking sample management practice on clinical and research outcomes.

5.2.2 The Pre- and Post-Acquisition Analytic Variable Relationship to EBBP

Pre- and post-acquisition analytic variables are believed to be relevant data points of interest to aid improved understanding of a biospecimen's molecular integrity during analysis. Improved understanding would likely increase accuracy and relevance of downstream interpretation of research results. So much so, that best-practice documents worldwide such as NIH/NCI and ISBER have formally recommended that these variables be included in biospecimen collection and sample management efforts. The US NIH and NCI OBBR 2010 Best Practices document formally recognizes that pre- and post-acquisition analytic variables are two crucial categories of data that affect quality and interpretation of research results. As per the NIH/NCI OBBR 2010 Best Practices, pre-analytic variables may be classified as they relate to three segments of the lifecycle: the physiology of the human research participant prior to study participation or 'patient,' specimen collection practices, and specimen handling practices prior to their inclusion in downstream testing [5].

It should be noted that defined data points and reference ranges for pre- and post-acquisition analytic variables are still being explored. Therefore only via more cogent definition, documentation, and tracking of these details can scientists foster learning of key factors related to integrity. Once this is accomplished, evidence-based quality improvements would then require implementation across research and resulting clinical sample management process chains. One practical hindrance to measuring integrity is that for many in the scientific arena 'biospecimen integrity' still remains a popular 'catch phrase.' The term '*integrity*' has yet to be formally defined, confirmed, and validated. This makes any intermediate interpretation still somewhat subjective and less relative. More objective parameters need to be explored and conclusively defined for integrity. Such parameters should aim to factor in effects of pre- and post-acquisition factor analytical variability. Meanwhile, scientists are collaborating to assist harmonization and real-time multi-center analysis of analytical variables in an attempt to more clearly understand these issues. One example is a recent effort by the ISBER's Biospecimen Science Working Group. This group has devised an initial pre- and post-acquisition analytic variable coding system referred to as 'Standard PRE-analytical Code' (SPREC) [6]. Current goals of the group are to promote biospecimens selection quality assurance schemes, assist identification of the critical points for biospecimen handling, and translate findings into standard formats to aid downstream analysis. This work, although still at an early stage, appears to be aiding aggregation and early understanding of biospecimen science-based research findings.

5.2.3
The Biospecimen Lifecycle Concept: a Framework to Aid EBBP Protocol Design

Many in the sample management field may still be unaware of the biospecimen lifecycle concept on which pre- and post-acquisition analytical variables are founded (Figure 5.1). The concept of the biospecimen lifecycle has been adapted worldwide over recent years in biobanking, sample management, and related industry circles. Much discussion has focused on increasing awareness of the value of biospecimen science as an area of study. More recent discussion has expanded to include exploration of related critical considerations. For ease of reference in this chapter we refer to the NCI model which may be viewed online at: *http://biospecimens.cancer.gov/researchnetwork/lifecycle.*

The biospecimen lifecycle's conceptual premise as depicted in Figure 5.1 [7] is the idea that biospecimens, having been living viable objects, have a resulting sample management-based 'course of care' to promote and optimize biospecimen integrity. The biospecimen lifecycle referred to here in Table 5.1 encompasses eight critical time points: patient, medical and surgical procedures, acquisition, handling and processing, storage, distribution, scientific analysis, and re-stocking used samples. The lifecycle represents a 'cradle-to-grave' approach to the biospecimen's lifespan; all temporal segments have their own separate but equally crucial pre- and post-acquisition analytical variable considerations. It should be noted that while pre- and post-analytical data considerations may be alluded to in this chapter to highlight the crucial concepts, guidance is not meant to be comprehensive. The intended message is that materials, methods, and logistics for daily sample and data management practice should be framed while keeping in mind such considerations.

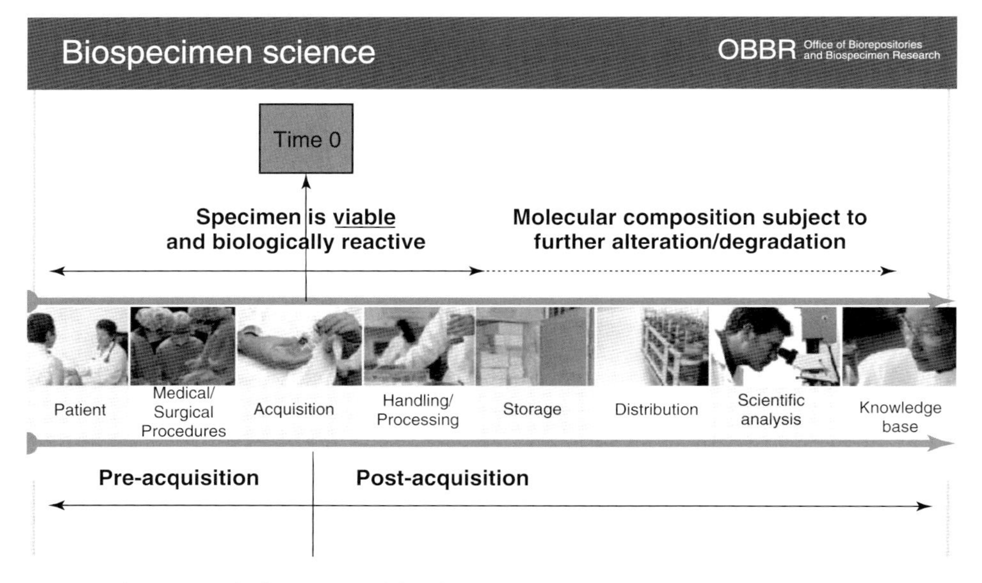

Figure 5.1 The biospecimen lifecycle as presented by NIH/NCI OBBR [7].

Table 5.1 Examples of biospecimen lifecycle time point-related technical and scientific considerations.

Time point 1: Patient

What lifestyle factors affect integrity? (e.g., age, dietary intake, overall health, medications, and treatment). What are the individual and combined effects on the biospecimens being sampled?

Time point 2: Medical and surgical procedures

How do type and length of anesthesia, and intra-surgical medications given affect sample integrity; Does ischemic window begin at 'cut time, clamp time,' or at time of excision'?

Time point 3: Acquisition

What is the ideal evidence-based approach to collection container selection?
By downstream analyte: vacutainer type, size, shape, material, and additive
By process: automated vs manual
By cryopreservation method: room air, fixed, and frozen

Time point 4: Handling and processing

How is EBBP baseline biospecimens quality defined and measured?
What are relevant biospecimen handling issues/data points across the biospecimen lifecycle?
What are the appropriate lags post-collection, during/after processing prior to preservation, transport, and storage?
Should molecules of interest be isolated prior to preservation?
What preservatives are required to optimize biopreservation? At what time point?
What are the key biopreservation concerns and best 'fixes' according to the data?
How do preservatives such as optimal cutting temperature (OCT) compound affect molecular integrity? What are the ideal type, volume ratio, temperature, and time of fixation for fresh tissues?
Which processing method should be chosen when more than one method exists? (e.g., should quantitative buffy coat (buffy coat procedure) processing be performed with or without trizol and be followed by immediate or delayed extraction and/or storage of ribonucleic acid (RNA) isolation?)
What is the effect of fixation time on RNA quality?
What is the effect of sample handling on specimen integrity?
What are the appropriate parameters for evidence-based quality control?
How can pathological review be carried out without putting the specimen integrity at risk?

Time point 5: Storage

What are ideal transport temperature- and transport-related handling conditions?
What is the effect of storage container type on quality and integrity?
What is the ideal storage container temperature per specimen, molecule, and/or organ type?
What is the maximum duration for which biopreservatives protect biospecimens?
What is the ideal length of storage per specimen, molecular, and/or organ type?
What is the time zero and rate of molecular degradation post freezing and its effect on specimen quality; Is 'freezer burn' a surrogate marker of quality? Relative risk?
What is the effect of humidity on FFPE specimens?

(continued overleaf)

Table 5.1 *(Continued)*.

Time point 6: Distribution and retrieval
What is the ideal retrieval method to maintain/optimize specimen integrity?
Manual vs automated: batched vs cherry-picking
Room air vs ice
Freeze/thaw vs cascading methods
What are the ideal freez/thaw and/or retrieval algorithms to reduce risk of compomising integrity? Aliquot single instance vs re-use; indefinite retrieval vs culling
Time point 7: Scientific analysis
What is the ideal thawing temperature and time?
Within what temperature range should a specimen be maintained so as not to alter integrity during time of analysis?
What are the desired quality markers and related reference ranges on which to frame a biospecimen's quality?
Time point 8: Re-stocking
What is the effect of re-use and re-stocking on specimen composition and integrity?
What specific artifact and changes occur as a result of the 'freeze/thaw' cycle(s)?

Ideally, this exercise should then be transitioned into formal workflows in a manner that complements the biospecimen lifecycle considerations in an effort to promote and optimize specimen integrity. To accomplish this goal it is recommended that sample management protocols be developed first with the analyte of interest in mind, a procedure referred to as '*fit-for-purpose*' planning.

5.3 Integrating Evidence-based Biobanking Practice into Sample Protocols

5.3.1 Protocol Planning for EBBP-based Sample Management

The initiation of evidence-based and 'fit-for-purpose' sample management involves factoring in both experiential data as well as evidence-based technical sample management design considerations ideally in tandem with prospective biobanking protocol planning. Historically, this has not been the case. Evidence-based protocol planning would best commence with utilization of a standardized format. The ideal template might begin with a description of the scientific rationale for biospecimen collection and utilization. This approach supports the belief that analysis should aim to be the driving factor in research methodological development. Therefore, one's protocol template should aim to include detailed yet reasonably standardized descriptions of evidence-based sample management rationale and related downstream analytical methodologies.

Using this approach, a proposed EBBP protocol template could aim to include but may not be limited to the following sections:

1) Scientific rationale for biospecimen utilization
2) Overview of quality control pilot design
3) Evidence-based sample management methods, techniques, and instrumentation
   a. Recruitment and collection
   b. Handling and processing
   c. Quality control
   d. Biopreservation
   e. Storage and retrieval
   f. Mode(s) and method(s) of transport.
4) Downstream analytical methodology, techniques, and instrumentation
   a. Overview of analytical objectives
   b. Overview of related testing and techniques in analysis
   c. Description of applied instrumentation
   d. Epidemiological applications and bio-statistical review
   e. Summary and conclusions related to biospecimen analysis.
      i. Report of confounders, resulting error, variability, and related artifact
      ii. Recommendations for revision of sample management methodology, study design, and/or downstream analysis.

It may be beneficial to elaborate as to the impetus for proposing inclusion of section two (above), which involves conducting a quality control-based pilot. In such a pilot, one would aim to collect a small number of samples under thoughtfully designed evidence-based and quality procedures. It should be noted that this is not currently a regular quality assurance step in most prospective biospecimen-based collection efforts, but such a step should aim to be included. Typically, quality assessments are conducted long after procurement targets are achieved. In fact, quality assessments are obtained at such a delayed stage of collection that it is often impossible to gauge the quality and relevance of the collections until research results are written up and analyzed for publication. This practice contributes serious risk to the value of collections and the operational viability of bioresources and projects in progress. A pilot-based approach may offer early opportunities to refine research protocols and process chain practices before irreversible mistakes are made. Procurement efforts would briefly be reduced or put on hold while the baseline assessments are being performed. Then when acceptable baseline 'benchmark' quality is ascertained, requirements defined and implemented into the sample management protocols, expanded procurement can then continue. Following the pilot quality control, interim, and post-procurement quality assessments can be performed at predetermined time points. The goal here is to mimic the traditional approach to pharmacogenomic planning of clinical trials and/or typical bench experimental design which factors scientific rationale ahead of any formal design considerations.

5.3.2 Crucial Scientific and Technical Considerations for EBBP Protocol Design

Once an evidence-based protocol template has been constructed, it is advisable that detailed processing workflows, standard operating protocols, and procedure methodologies be drafted and formally documented. One approach includes doing so with consideration of technical and scientific factors that may affect biospecimens and downstream biosample integrity. Initial procedural design should include consideration of factors throughout relevant biospecimen lifecycle time points, timing of feasible execution as well as specific ideal methodology. Collection protocol design should aim to identify requisite study sample size ('N') as well as individual physical amount of the biospecimen required for analysis. Consideration of relevant granular details may be dramatically important, for example, environmental temperature and container type should also be explored, especially when the analyte of interest has been predetermined and the ideal environment has been clearly defined. In many of these areas, these factors remain under exploration and, in some cases, speculation. Currently, no known complete list of scientific and technical considerations or extensive formal guidance exists. Therefore one approach may be to amalgamate multi-disciplinary technical and scientific team efforts in an attempt to identify and define such considerations prior to formal methods and materials development. Once the relevant considerations have been determined, it is recommended that a formal review of the literature be conducted to unearth related findings. To foster learning, educated debate and discussion are encouraged as part of the protocol planning exercise. For example, it may be helpful to understand how such considerations may also alter approach to procurement (prospective vs retrospective aggregation), what corresponding annotation is required in informatics infrastructure, as well as ideal parameters for histological review and molecular analysis.

While validated formats and pathways to evidence-based biobanking protocol development are still being developed and agreed upon, there is a process that may help to expedite real-time incorporation of evidence-based considerations into daily practice.

This process consists of five initial steps.

- **Step 1**, as briefly described above, includes definition of critical issues and corresponding pre- and post-acquisition variable-based data points that may compromise, alter, and affect quality and integrity of the biospecimen being sampled. Tables 5.2 and 5.3 provide an introduction to relevant issues of interest to consider regarding optimization of tissue biospecimen integrity.
- **Step 2** involves reviewing relevant scientific literature with the intention of obtaining guidance on how to draft evidence-based sample management techniques. Initial efforts should aim to keep in mind integrity and quality related issues and appropriately weigh guidance from both experiential and evidence-based data. Such guidance should aim to foster an evidence-based rationale for sample management protocols and procedures. Tables 5.2 and 5.3 provide a brief example of

Table 5.2 Example of relevant technical considerations related to FFPE tissue sample management (stage I EBBP level recommendations).

Technical consideration	Assumption/finding related to overall integrity	Literature/reference that may support this assumption
Level of sterility	Considerations should be made; RNA is sensitive; sterile container pre-transport, received in 'clean' area	Section J.8.110, ISBER 2008 Best Practices [8]
Collection transport method/temperature	Fresh on wet ice (0 °C) to decrease the rate of biomolecular degradation	Section J.8.110, ISBER 2008 Best Practices [8]; Leiva *et al.* [9]
Collection and related transport time	30 min to 2 h ischemic window ideal	NTRAC 2004 SOP Guidelines [10]; Mager *et al.* [11]
Tissue section size	Smaller size may promote better molecular preservation; 3–5 mm?	Leiva *et al.* [9]
Cutting table temperature	Chilled −20 °C	Experiential data
Blade Temperature	Chilled −20 °C	Experiential data
Cutting paper absorbency	Refrain from dissection on dry towel or absorbent material to prevent desiccation	Section J.8.110, ISBER 2008 Best Practices [8]
Type/volume ratio of fixative	10% buffered formalin vs70% ethanol: ethanol fixation and embedding demonstrated improved preservation of biomolecules over formalin fixation and embedding; 10 : 1 ratio with 10 mL of fixative for every gram of tissue	Leiva *et al.* [9]; Hewitt *et al.* [12]
Type/concentration of buffer for fixative	Phosphate-buffered formalin better for high quality RNA	Chung *et al.* [13]
Length of fixation time	16–32 h	Chung *et al.* [13]
Tissue position	To be determined	–
Length/time of processing	Longer tissue processing times are more ideal to obtain higher quality RNA; 12–24 h recommended	Chung *et al.* [13]
Amount of dehydration	'well dehydrated'	Chung *et al.* [13]
Section size	Weight $\leq$ 1 g; not larger than 1.5 cm × 1.5 cm × 0.4 cm	Hewitt *et al.* [12]
Type of storage	'Office like'; low humidity environment	Hewitt *et al.* [12]
Length of storage	To be determined	–

Table 5.3 Example of relevant technical considerations related to frozen tissue sample management (stage I EBBP level recommendations).

Technical consideration	Assumption/finding related to overall integrity	Literature/reference that may support this assumption
Level of sterility	Considerations should be made; RNA is sensitive; sterile container pre-transport; received in 'clean' area	Section J.8.110, ISBER 2008 Best Practices [8]
Collection transport method/ temperature	Fresh on wet ice (0 °C) to decrease the rate of biomolecular degradation	Section J.8.110, ISBER 2008 Best Practices [8] and Leiva *et al.* [9]
Collection and related transport time	30 min to 2 h ischemic window ideal	NTRAC 2004 SOP Guidelines [10] and Mager *et al.* [11]
Tissue aliquot size	Smaller size may promote better molecular preservation	Leiva *et al.* [9]
Cutting table temperature	Chilled −20 °C	Experiential data
Blade temperature	Chilled −20 °C	Experiential data
Cutting paper absorbency	Refrain from dissection on dry towel or absorbent material to prevent desiccation	Section J.8.110, ISBER 2008 Best Practices [8]
Ideal cryopreservative	'OCT is OK for DNA and RNA, bad for protein. and required for histology' (experiential feedback)	Experiential data (S. Hewitt, personal communication)
Rate and length of snap-freezing	Until the specimen is frozen to the core-exact measure TBD	Experiential data
Type of storage container	Container should be fit-for-purpose	NTRAC 2004 SOP Guidelines [10]; Experiential data
Temperature of storage container	Chilled slightly or same temperature as specimen? – TBD	Experiential data
Section size of quality control specimen	As per CAP guidelines	Experiential data
Ideal temperature of 'chucks and cryostat'	Pre-chill 'chucks' and cryostat at −22 °C?	Experiential data
QC slide storage environment	Gaseous nitrogen, vacuum, enclosed cases, · · ·	Chung *et al.* [13]
Type and length of storage for aliquot	Two years, five years, longer, · · ·	Experiential data

how one may commence this step in regard to processing tissue into fixed frozen paraffin-embedded and frozen specimens.

- **Step 3** includes integrating experiential and evidence-based guidance after discussion into design of sample management and biobanking related protocols using evidence-based frameworks such as the example presented in Section 5.3.1.
- **Step 4** focuses on utilizing such scientific knowledge to develop prescribed data sets relative to particular molecules of interest (RNA, DNA, proteins, ···) and/or disease of study in coordination with prospective clinical trial and even bench design. Initial data sets should aim to focus on pre- and post-acquisition analytical variables such as those referenced in the processing workflows in Figures 5.3 and 5.4 and discrete variables in Tables 5.2 and 5.3.
- **Step 5** occurs mid- and post-study when research methods and materials and related findings are incorporated into the scientific literature following recommended reporting and publishing guidelines. References to relevant reporting guidelines are provided in Table 5.4.

A simple example of a single-tissue biospecimen collected commonly in excess to be processed into a formalin-fixed paraffin-embedded (FFPE) block and a frozen tissue sample is shown in Figure 5.2 [9], Figure 5.3 [1], and Figure 5.4 [1]. Displayed in Figure 5.2 is an example of the initial basic workflow that might typically occur. It is recommended that all procedural design commence with a workflow draft

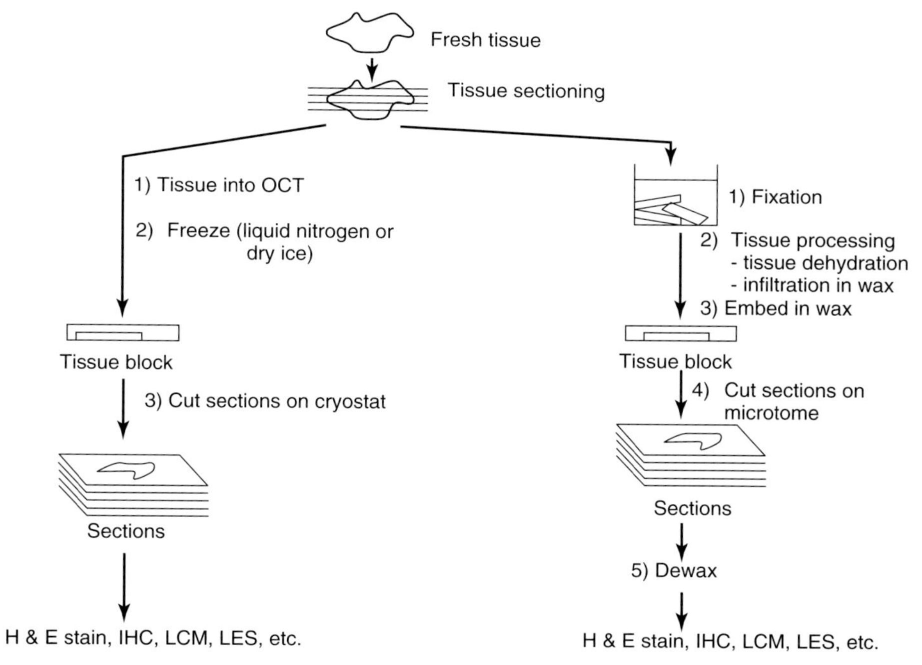

Figure 5.2 Processing workflow of fresh tissue collected for FFPE and frozen specimens. (Original diagram: Levia *et al.* [9].)

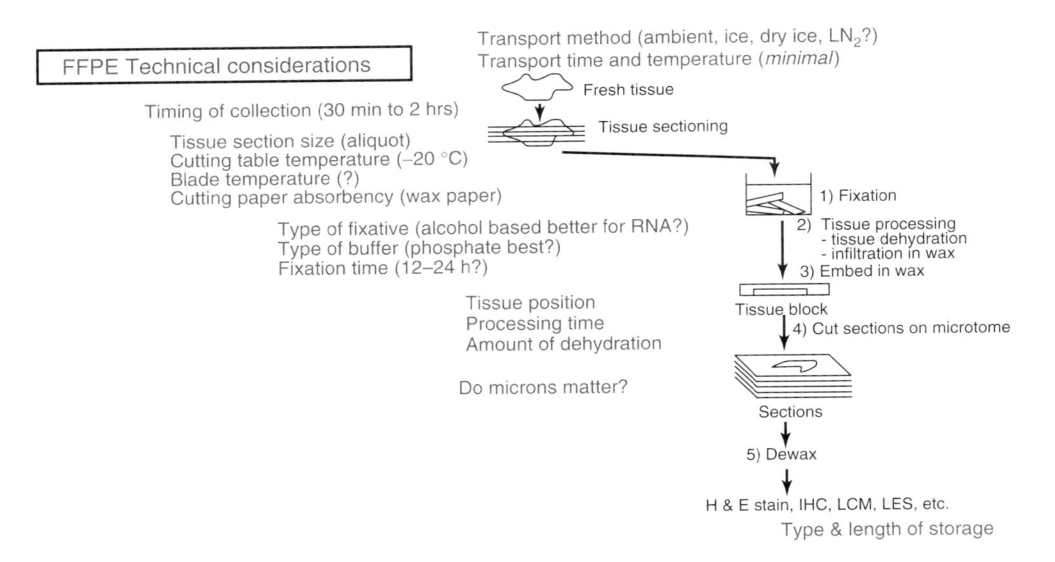

Figure 5.3 Example of a workflow-based approach to drafting relevant FFPE tissue sample management-based scientific and technical considerations. (Original diagram: Levia *et al.* [9].)

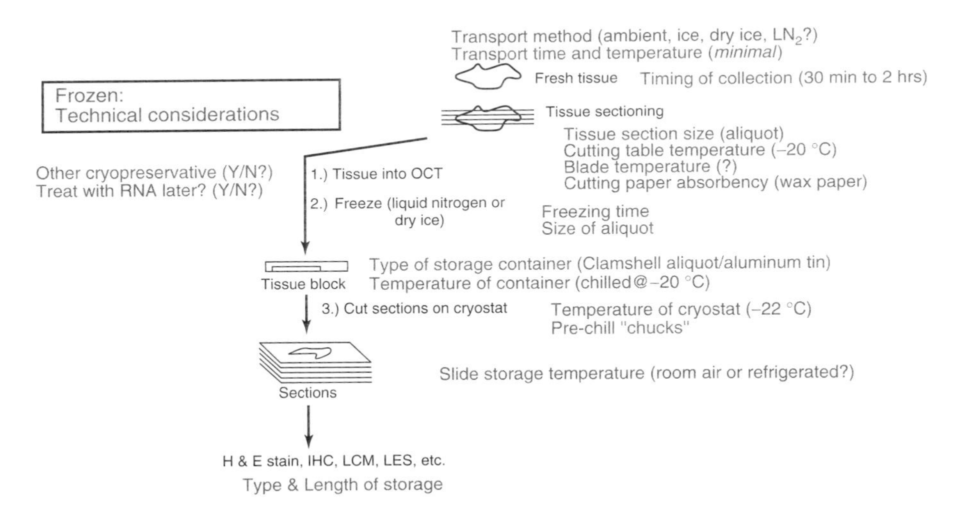

Figure 5.4 Example of a workflow-based approach to drafting relevant frozen tissue sample management-based scientific and technical considerations. (Original diagram: Levia *et al.* [9].)

to most accurately guide FFPE and frozen tissue Standard Operating Procedure (SOP) development. Figures 5.3 and 5.4 then detail the critical EBBP, scientific, and technical considerations commonly observed during FFPE (Figure 5.3) and frozen (Figure 5.4) specimen sample management workflows.

5.3.3 Utilizing Publication Reporting Guidelines to Guide EBBP Protocol Design

Early biospecimen science efforts have offered guidance aimed at refining the reporting of materials and methods via development of the pre- and post-analytical

Table 5.4 Example publication and reporting guidelines relative to promotion of evidence-based biobanking practice.

Guideline	Description/references
BRISQ: Biospecimen Reporting for Improved Specimen Quality	Description: Guidelines for reporting of pre- and post-analytical variables and related methods and materials in biospecimen-based research Related Reference: [14]
REMARK: Reporting recommendations for tumor MARKer prognostic studies	Description: Guidelines for reporting of tumor marker studies Related Reference: [15]
STARD: STAndards for the Reporting of Diagnostic accuracy	Description: The objective of the STARD initiative is to improve the accuracy and completeness of reporting of studies of diagnostic accuracy, to allow readers to assess the potential for bias in the study (internal validity) and to evaluate its generalizability (external validity) Related Reference: *http://www.stard-statement.org*
PRIMR: PResentation and Interpretation of Medical Research	Description: Group that aims to improve the design of studies, their presentation, interpretation of results and translation into practice Related Reference: *https://www.primr.org*
STROBE: STrengthening the Reporting of OBservational studies in Epidemiology	Description: The STROBE Statement is referred to in the Uniform Requirements for Manuscripts Submitted to Biomedical Journals by the International Committee of Medical Journal Editors Related Reference: *http://www.strobe-statement.org*
GPP: Good Publication Practice	Description: Guidelines that encourage responsible and ethical publication of the results of clinical trials sponsored by pharmaceutical companies Related Reference: [16].

focused publication reporting standards. One such set of guidance is the 'Biospecimen Reporting for Improved Study Quality' (BRISQ) publication [14]. This follows in the footsteps of preceding domain-specific standards such as 'REporting recommendations for tumor MARKer prognostic studies' (REMARK), 'STAndards for the Reporting of Diagnostic accuracy' (STARD), and 'STrengthening the Reporting of OBservational studies in Epidemiology' (STROBE) [17].

Publication and reporting guidelines were originally developed to facilitate standardized reporting of methods and materials used in research studies. However, reporting guidelines can also be utilized as helpful tools to aid prospective evidence-based protocol design and retrospective procedural refinements. In fact, this may be the only way to ensure 'fit-for-purpose' planning and offer a true comparison of collections. To assist these efforts some medical journals are now requiring compliance with all or some reporting guidelines. Information and references to these guidelines as noted in Table 5.4 below may be found on the 'Enhancing the QUAlity and Transparency Of health Research' (EQUATOR) [18] web site.

5.4 Final Thoughts and Recommendations

5.4.1 Proposed Staging System to Qualify EBBP Related Data

Currently there is no proven, formal methodology of how to qualify experiential and evidence-based data extrapolated from the scientific literature. As this may not be intuitive or straightforward, it would be ideal to have an algorithm to guide incorporation of such data into formal biobanking sample management protocols and procedures. One of the complicating factors is that there is no single, central reference compilation, registry, or source of experiential or evidence-based findings. This makes it difficult to locate, aggregate, review, and evaluate evidence-based data. Additionally, while early coding schemas and reporting guidelines exist, there is no universal consensus or full criteria, nor is there a standard procedure for integrating EBBP data. What exact conditions qualify relevant data as 'evidence-based' remains unclear. The potential contributing considerations to choose from are vast, which makes things even more complex. Ideally, evidence-based data evolve from experiential logic, with confirmation of reproducible research results in publication.

Consider the question: 'What is the desired storage temperature and length of storage time for serum ribonucleic acid (RNA) specimens to achieve the ideal RNA research integrity number (RIN)?' If data derived from experience, research and/or downstream experiments/analysis either in the laboratory or extrapolated from published literature demonstrate that storing specimens at $-86\,^{\circ}$C (versus -148 to $-196\,^{\circ}$C) for no longer than two years is the ideal approach to preserve RNA integrity, then this data may be a relevant evidence-based practice consideration. However, this may be an early finding that may need to exhibit an acceptable level of reproducibility and validation in reasonably standard or similar conditions.

Unequivocally establishing this would require a formal evidence-based procedural schema or framework; no known framework currently exists.

One approach to defining a formal evidence-based data classification schema could include offering an initial 'staged approach' to categorize evidence-based protocol-related data. A staging system may provide the baseline framework for categorical relevant comparison of evidence-based findings. One challenge to framing evidence-based data is that biospecimen science is young in its evolution. Therefore, while it can be assumed that baseline standards and parameters will be more clearly defined and confirmed, others will be in a period of flux. In these circumstances there will most likely be organic evolutionary shirts in thinking as to standards for EBBP. For ease of comparison of such standards, these periods of flux could be categorized in stages simply put as 'early, advanced, and mature.' These stages may in fact be a prerequisite to aid substantive development of EBBP based requirements. This concept could be explored, via collaborative staging related pilots, in biospecimens science-based research. If proven helpful, this concept could be expanded and refined to provide a legitimate framework to designate the level of evidence-based practice being conducted.

Three proposed stages to define potential EBBP maturation are described below.

- **Stage I: Early EBBP**: *Early EBBP* may be defined as the initial process of incorporating mined experiential and experimental data into biobanking protocols, processes, workflows, and procedures. This stage of practice may be conducted primarily to define baseline quality standards and scientific guidelines in a generalized fashion aimed at improving overall sample integrity. Bearing in mind current findings in biospecimen science, the exact procedures related to an analyte of interest may not necessarily be confirmed at this stage. Ideally one should aim to consider the analyte of interest in the most relevant manner to guide early EBBP procedural development and design.
- **Stage II: Advanced EBBP**: *Advanced EBBP* may be defined as the process of reviewing reproducible experimental and refined experiential biospecimen-based data utilizing standardized early EBBP recommended techniques. This stage of practice may be conducted by revisiting biobanking protocols, processes, workflows, and procedures to ascertain quality improvements and research outcomes based on integration of early EBBP practice. Lessons learned from downstream analysis would then be married with additional mined data and integrated into the sample management process chain aimed at validating the reproducible results and related findings. Optimization of sample integrity would, as always, remain a chief concern, with procedural development increasingly being driven by evidence-based data confirming parameters for ideal preservation techniques for the analyte of interest.
- **Stage III: Mature EBBP**: *Mature EBBP* may be defined as the process of reviewing aggregated validated reproducible biospecimen-based data utilizing standardized and proven advanced EBBP recommended techniques. Epidemiological biostatistical modeling techniques would serve to guide, support, and drive experimental and procedural design. Procedural revisions would be based on lessons learned

from the literature, updated mined data, review of operational and bench practices, updated quality metrics, and analysis of sampled collections as well as downstream biospecimen products.
Procedural parameters for integrity relative to the analyte of interest would be clearly defined at this stage. As such they would drive biobanking sample management protocols and procedures in a significant way.

Noted below is an example of how a technical and scientific consideration might be explored at the stage I for both frozen and FFPE tissue sample management as outlined in Figures 5.3 and 5.4. This exercise is aimed at briefly presenting how one might utilize the EBBP staging concept when drafting sample management protocols and procedures. In both examples the key issue under consideration is: 'What is the ideal ischemic biopreservation window ('time to biopreservation') to optimize biospecimen integrity for frozen and/or fixed tissues?'

1) **Two examples of a stage I/early EBBP consideration:**
   a. **Example 1: Fresh tissue for processing into FFPE blocks** In the case of 'time to fixation' of fresh tissue to be processed for FFPE blocks, Chung and colleagues propose the ideal time to fixation to be 12–24 h [13].
      'The optimal fixation period of 12–24 h in phosphate-buffered formalin resulted in better quality RNA. Longer tissue processing times were associated with higher quality RNA.' Therefore the recommended advanced EBBP fix includes the following:
      i. Optimize standards and workflows to achieve an average fixation period of not less than 12 and not greater than 24 h utilizing phosphate-buffered formalin to optimize RNA quality.
      ii. Establish initial RNA parameters and criteria such as ideal RIN ranges, which may be complementary for optimization of RNA integrity and integrate into operations in parallel as a research standard of care.
   b. **Example 2: Fresh tissue for processing into frozen aliquots** Relative to the consideration noted above, one may be aware that experiential data dictates that fresh tissues should be snap-frozen or fixed in formalin within a 30 min to 2 h timeframe. It should be noted that this is experiential data as recommended by the National Translational Cancer Research Network 2004 Report. SOPs can be adjusted to incorporate and track ischemic windows. Adherence to ischemic windows should be then reviewed as needed in downstream analysis to aid interpretation of related artifact.
2) **Examples of a stage II/advanced or stage III/mature EBBP consideration:** As biospecimen science is still in its early stages it may be difficult to predict an ideal example observed from daily practice of a stage II or III EBBP consideration. It is unknown whether at the present time this level of EBBP is occurring. Public, technical, and scientific presentations do not appear to make this evident. Therefore, while it is possible to offer a hypothetical example for stage II, it may be premature to attempt to offer a case-based example for stage III. As biospecimens science practice evolves, further research into what

defines stage III EBBP is advised. It is the author's opinion that stage III EBBP would be most commensurate with benchmark level practice.

Hypothetically speaking, building on the previous FFPE and frozen specimen stage I example, one could imagine how stage II EBBP could translate in practice. Stage II-based practice might commence with a meta-analysis of literature focused on review of processing methods and materials for FFPE and frozen tissue specimens. At this level of protocol investigation and design, it may be useful to ascertain crucial pre-analytic factors that affect integrity, the ideal window for 'time to biopreservation,' and the quality of research results from aggregated extrapolated data. The methods and materials review should aim to be viewed in standard frameworks, that is, BRISQ, SPREC, and other reporting guidelines noted in Table 5.4 to ensure an efficacious range of comparison. The framed data could then be compared to internal research results from similar procedures performed in tandem with scientific assessment of 'pilot data'-based quality assessments as referred to in Section 5.3.1. Ideal assessments would be based on recommended quality marker reference ranges (e.g., RIN range) for the analyte of interest. The sample size or 'N' of the data should be deemed statistically significant before any conclusions are generated and protocol recommendations offered.

5.4.2 Revisiting Crucial Considerations Related to Implementation of EBBP

So, one may inquire as to why evidence-based banking practice still exists on the periphery of biobanking and sample management operations? It may be timely to revisit the current cultural and logistical constraints that impede integration of real-time EBBP into bioresources and biosample-based laboratories today.

Firstly, in clinical settings from where the majority of biospecimens are obtained (and understandably so), there is a well known principle and best practice that 'clinical care supersedes research need'. While worthy and well understood, this historical requirement has a downstream affect on all sample management and biobanking operations. Ischemic windows are often extended and quality reviews delayed because of this issue. Interdisciplinary refinements will need to occur equally across both the research and the related clinical infrastructure for biospecimen science and evidence-based biobanking sample management practices to evolve. As this essential requirement cannot be changed, it will be necessary to develop a mechanism to improve the sample management process chain without impeding clinical care, keeping preservation of biospecimen integrity as a core objective.

Secondly, clinical biobanking in practice is less 'controlled' than bench or laboratory research. The inability to control the sample environment makes standardization of specimen processing and sample management environments challenging. It also makes it difficult to compare biospecimen science data derived from controlled experiments or bench laboratory studies with data derived from bioresource facility-supported specimens. As noted previously, many methods and materials are not well documented. As previously stated, documentation can be extremely helpful in aiding accurate review of pre- and post-acquisition variables-related artifact as well as relevant comparison of variance in processing and sample management. Daily operational performances of best and known quality practices are essential to provide a foundation for evidence-based adjustments.

Thirdly, there remains limited oversight, regulation, or formal policy, internal or otherwise, to confirm and evaluate attempts to integrate evidence-based biospecimen data into biobanking protocols, sample management procedure, and design. Regulatory review boards are not formally trained in the subtleties of biospecimen science to guide evaluation of protocol composition. There are also no regulatory requirements that state that protocols should contain some level of evidence-based data at time of design. As biospecimen science is still nascent, it may be too early for formal regulation; incorporation nonetheless should be encouraged.

Lastly, underlying all of these issues is the fact that the majority of bioresources and laboratories do not practice cost recovery. Absence of cost recovery practices dramatically limits requisite resources in all forms. In fact, the majority of resources have no formal financial infrastructure. This offers limited time to establish, review, and refine practices, acquire the requisite education and training, or develop proactively strategic biospecimen science-based research and development pathways.

5.4.3 Strategies to Optimize Real-Time Implementation of EBBP

Development and implementation of evidence-based biobanking and sample management practice-related strategy is crucial to ensure that EBBP becomes a reality. Such strategies if implemented would dramatically assist in optimizing and accelerating EBBP in daily practice and sample management-related operations. A few strategies are further described below.

Ideally, operating budgets should aim to earmark funds for integration of evidence-based biobanking sample management practices as well as biospecimen science focused bench-to-bedside research and development. For this measure to have the desired effect, it is essential that budgetary lines be created across the full process chain. This includes adjunct clinical and supporting departments or institutions. An appropriate portion of these funds should be allocated toward refinement of physical infrastructure, materials, and requisite labor to support the ideal evidence-based environment that facilitates refinement of biomedical research practices. This budget should also facilitate process improvements over time and be coordinated with grant, funding, and sponsorship review.

Next, integration of early evidence-focused practice should aim to assist development, adoption, and implementation of 'universal evidence-based biospecimen protocols' across the process and downstream product chains alike. Efforts should attempt to foster the evolution from best to evidence-based practice. Pre- and post-acquisition analytical variables and data points of interest should be documented. Procedural and protocol deviations should be noted and explained in project reports as well as annual Institutional Review Board protocol related updates. 'Specimen Biosketches' should aim to become a consistent biobanking practice detailing the historical life of a biospecimen from 'cradle to grave and beyond.'

When evidence-based practice standards and guidelines are conclusive and relative, appropriate methods and materials should be updated with related processes and adjusted accordingly.

As referred to in the previous section, current research, biobanking, and sample management methodologies should be assessed by review committees with oversight from governing stakeholders to ensure acceptable levels of best and evidence-based practice in research operations and adjunct clinical environments. Regulatory review boards should aim to acquire more education on the issue and perhaps include biospecimen science experts on review panels. Biospecimen utilization committees should aim to include evidence-based criteria in biospecimen research applications and review processes. Scientific Advisory Boards should aim to offer direct guidance, contributing to evidence-based protocol design and review both prior to initial regulatory submission and on any amendments or renewals. It should be noted that design by L. Miranda and colleagues of an IRB EBBP-focused protocol evaluation tool is currently in progress.

Once research results are available, biospecimen science-based lessons learned should be disseminated via technical and best practice-related guidance in a reasonably timely fashion. Good tissue practices should require that such data when not restricted by proprietary constraints be publicly available. Utilizing the staging system, categories of EBBP data could then be flagged to qualify the current state of the research findings. As a result, gradually best practices should then evolve into robust evidence-based practice guidelines. Future technical guidance should also aim to incorporate guidance as how to best implement EBBP recommendations into formal operations as well as research and development.

Relevant stakeholders and players in the biobanking sample management and biomedical research process should aim to keep abreast of evidence based guidance related developments. Interdisciplinary collaboration would be advised to ensure lessons learned long ago in one area are not reinvented by a second area.

Lastly, integration of evidence-based design (EBD) should be a prime focus in advancing biotechniques, biotechnology pipelines, consumables, and related instrumentation and infrastructure alike. As biospecimen science and evidence-based biobanking and sample management practices mature, vendors should aim to validate technology in accordance with definitive evidence-based practice standards and design requirements.

References

1. Miranda, L. (2009) Workshop: scientific and technical considerations for development and management of biobanking protocols. Cambridge Health Institute's: The Science of Biobanking Conference, Crown Plaza Hotel, November 15, 2009, Philadelphia, PA.
2. Technical article: Miranda, L. (2009) The Evidence-based Biobanking Movement. Biopreservation Today,

Volume 1, Issue 3, Summer 2009. Biopreservation Today archives online publication: *http://www.biolifesolutions.com/index.php/biozone/biopreservation-today-newsletter/* (accessed 2009).

3. Miranda, L. (2011) The BRIF: success factors and crucial considerations for ensuring impact. INSERM, BBMRI and Gen2Phen BRIF Workshop, Novotel Campans Cafarelli, January 18, 2011, Toulouse, France.
4. Cambon-Thomsen, A. (2003) Assessing the impact of biobanks. *Nat. Genet.*, **34**, 25–26.
5. National Institutes of Health, National Cancer Institutes Office of Biorepositories and Biospecimen Research Best Practices, v 2010. *http://biospecimens.cancer.gov/practices/2010bp.asp* (accessed 2010).
6. Betsou, F., Lehmann, S., Ashton, G., Barnes, M., Benson, E.E., Coppolo, D., DeSouze, Y., Eliason, J., Glazer, B., Guadagni, F., Harding, K., Horsfall, D.J., Kleeberger, C., Nanni, U., Prasad, A., Shea, K., Skubitz, A., Somiari, S., and Gunter, E., International Society for Biological and Environmental Repositories (ISBER) Working Group on Biospecimen Science (2010) Standard preanalytical coding for biospecimens: Defining the Sample PREanalytical Code (SPREC). *Cancer Epidemiol. Biomarkers Prev.*, **19** (4), 1004–1011. [Epub 2010 Mar 23].
7. Carolyn, C., Compton M.D. (2010) Developing Common Biorepository Infrastructures. *IOM Workshop: Developing Precompetitive Collaborations to Stimulate Genomics-Driven Drug Development* , Washington, DC.
8. Pitt, K., Campbell, L., Skubitz, A., Somiari, S., Sexton, K., Pugh, R., Aamodt, R., Baird, P., Betsou, F., Cohen, L., De Souza, Y., Gaffney, E., Geary, P., Grizzle, W.E., Gunter, E., Horsefall, D., Kessler, J., Kaercher, E., Michels, C., Morales, O., Morente, M., Morrin, H., Petersen, G., Riegman, P., Robb, J., Seberg, O., Thomas, J., Thorne, H., and Walters, C., International Society of Biological and Environmental Repositories (ISBER) (2008) Best practices for repositories: collection, retrieval, and distribution of biological materials for research. *Cell Preserv. Technol.*, **6** (1), Spring, 3–58.
9. Leiva, I.M., Emmert-Buck, M.R., and Gillespie, J.W. (2003) Handling of clinical tissue specimens for molecular profiling studies. *Curr. Issues Mol. Biol.*, **5**, 27–35.
10. Mager, R., Ratcliffe, K., and Knox, K., of the National Translational Cancer Research Report Network, NTRAC (2004) The NCRI National Cancer Translational Research Network Report: Developing an Operational Framework: Standard Workflows, Operating and Quality Control Procedures for the Collection, Storage and Distribution of Frozen and Paraffin Embedded Tissue and Blood, May 2004 (online public source).
11. Mager, S.R., Oomen, M.H., Morente, M.M., Ratcliffe, C., Knox, K., Kerr, D.J., Pezzella, F., and Riegman, P.H. (2007) Standard operating procedure for the collection of fresh frozen tissue samples. *Eur. J Cancer.*, **43** (5), 828–834.
12. Hewitt, S.M., Lewis, F.A., Cao, Y., Conrad, R.C., Cronin, M., Danenberg, K.D., Goralski, T.J., Langmore, J.P., Raja, R.G., Mickey Williams, P., Palma, J.F., and Warrington, J. (2008) Tissue handling and specimen preparation in surgical pathology: issues concerning the recovery of nucleic acids from formalin fixed, paraffin-embedded tissue. *Arch. Pathol. Lab. Med.*, **132**, 1929–1935.
13. Chung, J-Y., Braunschweig, T., Williams, R., Guerrero, N., Hoffman, K.M., Kwon, M., Song, Y.K., Libbuti, S.K., and Hewitt, S.M. (2008) Factors in tissue handling and processing that impact RNA obtained from formalin-fixed, paraffin-embedded tissue. *J. Histochem. Cytochem.*, **56** (11), 1033–1042.
14. Moore, H.M., Kelly, A.B., Jewell, S.D., McShane, L.M., Clark, D.P., Greenspan, R., Hayes, D.F., Hainaut, P., Kim, P., Mansfield, E.A., Potapova, O., Riegman, P., Rubinstein, Y., Seijo, E., Somiari, S., Watson, P., Weier, H.U., Zhu, C., and Vaught, J.J. (2011) Biospecimen reporting for improved study quality

(BRISQ). *Cancer Cytopathol.*, **119** (2), 92–102. [Epub 2011 March].

15. McShane, L.M., Altman, D.G., Sauerbrei, W., Taube, S.E., Gion, M., and Clark, G.M., Statistics Subcommittee of NCI-EORTC Working Group on Cancer Diagnostics (2006) Reporting recommendations for tumor MARKer prognostic studies (REMARK). *Breast Cancer Res. Treat.*, **100** (2), 229–235.
16. Graf, C., Battisti, W.P., Bridges, D., Bruce-Winkler, V., Conaty, J.M., Ellison, J.M., Field, E.A., Gurr, J.A., Marx, M.E., Patel, M., Sanes-Miller, C., and Yarker, Y.E., International Society for Medical Publication Professionals Research Methods and Reporting (2009) Good publication practice for communicating company sponsored medical research: the GPP2 guidelines. *Br. Med. J.*, **339**, b4330.
17. Simera, I., Moher, D., Hoey, J., Schulz, K.F., and Altman, D.G. (2010) A catalogue of reporting guidelines for health research. *Eur. J. Clin. Invest.*, **40** (1), 35–53.
18. EQUATOR Directory of Reporting Guidelines. Website Reference: *http://www.equator-network.org/resource-centre/library-of-health-research-reporting* (accessed 2011).

6
Thinking Lean in Compound Management Laboratories

Michael Allen

6.1
The Emergence of 'Lean Thinking'

'Lean thinking' is a phrase used to describe a production philosophy that seeks to progressively eliminate waste while preserving the essential value (as judged by the customer) in a process. It also focuses on 'steady flow' of processes so that work moves seamlessly from one step to the next with no waiting between process steps.

The phrase 'Lean thinking' is a relatively recent term, coined by Krafcik in 1988 [1] and popularized by Womack and Jones in 1990 [2], who described the production philosophy used by Toyota. The philosophy of '*Lean,*' however, builds on a long history of continuous process improvement and 'Just-In-Time' production. The term is sometimes used interchangeably with the 'Toyota Production System' as described by Ohno [3] and Liker [4], though the term 'Lean' may be used within and outside of manufacturing environments. In this chapter we will examine the essential components of any Lean system and examine their application in R&D, laboratory, and compound management systems.

6.2
The Application of 'Lean Thinking'

A team looking to apply Lean thinking in their environment will first examine their processes to identify areas of waste. Waste is any activity that does not directly add value to the product or service. One way of thinking about this is to consider whether a customer would consider paying extra for this activity. For example, in the pharmaceutical industry a customer may be willing to pay extra for the coating on a pill if it means that they only need to take the pill once a day instead of four times a day. A customer, though, is unlikely to be willing to pay extra if the only difference between two competitors' pills is that one company wants to charge more because they have more paperwork and bureaucracy within their company. Waste may often be, or appear to be, essential in a process, which may make it hard for

Management of Chemical and Biological Samples for Screening Applications, First Edition.
Edited by Mark Wigglesworth and Terry Wood.

people to spot it as waste. For example, a process may have multiple quality control (QC) steps to ensure a good quality product. But these QC steps do not directly add value to the finished product; they simply eliminate defected goods. If the process could be made so that it could not make defective goods, then the QC steps may be reduced or eliminated without changing the value of the goods offered to the customer.

As an aid to identifying waste, Ohno produced a framework of seven wastes:

1) Transportation
2) Inventory
3) Motion
4) Waiting
5) Over-production
6) Over-processing
7) Defects.

Processes may be analyzed to identify (and then reduce/eliminate) all of these.

- **Transportation** is the waste of moving materials. Generally speaking the larger the distance between process steps the greater the waste (both in costs of transportation and costs of slowing the process down). Distance between process steps often encourages batch work as well, building more waiting into the system.
- **Inventory** is the waste of holding material. The costs of this waste are in the value of the inventory held, the costs of storing it (space, specialized storage facilities, and staff), and the potential loss of material because it is not being used before it is no longer valuable.
- **Motion** is the wasted time associated with movement of staff. This includes travel of staff, movement of staff around the workplace, and the movement of arms/hands required when in one position in the workplace.
- **Waiting** is the wasted time between process steps. This may be due to queuing within the system, batch work (where one item of work cannot move to the next process step until the remainder of the batch has been processed), or scheduling problems.
- **Over-production** is waste associated with producing beyond customer requirements. This may be producing excess stock or may be over-engineering the product (e.g., producing excess data that is not valued by the customer).
- **Over-processing** may also be thought of as 'bureaucracy' and would include excess management processes (meetings, e-mails, and forms) and excess quality inspection (after all, processes that are made mistake-proof do not require inspection).
- **Defects** include waste due to poor products (such as impure molecules) and waste due to defects in the process (such as work being delayed because the required resources, tools, or stock were not available). There are frequently hidden costs of poor quality production including the costs of inspection which attempts to prevent the defect reaching the customer, the cost of re-work to replace the defect, and the cost of letting defective goods/data through to the following process steps.

Where goods have reached the next process step or customer significant rework is often required, and there may be a loss of trust/confidence by the customer risking loss of custom.

In addition to the aim of reducing (and ultimately eliminating) these seven wastes, Lean thinking aims for smooth flow in any system, so that work may move between process steps with no waiting between these steps. In order to achieve maximum flow there must be evenness, steady work arrival, and transfer between steps in a system. Variability in work arrival disrupts flow [5], and any reduction in that variability will enhance flow. The problem of variability may be easily observed in shops. Sometimes a queue may have developed at a till in a shop, but come back 30 min later and the queue has gone. Simply by chance several people may have arrived at the till in rapid succession causing the queue to develop. Similarly just by chance later on nobody has arrived at the till and the shop assistant has nobody to serve. The variability has meant that even though the shop assistant is not busy the whole time sometimes people are having to wait to be served.

When looking for variability within a system we must look for both random and systematic variability. Random variability may be the hardest to control, but systematic variability is frequently the major cause of unevenness in production systems. Systematic variability may frequently be seen as 'scheduling,' but closer examination may reveal it is the major cause of pulses (or batches) of work running through the system. For example, in our shop example, there will be a predictable surge in demand at weekends. The shop then has two possibilities. It may increase the number of tills and staff at weekends (though it may struggle to afford the tills or the space needed for them) or it may make some efforts to smooth the demand throughout the week, for example, by tempting people to shop on weekdays by offering a small discount on some weekdays to regular customers. Either approach may be appropriate – it may sometimes be better to be able to flex resources to match demand patterns and at other times it may be better to try to smooth demand/flow. A combination of the two approaches may be best.

Examples of systematic causes of variability in R&D might include yearly targets (with all projects trying to reach the same point at the same time of the year, thus requiring access to the same limited resources at the same time of year), schedules where work is only allowed to be performed intermittently rather than on demand, or work processes that require a minimum amount of work to accumulate before processing. When applying Lean thinking, these sources of systematic variability are targeted for removal (for example, Toyota sales staff were not given sales targets that were driving uneven demand in other car manufacturers) or progressively reduced (for example, if batch work is currently scheduled to take place just once a month then the process should be improved such that it may be run just as efficiently twice a month, with a smaller batch size).

Batch work in particular can significantly slow a process. Once one part of a process decides work is most effectively performed in a batch then that batch of work becomes passed onto others and the whole system steadily becomes a batch system. The frequency that work flows through the system can easily become set by the step that chooses to work in the largest batch sizes. Identification of these bottleneck steps restricting the flow in the system becomes a priority in achieving faster flow.

The progressive reduction in batch sizes requires a different mind-set than the one that is always looking to be able to process ever larger batch sizes more efficiently. A primary aim in designing the process (or the automation) becomes very rapid set-up times, rather than just focusing on the speed of processing once the batch has started. The ultimate aim of Lean thinking is to be able to efficiently process work in 'single piece flow.' For example, in assaying compounds in drug discovery, the process should be able to efficiently and rapidly deal with a single compound if required. Though this may at first seem an impossible or inefficient aim (especially in an industry which has spent the last decade focusing on scale of production through ever larger automation and batch sizes), such single-piece systems are commonplace in clinical laboratories where speed of reporting back diagnostic data is critical and where processes have therefore been designed to allow efficient and rapid analysis of single samples. Lean thinking aims to combine high productivity with rapid flow. Though some systems may never run as single-piece flow, as batch size is reduced (or the frequency of running any particular activity is increased), queuing within the system will decrease and speed through the system will increase [5].

6.3
Lean Thinking in Drug Discovery

Successful implementation of rapid Lean processes should run throughout drug discovery. In a typical pharmaceutical project many disciplines are involved (chemistry, physical chemistry, compound management, pharmacology, drug metabolism and pharmacokinetics (DMPK)). Figure 6.1 shows a schematic of a typical drug discovery process. Some parts of the process, such as high-throughput screening (HTS), run through generally linear processes, while other parts of the process such as lead optimization rely on iterative cycles (Figure 6.2). Compound management processes must support both linear and iterative parts of the process.

For those involved in iterative work, any reduction in cycle time (the time for one complete cycle of the design-synthesize-test-analyze cycle) will be multiplied across as many cycles as the team goes through [6]. For projects to run faster, all involved disciplines must therefore learn to work to support a faster rhythm (though with less work done on each beat of the drum). Petrillo [7] reported that

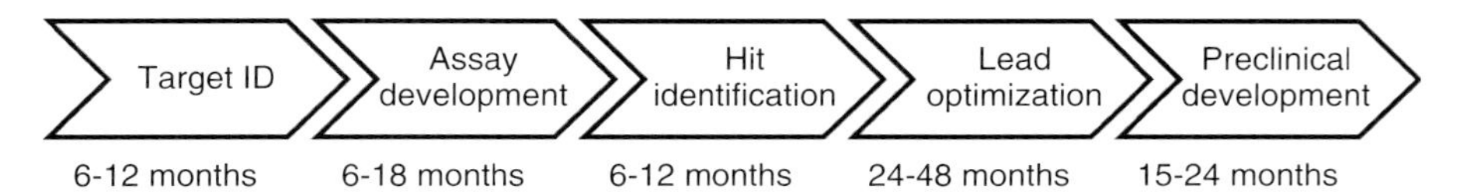

Figure 6.1 Outline of typical drug discovery process. (timings from Ullman and Boutellier [6]).

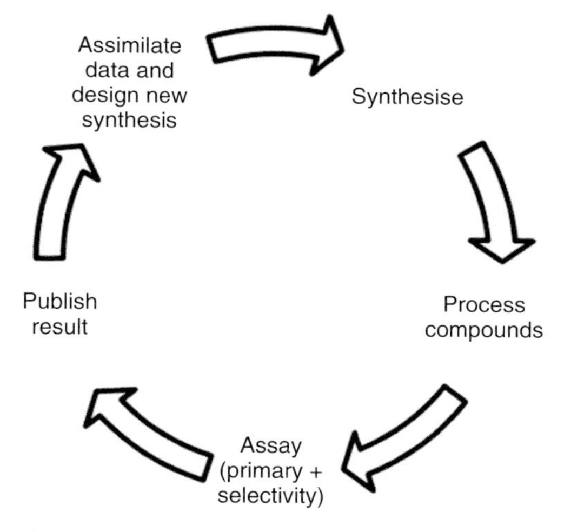

Figure 6.2 A typical lead optimization cycle.

at Bristol-Myers Squibb, improving the speed of processing of samples, assaying compounds, and publishing results would improve the speed of the whole iterative lead optimization cycle. He found that while primary assays were generally keeping pace with chemical design and synthesis, production of secondary selectivity data was delaying the cycle. Compound management systems must be at the heart of, and be able to support, this faster rhythm of work.

Though in this chapter we will focus on compound management processes, previous experience has shown that Lean methodology can reduce compound library synthesis time by half [8] and can reduce the assay cycle time by more than half while more than doubling output [9].

6.4 A Lean Laboratory Toolbox

The full range of Lean tools has been excellently described by Bicheno [10]. We will focus here on tools that are most relevant to laboratory work, and place them in the context of drug discovery and compound management.

6.4.1 Defining Value

Any Lean journey must start with understanding what the customer values. When dealing with in-house processes, the customers are other steps (usually downstream) within the process chain. Each process step should work so that the output can move smoothly to the next step; this requires a clear understanding of downstream processes and customer requirements. In practice there may be compromises to be made, but the ultimate aim of smooth flow between process steps should always be kept in mind. The aim for any compound management department should be for the compounds to be assay-ready (if that is the next step), with no reformatting required by the customer. The customers should define what 'assay-ready' looks like.

6.4.2 Understanding the Current Process – Process Mapping

After understanding what the customer requires, the next step in a Lean journey is to understand the current process as it really is [11]. Two common methods are employed, having their own different strengths:

- **'Brown paper process mapping'**: the team (all those involved in a process) is gathered together in a room, and one side of the wall is covered with a long piece of brown wrapping paper. The team members then discuss the process and describe all the steps. These are written down onto sticky 'Post-It' notes, which are attached to the brown paper (allowing the process steps to be moved around). The final process may then be captured by camera. This method has the advantage of leading to a rapid understanding of the process which is shared within the team (who, as individuals, may only see one part of the whole process). The disadvantage is that often only the value-adding steps are remembered.
- **'Gemba'**: this is a Japanese term meaning 'the actual place.' The process is mapped by following the work and the people. This method has the advantage of describing the process in much more detail, capturing times, movement, and waste. It may reveal the 'hidden factory' of multiple small steps that people think are too insignificant to mention during brown paper mapping. The disadvantages are that it is time consuming (requiring someone to follow the process several times) and that it may worry staff who feel they, rather than the process, are the subject of examination.

Both of the above methods are valuable in process mapping, and should be combined wherever possible.

6.4.3 Identifying Waste

As the process mapping continues, attention should be paid to separating out value-adding activities from the seven wastes. Table 6.1 gives examples of the seven wastes in laboratories, especially in compound management laboratories.

Table 6.1 Example of Ohno's seven wastes in laboratories.

Waste	Examples in laboratories
Transportation	Movement of samples within the laboratory
	Movement of samples between laboratories/sites
	Movement of consumables
Inventory	Queuing of samples to be processed
	Space used to store consumables
	Consumables (which may not be used before expiry)
Motion	Movement of people between laboratories
	Movement of people within the laboratory
	Movement of people to obtain consumables
Waiting	Work waiting for required resources before processing
	Samples waiting for next schedule
	Samples waiting for others in batch to be processed
	Waiting for approval for work
Over-processing	Paperwork/electronic processing
	Meetings and phone calls
	E-mail
	Redundant/repeated steps in process
	Reformatting of results data
Over-production	Dispensing of excess volume/weight of compound
	Supply of unwanted/unused data
	Quality inspections
Defects	Poor-quality compounds dispensed
	Incorrect compound identification
	Wrong volume dispensed
	Re-work (repeat of failed assay)
	Poor-quality assay giving inaccurate data

6.4.4
Standardized Work, Future State Mapping, and Continuous Improvement

'Where there is no standard, there can be no improvement' (Taiichi Ohno, Vice-president, Toyota Motor Corporation). Standard work forms the basis of all process improvement techniques. Standard work does not imply that all types of work must be done in the same way, but implies that one particular type of work is always done in the same way, regardless of operator; individual preferences are eliminated. Standard work also does not imply that it can never be changed (that would be a barrier to continuous improvement), but rather standard work implies 'this is the best way we know how to do this type of work today.' The standard may then be continually improved based on evidence or data (Figure 6.3).

As processes are mapped it may become apparent that different people follow slightly different procedures. Once these are identified, the team should first come to a consensus on the first standard process. As part of this it may be decided that

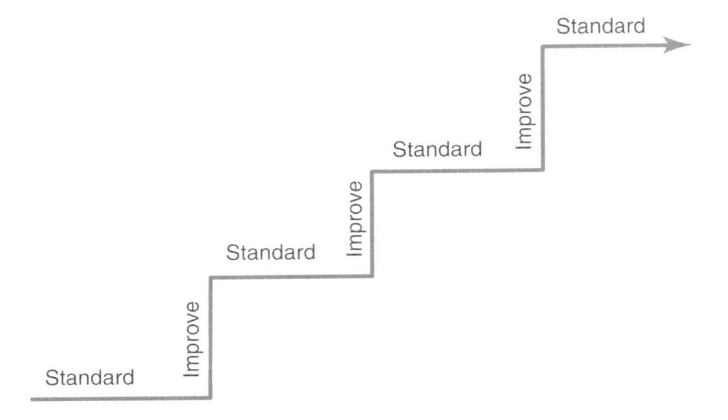

Figure 6.3 Combining standard work with continuous improvement.

some of the different ways of working need investigating. If these can be compared quickly and easily, then this can be done before the first standard is constructed. Otherwise a consensus must be reached, and the differences are recorded for future investigation.

The definition of the new standard is called '*future state mapping*.' Once a process is understood and the wastes are identified, then a new standard way of working is designed and defined by the future state map. It may often be that the team cannot move directly to this new future state (for example, if new equipment or information technology is needed), and so interim standard procedures may be designed to take the process from the current state to the preferred state in a number of smaller steps.

6.4.5
Batch Size Reduction, Changeover Time Reduction, and Workload Smoothing

When dealing with repeated work (such as lead optimization cycles) the rhythm may be driven by a schedule of work. A Lean ideal is to run every product every day. Ideally, in lead optimization the compound management team and the compound profiling teams would work so that every assay could be run every day (as many clinical or pathology laboratories must do with time-critical diagnostic tests). Many pharmaceutical compound profiling departments are not set up to easily perform that task, as in recent years the focus has often been on being able to test larger batches of compounds rather than to test more frequently. Departments may have switched to higher-density plate formats (96-well, then 384-well or even 1536-well), which make testing small numbers of compounds inefficient. As scale has increased, many companies may have seen the speed of the iterative drug discovery cycles actually slow down, despite large investments in the technological infrastructure. Few chemists are now able to dry a compound overnight and hand it to the biologist in the morning for results by lunchtime, as they may have been able to before the advent of the sample management/centralized screening infrastructure.

Manufacturing went through this same problem – economy was built on running large batches of identical work – 'Any customer can have a car painted any color that he wants so long as it is black' remarked Henry Ford about the Model T in 1909. The 'economic batch size' became a favored way of calculating when to run a production batch, divorcing production control from the direct pull of the customer. Customer pressure since then has demanded more choice and options, with the automotive industry having to produce a larger range of products on a regular basis. Lean manufacturers realized that what was creating the pressure for larger batches was the change-over time between different production runs – it was senseless to spend a day retooling the line to only run it for a couple of hours in that configuration. But rather than take the approach of 'therefore we need to run large batches,' Lean manufacturers sought to continually reduce the changeover time to make running ever smaller batches more efficiently. This approach became known as SMED, or Single minute exchange of die, as the target was to be able to retool any part of the line is less than 10 min.

As scale of work in pharmaceutical laboratories has increased (for example, by the use of combinatorial libraries) the focus has often been on the speed of technology per sample rather than on the set-up times. Spending 20–30 min setting up an automated assay station is quite acceptable if there are several hundred compounds to assay, but few people would be willing to spend the same time (and perhaps use the same required dead volumes of expensive reagents) for a handful of compounds. This is not to say that automation should not be used, but that automation should be designed such that it enables rapid processing of a single compound if required.

Two methods are generally used to speed up set-up times. The set-up process is analyzed and divided into 'internal' and 'external' activities. Internal activities must be done at the time, whereas external activities could have been prepared beforehand. External activities should be maximized and taken out of the process, and thought should be given to how internal activities may be speeded up. For example, an automated compound-processing platform may need to be loaded with new work between runs, making the changeover 'internal.' Here, thought should be given to how new work could be loaded onto the platform while the automation is running the previous job, for example, by having safe loading of plates and compounds into a hotel while still being protected from moving parts. This has changed the internal changeover to an external one, and the automation can then move rapidly from one job to the next.

A particular challenge for high-speed but low-throughput work is the choice of high-density screening formats. 384- and 1536-well plates are efficient means to screen large numbers of compounds but may be wasteful when screening small numbers of compounds. Even if assay set-up times can be contracted, an assay scientist may be reluctant to increase the frequency of screening if only a single compound has to be screened each time. Running very low throughputs efficiently in the same team as higher throughput work is likely to take some further development of the standard microtiter plate format, though low-volume lower-density plates are available.

6.4.6 5S and Kanban

A basic premise in Lean thinking is that all available equipment and consumables should be readily on hand. Time should not be lost having to search for (or worse, order) stock because stock has not been organized and maintained well. The Lean laboratory will not want to ensure adequate stock by always having large inventories, as this leads to loss of precious laboratory space to storage and runs the risk of material not being used within its use-by date. A system is needed that keeps the stock (and the workplace generally) well organized and replenishes stock as necessary. We will look to the Lean techniques of 5S and kanban to provide the required solution.

5S may be best summarized by saying 'everything has a place and everything is in its place.' The 5S describes five steps to creating that organized workplace:

1) **Sort**. The first step is to remove clutter from the workplace. In a shared workplace, equipment and material may accumulate over time and no-one is quite sure who owns it or when it is used. One way of dealing with this is a 'red tag' exercise. Place a red tag on all equipment and boxes of consumables. Ask people to remove the tag if they use the equipment/material. After one month it is easy to identify all those items not used in the last month. Ask people to identify any items that they know are definitely used. For the remaining red-tagged items, decide whether they can be disposed of immediately or whether they should be placed in a quarantine area for a further period (but move them out of the daily work area). Excess inventory should also be removed – how many laboratories seem to have an almost inexhaustible supply of gloves and tissue paper cluttering up the workplace?
2) **Straighten, set in order, or simplify**. Everything should be given a designated place close to where it is used. Ensure the place is very clearly labeled (shadow boxes may be used to mark where equipment is to be placed). Labels such as 'miscellaneous' should be avoided. Where possible supplies and suppliers should be simplified – avoid unnecessary duplication of supplies or suppliers (discounts usually get better as the number of suppliers is reduced).
3) **Sweep or shine**. Thoroughly deep-clean the workplace.
4) **Standardize**. Everybody should use the workplace in the same way. Use training and clear signage to remind people how they should be using the workplace.
5) **Sustain**. The first three to four items may be typical of any spring clean. But 5S must continue. Standard practice should be to ensure the laboratory area is properly clean and put back in order after all work. It may be useful to display a photograph of what the area should look like. Deeper cleaning should be regularly scheduled. Some type of auditing system should be put in place (which may be encouraged by reward for the best areas, and display of all scores). Random checks may be used to avoid areas being cleaned only before

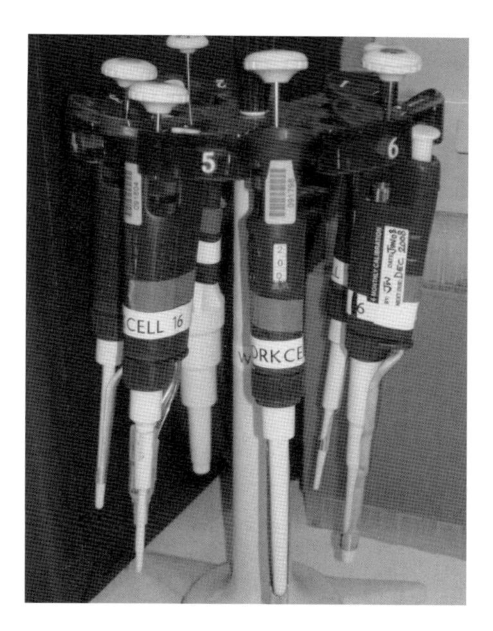

Figure 6.4 Color coding equipment allows rapid identification if it is moved out of its home area. (Photo courtesy of GlaxoSmithKline.)

a scheduled inspection. A successful 5S system should permanently replace the 'old 5S' of Stash, Search, Scrounge, Steal, and Stall!

One challenge that a laboratory may face is that when one area becomes very well organized, other people may sometimes 'borrow' equipment from it as they know how to find what they want quickly. One simple way to help keep equipment in the right place is to color code it (see example in Figure 6.4) so that equipment that it outside of its 'home' is rapidly identifiable.

5S creates an ordered workplace in laboratories or offices. The steps however may also be usefully applied to computers and e-mail (where much of our clutter these days may be hidden).

Once the workplace is organized, a simple and reliable system is required to maintain stock levels without excess inventory being used. A regular check and re-order may be used, but an alternative is to use a kanban (or 'signal') system.

The simplest, and most common, type of kanban is known as a *two-bin kanban* (Figure 6.5). Stock is divided into working and reserve. The reserve stock has a kanban card attached which contains details of the identity, location, and quantity of the stock. When the working stock is exhausted the reserve stock is started. The kanban card is removed and this triggers resupply of the stock (a daily collection round may be used to collect the cards, or a more high-tech electronic system may be used to immediately signal the need to replenish the stock). When the stock

1. Stock is drawn from working stock. A 2nd stock (with kanban card attached) is kept in reserve.

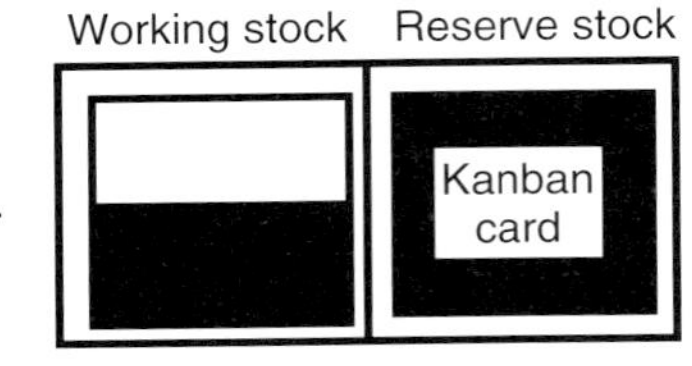

2. When the working stock is depleted, the reserve stock now becomes the working stock. The kanban card is removed and this triggers resupply.

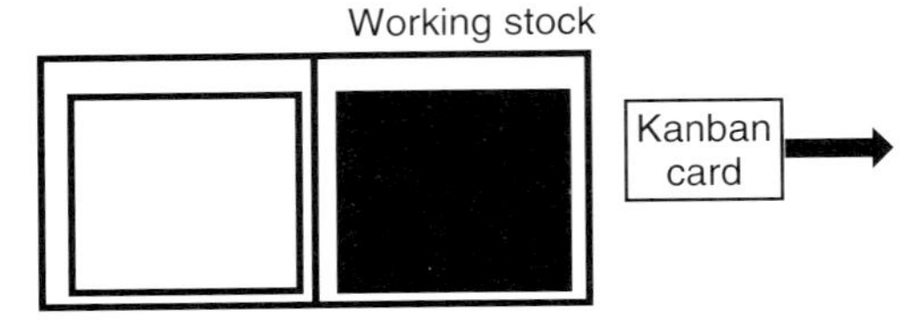

3. The resupplied stock becomes the new reserve stock and the kanban card is re-attached to this reserve stock.

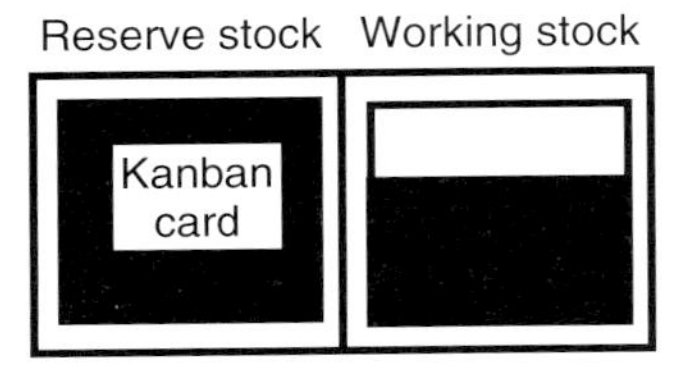

Figure 6.5 A simple two-bin kanban system.

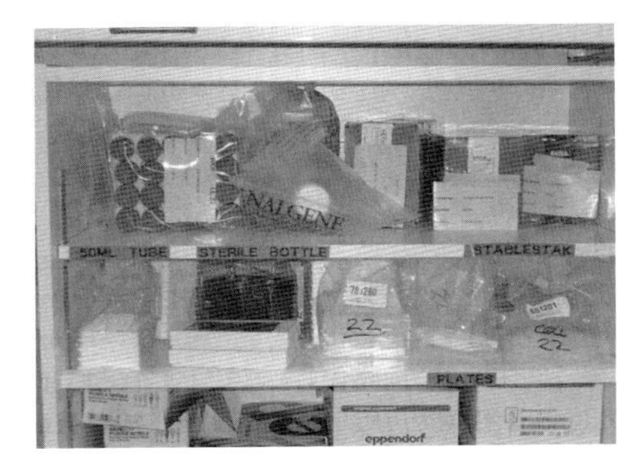

Figure 6.6 A clearly labeled open cupboard (removing doors provides clear visibility of contents). Note the kanban cards on items – when the item is removed for use the card is placed in a kanban box in the workcell and laboratory services replenish the item. (Photo courtesy of GlaxoSmithKline.)

is replenished this new stock becomes the reserve stock and the kanban card is re-attached. Stock levels should be chosen to ensure resupply within the period that the reserve stock is used (it is generally best to start with high stock levels and then progressively reduce as confidence is gained in the system). Figure 6.6 shows an example of a simple kanban system in operation.

6.4.7
Lean Layouts and Flow

Our Lean system must be designed to allow flow of work rapidly from one process step to the next. A common barrier to achieving this flow is a layout that physically separates process steps. Physically separating steps frequently impedes flow by creating handover or queuing points between the process steps. The greater the physical separation between steps the greater is the tendency to introduce more and larger batch work, and the greater is the tendency to create handover points between steps. The addition of these waiting and transports steps can cause processes that should take minutes to take hours or even days, as large batches of work are handed between the centralized functions in the laboratory.

A common approach to designing laboratories is to centralize functions and equipment (e.g., putting all weigh stations together in a compound management laboratory). Where a centralized and function-based layout is used, it is common for individuals to be tied to one step of the process, and batch work is common with large batches of work moving between the functions.

A Lean layout breaks up these centralized functions and disperses the components into multiple cells (see Figure 6.7). The aim of the cell-based approach is to allow much smaller batches of work to move rapidly between process steps. The cell is kept small to allow one or two people to reach all parts of the process easily within a few steps. Most commonly, cells are U-shaped, which reduces the walking distance to cover all parts of the process. Process steps are then ordered around the cell. Minimal space should be left between the process steps in the cell, as spare space often allows queues to build up between process steps. Cells should be made so that they may be rapidly reconfigured as needs change – where possible all equipment and cupboards should have retractable wheels so that they

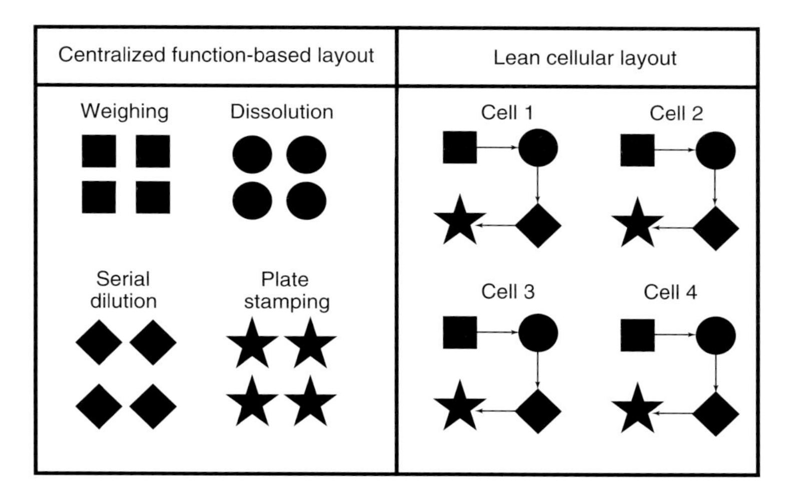

Figure 6.7 Comparison of centralized and lean layouts.

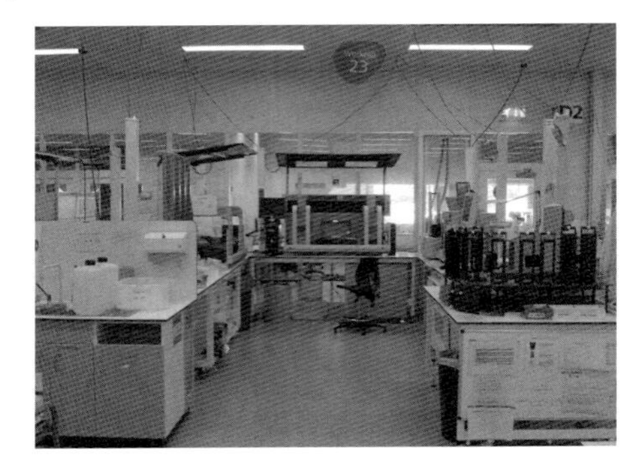

Figure 6.8 A compound processing cell. Stackers have been removed from Beckman FXs to speed processing. (Photo courtesy of GlaxoSmithKline.)

may easily be moved. Figure 6.8 shows an example of a compound processing cell with all equipment easily accessible within a couple of steps.

One more radical step that may be considered during the move from centralized functions to a Lean cell-based layout is removing stackers and hotels from equipment. Stackers require extra set-up and clean-down steps, encourage batch work, and are often significantly slower than use of human hands. A careful study should be undertaken to look at the advantages and disadvantages of stackers or other automation in any operation. A worst-case scenario is where stackers encourage unnecessary batch work and slow the process down due to the slower movement of automation compared with hands, where the operator simply waits for the operation to complete. Frequently it may be possible to remove stackers, speed up work, and not harm the throughput in any way. Reducing the quantity of automation may also improve quality, as the operator is in constant touch with the process and can detect and correct errors as they occur. A reduction in the level of automation may often seem counter-intuitive to staff (especially those brought up in a culture where automation is heavily promoted), and in such circumstances it may be useful to run the process with and without stackers as a trial.

When moving from centralized functions to Lean cell-based layouts, care needs to be taken to balance work between cells. Staff should be also trained so that they can move to wherever the current workload is. One tool that may be used to assist this is a skills matrix (see Figure 6.9) where the level of training and experience of different staff members in different cells is drawn out on a matrix to plan training and ensure there is good cover for all cells.

6.4.8 Total Productive Maintenance

Lean cellular layouts require high equipment reliability. An advantage of centralized layouts is that there is often duplication of equipment so that if one piece of equipment fails another identical piece close by may be used, but in a Lean layout

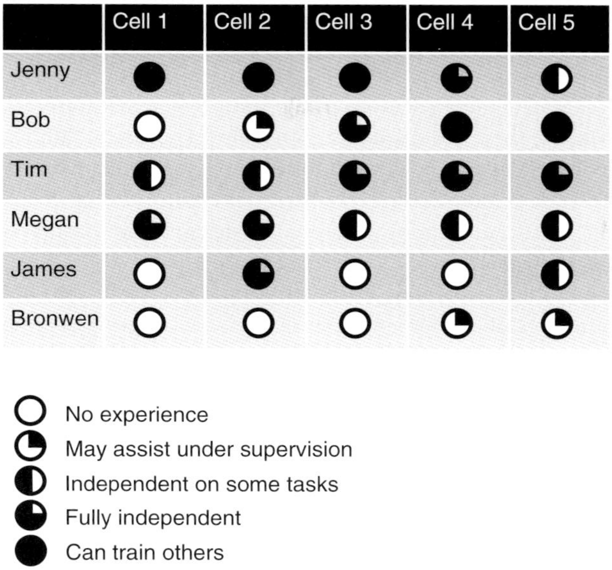

	Cell 1	Cell 2	Cell 3	Cell 4	Cell 5
Jenny	●	●	●	◕	◑
Bob	○	◔	◕	●	●
Tim	◑	◑	◕	◕	◕
Megan	◕	◕	◑	◑	◑
James	○	◕	○	○	◑
Bronwen	○	○	○	◔	◔

○ No experience
◔ May assist under supervision
◑ Independent on some tasks
◕ Fully independent
● Can train others

Figure 6.9 A typical cross-training matrix.

equipment is dispersed between cells so a spare may not be so easily accessible. Additionally in a Lean 5S layout there should be one set of equipment that is trusted by the whole team (rather than individuals having their own preferred pipettes or other equipment).

High equipment reliability is a hallmark of good Lean environments. Some simple steps may be followed to develop and sustain this high equipment reliability:

- **Routine maintenance.** All equipment should have clear preventative maintenance and calibration routines where necessary. This is likely to be composed of user maintenance and specialist maintenance. The required user maintenance should be clearly displayed next to the machine and a simple visual tool used to show when the maintenance is due and whether it has been completed. For example, if equipment requires daily cleaning and inspection, a sign indicating that it has been done should be displayed. The Lean laboratory will not see maintenance routines as fixed in stone: the team will learn from data collected to continually refine these preventative maintenance routines.
- **Monitor all equipment failures and time taken to repair.** In order to continually improve equipment reliability it is important to know all the failures that occur. Over time a database can be built up of what fails on equipment and how often. One simple measure that may be taken is for each piece of equipment to have a failure/repair card placed on it. Should the equipment fail, the nature of the failure and the type of repair required is recorded on the card, and the information is entered into a breakdown/repair database. If 'T-cards' are used,

the cards can be displayed between breakdown and repair so that all may quickly see how many pieces of equipment are currently out of operation.

- **Refine maintenance based on an equipment failure database**. The preventative maintenance routines should be adjusted to anticipate failures or, as a minimum, the laboratory should ensure there are spare parts available for common failures and that there is someone available who can rapidly fix these errors. Equipment failures may also be fed back to manufacturers to help them in improving scheduled specialized maintenance and in improving product design (manufacturers may be ignorant of failures that are fixed by end-users).
- **Monitor and display equipment reliability**. From the database, each cell should be able to print the number of equipment failures and the amount of time during which equipment is unavailable. Generally, equipment reliability should improve (as equipment maintenance becomes better targeted and as any new equipment settles in), and this improvement should then be maintained. A gradual worsening of reliability of any piece of equipment is then a predictor of senescence, and replacement should be planned.

As with much of Lean, the culture is at least as important as the procedures used. A culture must be developed that takes equipment reliability and maintenance very seriously. Users should become used to listening and looking for signs of a change in the equipment (new squeaks or rattles should be investigated). When a team is under a heavy workload, time for preventative maintenance must be preserved (equipment reliability can often be at risk at times when it is needed most). Teams should be rewarded for maintaining a strict preventative maintenance routine.

Figure 6.10 shows two examples of total productive maintenance (TPM) in operation – all equipment has T-cards that are attached to monitor all breakdowns and repairs, and all equipment has the current QC status.

(a)

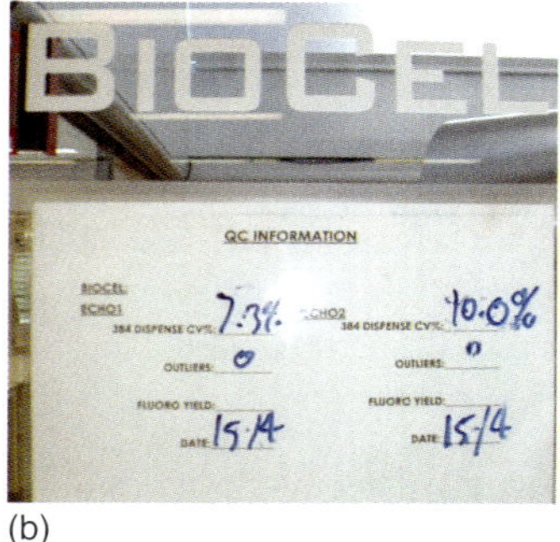

(b)

Figure 6.10 A total productive maintenance system. (a) Equipment has T-card (with barcode to allow rapid logging in database) that is used to report any fault. The type of fault is recorded as well as the repair details. A fresh card is replaced on the equipment once repaired. This forms the basis of a database used to monitor and improve equipment reliability. (b) Current QC information is displayed on all equipment. (Photos courtesy of GlaxoSmithKline.)

6.4.9
Theory of Constraints (TOC)

The Theory of Constraints (TOC) is a production philosophy introduced by Eliyahu Goldratt in his 1984 book entitled *The Goal* [12]. The principle of TOC is that at any one time the productivity of a company is limited by a few critical constraints. The aim of TOC is to identify these constraints and to lever them to a higher level (exposing another set of constraints). One potential constraint a company or a department faces is demand. Internally within a company there needs to be a discussion about what happens when a department is limited by demand for its services – should it compete with other similar departments so that in time fewer resources are needed overall within the company to provide this service, should it try to broaden the range of services it can offer, or should it be left alone while it is not a constraint?

There are five general steps when following a TOC approach:

1) **Identify the constraint**. Do you know what currently limits productivity? Is it staff, budget, equipment, demand, or management systems (e.g., work schedules)? Identifying the key constraint may not be a trivial task, and, as with any Lean journey, the first significant task is to understand the current system well.
2) **Exploit the constraint**. The constraint is the one place where it may be appropriate to engineer a buffer of work. This buffer of work is there to keep the constraint busy as time lost at a constraint may never be regained. Also, work that can be done elsewhere should be removed from the constraint (even when it is slower elsewhere, work should be rebalanced away from the constraint).
3) **Subordinate everything else to the constraint**. The constraint becomes the most precious part of your department (even though it may not be the most expensive part). If the constraint is a piece of equipment, then it must have first call on staff to run it or maintain it. If the constraint is a particular subgroup of staff (with the required skill) then other staff should work to help ensure the specialized staff are always able to focus on the constraining work type.
4) **Elevate the constraint**. Increase the capacity of the constraint. If the constraint is a piece of equipment, can the operating hours be extended? If the constraint is a particular skill set held by limited staff, can other staff be trained with the same skills? Eventually investment in more/new equipment or staff may be required.
5) **Return to step 1**. There is always a constraint. As you alleviate one, another will be revealed – companies need to continually identify and elevate constraints in order to increase productivity or profitability of the company.

As an example, dispensing compounds onto plates may have been identified as the key constraint in an HTS process (this would have been identified by studying the process in step 1). Having identified that this activity is limiting the throughput of the HTS process, priority is given to keeping the plate dispensing equipment busy (step 2). This might mean, for example, that staff choose to load the equipment with new job lists before engaging in any other of their activities that are not directly related to the bottleneck. The HTS plate dispensing now becomes the highest priority for any service or repair staff (step 3) – other work is interrupted if the plate dispenser fails. The plate dispenser should have a higher level of contract cover (ensuring rapid repair) than other equipment (non-bottleneck service contracts may be reduced in cover if needed to fund for the higher cover of the bottleneck equipment). Ways of increasing capacity of the current equipment are also investigated (step 4). Can any of the work be moved to other equipment? Can the operating hours of the equipment be increased (for example, by introducing shift work to cover the machine for more hours per week)? Can the equipment be refined in any way to improve speed? Only when ways of maximizing the equipment capacity have been investigated should purchase of additional equipment be considered. The final step then takes us back to looking at the process again in the pursuit of continual improvement.

6.4.10 The Visual Workplace

A common theme that runs through all Lean thinking is making things visual and displaying information where the work is performed. 5S relies on clear signage of where equipment/consumables should be kept so that anyone can instantly find what they need in the laboratory. Total productive maintenance relies on servicing QC schedules and information being clearly displayed along with the current QC status of the equipment. The quality of the product (both material quality and service quality such as percentage of plates delivered on target) should be displayed at the site of the work. Operators are then continually confronted with any quality issues and will strive to correct them, often becoming proud of their achievements, while managers are encouraged to monitor quality by walking the workplace; any quality issues may then be discussed at the site of production rather than in a distant meeting room.

The visual workplace works to help maintain a high standard of work and monitor quality, but it may also be used over time to monitor long-term improvements. Teams should be encouraged to display long-term graphs on key performance indicators (including quality, throughput, and speed) to help foster a culture of continual improvement.

6.4.11 Engaging Staff

Staff involvement is critical for process improvement; without good staff engagement, process improvement initiatives will fail. Carleysmith *et al.* [13] identified key factors in implementing a successful Lean program in an R&D environment:

- **Sponsorship**. Senior managers must endorse the activities of the staff. A compelling case for change should be made by senior managers.

- **Experience**. Experience in process improvement techniques should be brought in. This may be in the form of secondments from an area of the business which has been through these changes before. Allowing some staff to move to full-time facilitation of process improvement builds experience within the company and facilitates knowledge transfer around the company.
- **Training**. All staff should have some basic training on the aims of process improvement and on an overview of the tools that are used. (One-day courses are usual for the workforce, with managers having two to three days of training and experts having two to four weeks training.) Training should be customized to the audience. Staff may then gain experience of the tools most appropriate to them so that they may embed the tools in their own environments.
- **Build on success**. Start by streamlining the most common repetitive tasks. The results of projects should be communicated and the complexity of projects may be increased as experience is built up. Use facts and data to show improvements (converting time saved into full time equivalents (FTEs) may be used to help put improvements into a common currency).
- **Knowledge sharing**. Share what has worked well and what has not. Not all tools and approaches will work equally well in all environments, so openly share what works well where.

6.5 Streamlining Compound Processing – An Example

Allen and Wigglesworth described a Lean project to speed up the compound profiling process at GlaxoSmithKline [14].

The improvement process began by defining the goal. The aim was to reduce the required latency time of compounds within sample management for compounds being processed for structure–activity relationship (SAR) screening. This was to be done with minimal investment.

The first step was to understand the current process through observation and measurements of that process. Measurements were based on the value stream mapping technique. Compounds sent from chemists were tracked through the process. At each process step, the required resource was noted along with process times broken down into changeover time (the fixed time required before the first compound can be processed through a step, e.g., equipment setup, preparation of the work area) and cycle time (the time required for each unit of work processed; the unit may be a compound or a plate of compounds depending on the stage of the process). A 'current-state map' was drawn up to represent the process (Figure 6.11). It was found that the process took 48 h, but of those 48 h the compounds were only being processed for 150 min (5% of the total lead time). A significant proportion of the raw process time is non-value-adding, that is, time that is not directly contributing to the physical processing of the compounds (e.g., compounds are waiting while other compounds in the batch are being processed, compound is transferred into two intermediate containers and one intermediate

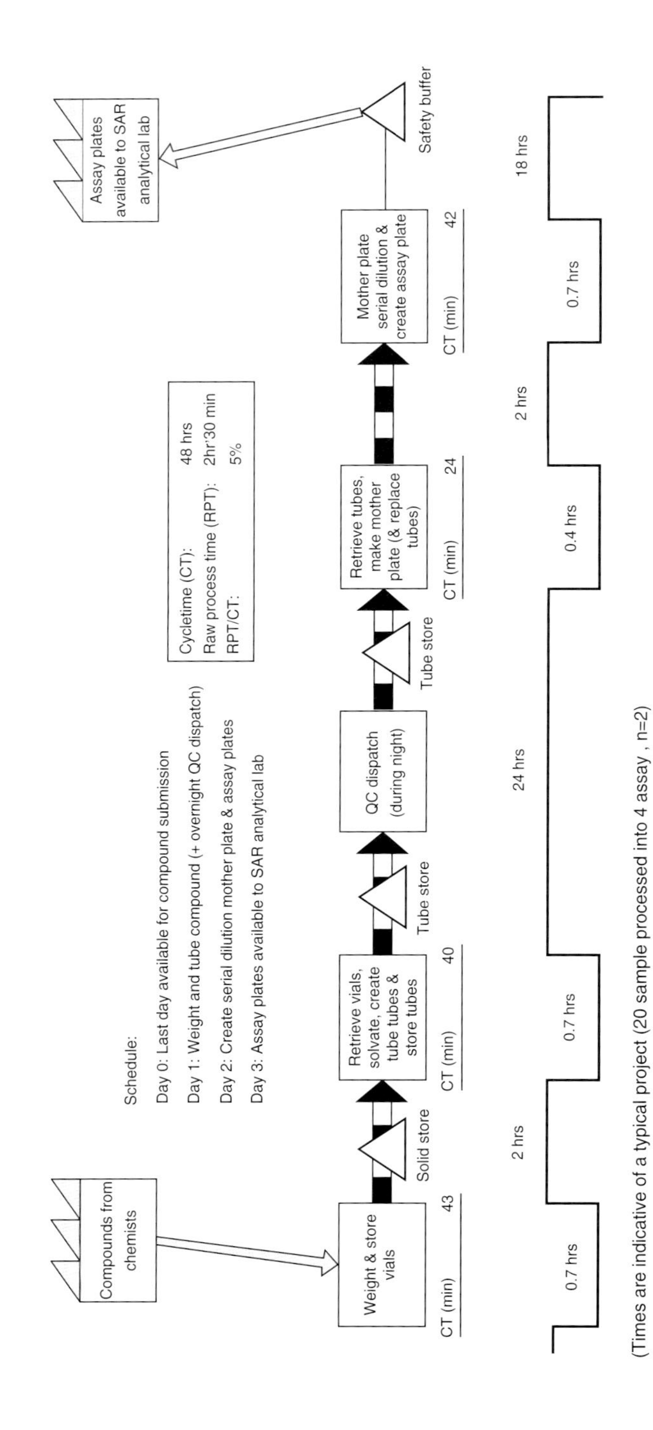

(Times are indicative of a typical project (20 sample processed into 4 assay , n=2)

Figure 6.11 Current-state map for compound processing.

plate before transfer to the assay plate, the compounds are transferred three times into and out of stores during the process, and the compound is frozen and thawed during the process). Of this 150 min total process time, there was significant non-value activity (e.g., waiting for other compounds in the batch to be dissolved). The value-adding time (the time an essential physical process is occurring) for a single compound to a single assay was 11 min: weigh: 1.9 min, dissolve and sonicate: 0.5 min, serial dilute: 8 min, and transfer to plate: 0.5 min. For a single compound, less than 0.5% of the time in the compound management laboratory was value adding. These value-adding proportions are quite typical in processes, and if this was reflected across the drug discovery process there may be potential to speed drug discovery by more than 100 times!

As samples could not be assayed independently, the process was redesigned to move a single batch of compounds faster through the process. Several points in the process could be significantly improved: (i) Chemists were already weighing compounds to transfer the appropriate quantity to compound management; rather than reweighing, chemists were asked to weigh out within a tighter range and record the weight. This eliminated reweighing in the compound management laboratory. (ii) Compounds had previously been loaded into a solid store and then removed; these steps were unnecessary. (iii) Compounds had to wait for an overnight QC dispatch before being dispensed into assay plates. The results of the QC were not available at the time of the assay so delaying the process for the QC dispatch process added no value. The QC dispatch was moved out of the critical path of compound processing. (iv) The assays were scheduled to run the day after the compound processing (with the plates frozen overnight to avoid evaporation of low-volume dispenses). This safety margin was excessive and could be reduced to 2 h and the freeze/thaw step removed. A planned 'future-state map' was drawn up to illustrate the new planned process (Figure 6.12). These changes were relatively simple to make (some IT needed changing but no investment in equipment or staff was required) and reduced the processing time from 48 to 4 h and raw process time from 150 to 110 min. The new process was established as a trial over a two month period; the immediate processing trial evolved a process that could take solid compounds, generate assay-ready plates, screen these compounds, and post data within the same day. This trial demonstrated a greatly improved solid-to-assay-ready plate cycle time of 2 h and 19 min (standard deviation (SD) 0.44 h, $n = 10$). Processing compounds faster also eliminates intermediate storage steps, so this streamlining of the process may also improve quality as storing compounds at 4 °C and rewarming to room temperature may cause loss of purity of some chemotypes [15].

6.6 Summary

Lean thinking can help to significantly improve laboratories and can increase speed and efficiency in compound management laboratories. Speed increases of an order

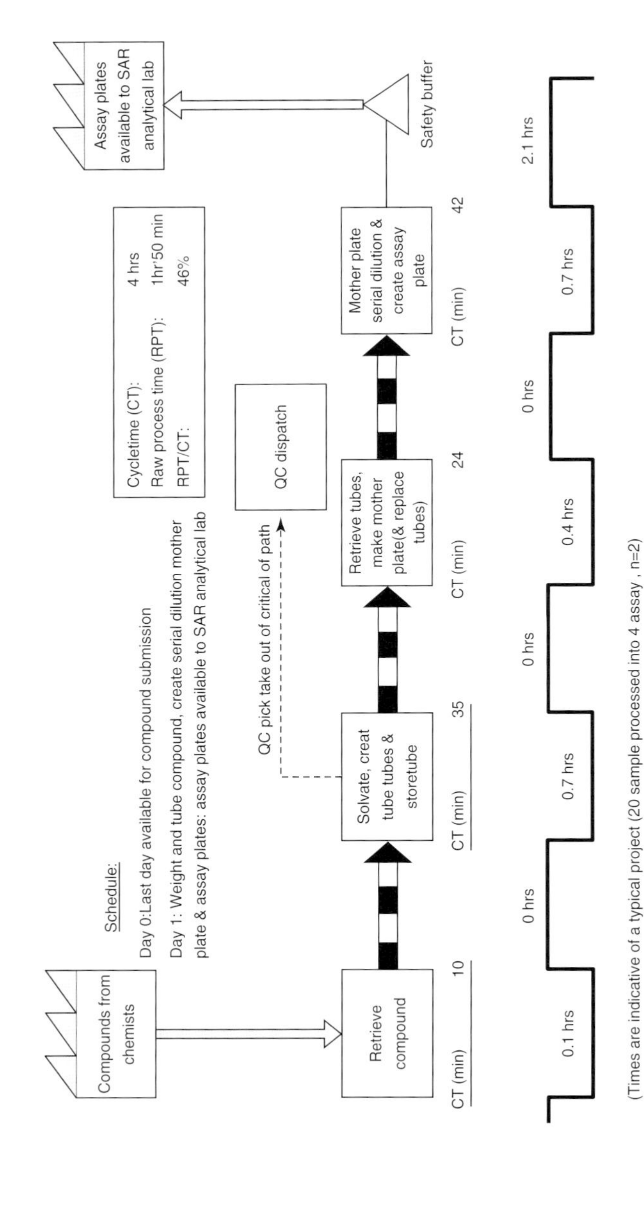

Figure 6.12 Future-state map for compound processing.

of magnitude may frequently be made, costs may be reduced, and quality may be improved. Key concepts from Lean thinking that are useful in laboratories are:

- Understand your current process, and know what limits your speed and throughput.
- Break up centralized functions to remove handover points in systems.
- Co-locate process steps in small cells.
- Know who can do what, and plan training to ensure you have a flexible workforce that can move to where the work is.
- Have a place for everything and have everything in its place.
- Replenish stock only when necessary.
- Monitor all equipment failures – use data to improve routine maintenance.
- Use a visual workplace to give people the information they need to work and to monitor quality, quantity, and speed over the short and long term.
- Equip and involve everyone.
- Permanently change the culture of the research laboratory.

References

1. Krafcik, J. (1988) Triumph of the lean production system. *Sloan Manage. Rev.*, **30**, 41–52.
2. Womack, J.P., Jones, D.T., and Roos, D. (1990) *The Machine That Changed the World*, Rawson Associates, New York.
3. Ohno, T. (1995) *Toyota Production System: Beyond Large-scale Production*, Productivity Press Inc.
4. Liker, J. (2003) *The Toyota Way: 14 Management Principles from the World's Greatest Manufacturer*, McGraw-Hill.
5. Hopp, W. and Spearman, M. (2000) *Factory Physics*, 2nd edn, McGraw-Hill/Irwin.
6. Ullman, F. and Boutellier, R. (2008) A case study of lean drug discovery: from project driven research to innovation studios and process factories. *Drug Discov. Today*, **13**, 543–550.
7. Petrillo, E.W. (2007) Lean thinking for drug discovery – better productivity for pharma. *Drug Discov. World*, Spring, 9–14.
8. Weller, H.N. *et al.* (2006) Application of lean manufacturing concepts to drug discovery: rapid analogue library synthesis. *J. Comb. Chem.*, **8**, 664–669.
9. Sewing, A. *et al.* (2008) Helping science to succeed: improving processes in R&D. *Drug Discov. Today*, **13**.
10. Bicheno, J. (2004) *The New Lean Toolbox: Towards Fast, Flexible Flow*, 3rd Revised edn, PICSIE Books.
11. Rother, M. (1999) *Learning to See: Value Stream Mapping to Add Value and Eliminate Muda*, Lean Enterprise Institute.
12. Goldratt, E.M. and Cox, J. (1992) *The Goal: A Process of Ongoing Improvement*, North River Press.
13. Carleysmith, S.W., Dufton, A.M., and Altria, K.D. (2009) Implementing Lean Sigma in pharmaceutical research and development: a review by practitioners. *Res. Dev. Manage.*, **39**, 95–106.
14. Allen, M. and Wigglesworth, M.J. (2009) Innovation leading the way: application of lean manufacturing to sample management. *J. Biomol. Screen.*, **14**, 515–522.
15. Matson, S.L. *et al.* (2009) Best practices in compound management for preserving compound integrity and accurately providing samples for assays. *J. Biomol. Screen.*, **14**, 476–484.

7
Application of Supply Management Principles in Sample Management

Paul A. Gosnell

7.1
Introduction

The past decade in sample management has been an extremely dynamic period of development and growth. During this period, all of the major pharmaceutical companies and many of the larger biotechnology companies invested heavily in their sample management capabilities. Much success was achieved, and, through a partnering with technology vendors, sample management has become an integral part of successful drug discovery. This period of time heralded rapid change, and a myriad of paths were pursued in drug discovery, which resulted in a requirement to rapidly evolve sample management in response to the emerging discovery paradigms, sometimes with a very short perspective of the long-term strategy. Likewise, as organizations expanded their discovery operations, sample management replicated systems and capabilities across the organization, often in a 'One Size Fits All' fashion, to rapidly provide much-needed services and to contain costs. However, in doing so, some sample management groups also duplicated their design flaws and further complicated their weaknesses. Moving forward, sample management must take a new approach and avoid falling victim to the same pitfalls created in the past decade.

This article presents supply management and logistics principles commonly applied in non-sample-management supply chains [1] and discusses how these principles can be applied in sample management.

7.2
Common Pitfalls of Sample Management

There have been several obstacles that have hindered success in sample management. These have included budget constraints, cultural resistance to change, an inability to obtain reliable demand forecasts, failure to foresee a major shift in direction within the customers' business, and attempting to force a 'One Size Fits

Management of Chemical and Biological Samples for Screening Applications, First Edition.
Edited by Mark Wigglesworth and Terry Wood.

All' approach in problem solving. The last mentioned obstacle is the most serious and has been self-inflicted.

In an environment of budget and resource constraints, it is common practice to deliver standardization and reuse as much as possible the processes and methods successfully implemented. However, by focusing too much on the standardization and reuse aspect, the end customer's requirements can be overlooked. From database management systems [2] to project management approaches [3] to Six Sigma [4], 'One Size Fits All' is a practice that should be avoided and has outlived its time.

7.2.1 One Size Does Not Fit All

To avoid the 'One Size Fits All' pitfall one must examine sample management as it really is: a logistics service supporting the discovery supply chain. Logistics is a word often used in sample management discussions, but it is a term that is often misapplied. Most commonly this term was misused as a 'bucket' for all of the things that were not science. In doing so, sample management failed to understand that logistics is a science. As a science it can be observed, studied, and measured in a systematic way to determine its nature and principles. The Council of Supply Chain Management Professionals, formerly the Council of Logistics Management, defines *logistics* as: 'that part of the supply chain process that plans, implements, and controls the efficient, effective flow and storage of goods, services, and related information from the point of origin to the point of consumption in order to meet customers' requirements' [5]. Using that definition, *sample management* can be defined as the logistics within the drug discovery supply chain.

7.3 Sample Management and Supply Chain Concepts

From this point forward in this chapter, the discussion will focus on sample management and the application of supply chain strategies as they relate to drug discovery, from a consumer-oriented perspective.

Sample management will be defined as those processes required for transferring, storage, and routing of substances along the supply chain from the point of creation to the point of results. This supply chain comprises three contributors: (i) those that produce, (ii) those that distribute, and (iii) those that consume.

- **The producer** is any organization or process that produces, synthesizes, extracts, or acquires substances for the purpose of drug discovery. Typically, this includes medicinal chemistry teams, combinatorial chemistry teams, natural products chemistry, and compound acquisitions efforts.
- **The distributor** is sample management, previously defined as those processes necessary to move a proprietary substance along the supply chain. Distributors manage the flow of substances and are responsible for coordinating the logistics

and quality control processes necessary to move product and product data to the consumer. Because sample management is a distribution service it has many opportunities to identify much of its strategy by examining the practices developed in other industries and avoid the temptation of reinventing the wheel. An excellent reference is Tompkins' Distribution Management Handbook [6].
- **The consumer** is any organization that uses the samples and supporting data to produce results, including therapeutic areas, discovery chemistry, and discovery biology organizations.

7.3.1 Goods and Services – Classification and Strategy

The goods and services that the sample management distributors provide can be placed into two broad classifications, information and inventory.

- **Information**
  - Sample registration and informatics
  - Sample history
  - Consumer requests management and inventory availability
  - Restrictions management.
- **Inventory**
  - Sample preparation, reformatting, and distribution
  - Sample storage
  - Corporate archive services
  - Acquisition and receipt of third party/alliance samples
  - Receipt of internally produced samples.

Next, evaluate the sample management team's goods and services from the perspective of the end user, determining whether they are core or non-core to their business process and whether the customer need is fairly standardized or requires customization to support the customer's objectives [7].

- Non-core goods/services are not directly tied to the consumer's process, but add significant supporting capabilities to the consumer's business.
- Core goods/services are those which directly add value to the consumer' business.
- Standard goods/services are those that serve all or a majority of the consumer base without the requirement for customization.
- Custom goods/services typically require specialty solutions and are directly tied to the support of a specific consumer niche.

Using the Riggs model presented in Figure 7.1, further classify and determine the positioning of the goods and services based on their relationship to the consumer. The Riggs model goes on to assign rules for identifying strategies based on positioning within the model.

- **Quadrant 1: Non-core/custom**, represents those goods and services not directly connected to the end customer but contributing significant supporting capability

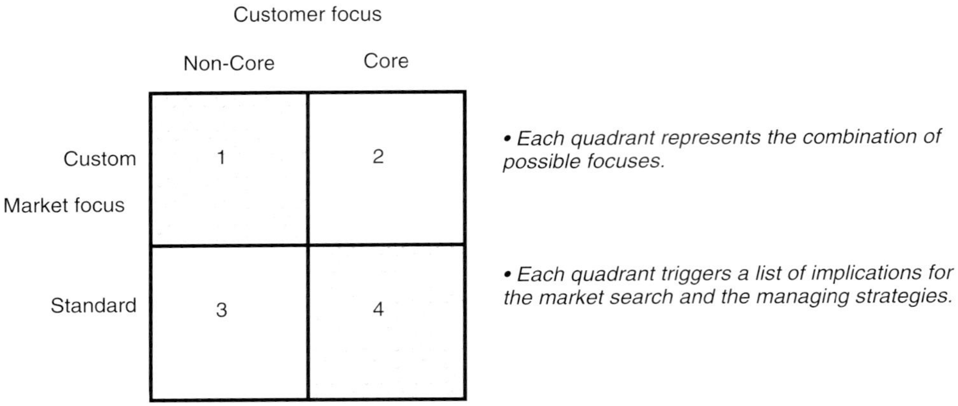

Customer focus

Market focus	Non-Core	Core
Custom	1	2
Standard	3	4

- *Each quadrant represents the combination of possible focuses.*
- *Each quadrant triggers a list of implications for the market search and the managing strategies.*

Figure 7.1 Positioning of goods and services. Each quadrant represents the combination of possible focuses. Each quadrant triggers a list of implications for the market search and the managing strategies. (D.A. Riggs, p. 119 [1].)

to the customers business. Discovery teams may vary from site to site; therefore a local decentralized strategy is recommended. Solutions for services in this classification can best be found by seeking out a low-cost supplier and deploying industry-proven technologies. Sample management activities in this quadrant might include: sample registration and identification, management of restrictions, and specialized distribution services such as screening follow-up support and external collaboration support.

- **Quadrant 2: Core/custom**, represents those goods and services that are directly connected to the discovery team and contribute significant creative capability to their business. Consumer requirements may vary from site to site, and again a local decentralized strategy is recommended. However, this time, since creative capabilities are required, solutions for services can best be found by seeking out the innovative supplier that can provide leading-edge technologies. Sample management activities in this quadrant are starting to push the envelope of what is commonly thought of as sample management, and they include specialized distribution services for niche customers. This is not to imply that industry-standard solutions and equipment cannot be used to support this niche, rather that the implementation will require customer-specific focus.
- **Quadrant 3: Non-core/standard**, represents those goods and services that are not directly connected to the discovery team and contribute essential support capability with consumer requirements that do not vary site to site. For these goods and services, a centralized approach is the best model, and solutions can be found by seeking out a low-cost supplier and trailing-edge technologies. Activities in this quadrant include standardized storage and distribution services such as the corporate compound archive services, sample tracking, and history data (services that mitigate the risk of property loss), preparation of screening

decks, compound acquisition, and compliance with regulatory agencies as well as import/export compliance.

- **Quadrant 4: Core/standard**, represents those goods and services which are directly connected to the end customer and contribute significant competitive capability, with consumer requirements that do not vary site to site. Because the activity is directly connected to the end customer, a local, decentralized approach can provide the best coverage. Best solutions can be found by seeking out an innovative supplier that provides leading edge technologies. Activities include support of post-lead generation groups including lead evaluation biology, support of activities within the therapeutic areas and discovery working groups, preparations of assay-ready plates containing both research samples and assay controls and references, and new technologies such as nano-dispensing technologies.

The results of this model will lead toward the development of a sample management strategy that centralizes services, where centralization and standardization will achieve lower cost success, decentralizes services where flexibility is needed in meeting the partners' business requirements, and opens opportunities to deploy sample management processes beyond the traditional boundaries of the sample management organization. As a result the company's sample management strategy will become one that no longer forces a 'One Size Fits All' constraint on discovery.

7.4 Implementing the Sample Management Strategy

The successful implementation of any sample management strategy will depend upon the sum of its parts. These include: the sample management organization, sample management informatics, robotic storage and sample processing, identification of opportunities to expand process boundaries, the positioning of inventory, and the routine application of efficiency metrics and standardized quality measures. Any successful implementation must also be founded on organizational behaviors and the organization's culture. This culture must promote a true partnering between the sample management team, its scientific partners, and its equally important partners in automation and information technology. Failure to establish an open and transparent culture that spans discovery will result in much wasted time and money with little benefit to the drug discovery supply chain. Throughout the design and implementation phase, avoid the 'One Size Fits All' pitfall.

7.5 Sample Management Organization

In evaluating the goods and services of the sample management organization, it may be seen that a mixture of classifications results. This is normal and demonstrates that the sample management organization must support a broad spectrum of customer requirements that covers non-core/standard services, such as a centralized collection archive, to core/custom services, such as geographically dispersed lead optimization satellites. The question as to how this mixed approach can be managed is not new to sample management and is seen at multinational organizations such as Novartis, where their response has been to create a network of centralized and decentralized efforts across multiple sites and teams [7]. At Novartis, they met the challenge of managing the support of eight research locations, centralizing their solids collection at Basel and supporting their various sites through an integrated network of local stores dedicated to site specific activities. Similarly other companies have implemented international networks for sample management.

Some were implemented as a single organization, with each site reporting to a central leadership team, while others were implemented as a federated organization, with local teams under the leadership of their customer. Though these models differ, both have been successful, suggesting that organizational structure is not the key to success. Advantages and disadvantages can be identified in either approach. Regardless of whether a traditional pyramid structure or a dispersed matrix is created, the key to success is not how it is structured but rather how it behaves. Sample management, like most supply management organizations, needs to focus its efforts on the 'work tasks and role definitions, processes and organizational mechanisms and competencies that work together to span functional groupings, and geographic and business locations' [8]. However, the internal closed market of a drug discovery operation presents the challenge that the primary customers, chemistry and biology can play the role of both the producer and consumer, and they themselves are part of the supply chain.

The sample management leader's role is to provide focus on the needs of the both the producer and the consumer, conducting an analysis of the 'customers' market,' performing cost analysis, managing resources, and benchmarking new technologies. Leadership needs to be provided at two different levels. The first is at the location, focusing the teams' efforts on the local customer, the second being at the higher supply chain level, developing and implementing strategies across the length of the total supply chain.

The sample management team, the distributor, can itself take on the roles of producer and consumer. In relation to the chemistry groups, sample management is a consumer, in that it consumes the data that chemistry provides, while, at the same time, sample management reformats and distributes the chemistry product to accelerate its flow into biology, taking the role of producer. At times, these roles seem to collide with each other, causing sample management to focus on internal processes, losing sight of customer's need. To prevent this from happening, one should recognize that both external and internal processes are equally important. From the supply chain perspective, an ideal sample management organization addresses both needs, dividing their teams into sub-teams. The first is the operations team that manages the inventory goods and services – the physical production processing of sample work orders – and physically manages and maintains the operations of equipment and inventory. The second is the order management team that manages the information goods and services – the tactical communications with customers, the customer request processes and the collection of demand forecasts – and serves other areas that require sample management data.

7.6 Sample Management Informatics

The successful management of the sample management operations relies greatly upon the design and success of the underpinning informatics. For this reason, whenever possible it is important to lay this foundation prior to building the organization and production capabilities. The requirements and functionality of sample informatics systems are laid out in Chapters 13 and 14. Missing the opportunity to begin services with quality informatics can result in mixed success and a high degree of customer dissatisfaction, let alone an inability to grow and change the sample management efforts in parallel with the changing discovery paradigms. Very few sample management organizations have had the opportunity to begin operations from a clean informatics slate. Likewise, several sample management organizations have been stalled by dismal software projects in efforts to establish multi-site and global integration. A prime contributor to these failed projects has been scope creep. This is mainly because most companies did not start their evolution of sample management on a multi-site or global level; it began on several local levels simultaneously and was then exported to other sites or brought together under one global umbrella, resulting in a multitude of local systems that did not speak to each other. The scope creep in developing the customer ordering systems and the global integration level is the typical 'Panacea' requirement phase, attempting to include everything possible into the inventory request and search utilities, aggravated by a 'throw it over the wall' approach in managing the project by both IT and the business.

To avoid this pitfall, each informatics project should be started with a detailed assessment of the situation and target before identifying the proposal. A full understanding of the user specifications, including both operational and customer users, is highly important. Each requirement identified should be evaluated to understand the benefits (both tangible and intangible) delivered and ranked as to their priority and impact.

Next, evaluate options to deliver the benefits, including in this evaluation options for both building and buying a solution.

It has only been in the past few years that sample management has had the option to purchase 'industry standard' sample management applications. During the preceding decade the only real software choices that were available for sample management were to develop it internally or outsource it as a custom development. Today there are quality software packages commercially available, with a growing number of companies using the same sample management software as the basis of their informatics. Commercial options include Titian's Mosaic, LabVantage's LIMS, and Xavo's Lab Logistics. These off-the-shelf, semi-customized offerings point to the fact that sample management software has become a commodity. As such, these tools provide very important enabling capabilities, but no longer present the same competitive edge as represented in the early years of high-throughput screening.

However, if the target objectives truly represent a competitive advantage and cannot be met by the commercial tools, so that internal development or external customization is the choice, then keep it simple and apply this basic rule – sample management is a supply management process. Examples can be found outside of sample management and drug discovery. Those internally developed systems that have demonstrated longevity and success have taken this approach [9].

- Start by reading a good reference for supply management informatics such as Tompkins' Distribution Management Handbook [10]. Then, move on to case studies for companies like Walmart, UPS, and Amazon.
- Next, divide the project into sub-teams that will focus on each of the system modules.
- Develop the sample management informatics as a modular and object-oriented solution, comprising:
  - Sample submission
  - Shipping and receiving
  - Customer ordering
  - Work order management
  - Inventory management database, as the core.
- Keep the sample management system separate from the results management systems, so that each can evolve separately, but aligned.
- Provide the peripheral systems access via a reliable application programming interface.
- Keep the control of robotics and workstations separate and managed by their own software.
- Plan for significant data cleanup and data migration.
- Validate the systems as though they were working under the auspices of good laboratory practices. Test plans should not only test the best case scenarios, but also the worst case and the stress loads of the processes.

7.7 Avoid Monolithic Silos of Excellence

One of the most apparent examples of where sample management has forced the 'One Size Fits All' approach can be seen in the selection and implementation of robotic storage systems during the early years of sample management growth. These systems represented significant engineering advances when they were initially designed, but as newer technologies became available, the older systems quickly fell behind. Most of these systems were bespoke solutions that were developed in support of a specific customer niche, such as high-throughput screening, and were only designed to manage a single container type: tubes only or plates only. To change these systems would require significant costs and reengineering, and this change would take on a life of its own, most likely coming to a conclusion about the same time that the business partners changed toward a new discovery direction. Such was the history of Eli Lilly and Company's FLEXSTORE at Research Triangle Park in North Carolina, and, as described by that team, the system had become a monolithic silo of excellence [11].

To avoid monolithic silos, start by selecting solutions that can be implemented quickly: tools that are reusable and can change or are less affected by change in the end customer's business. The systems should be capable of managing a variety of container types – vials, tubes, and plates – with little or no re-engineering. Next, adapt to the tool rather than trying to build a custom tool.

A distinct advantage is held by those companies just beginning the process of establishing sample management because they are not faced with the challenge of managing the transition from legacy systems to the new system. Maintaining support to the customers (who depend upon sample management without interruption) while covering the conversion costs and the additional staffing requirements during the transition all present major problems. The solution may not be cutting the losses and abandoning the current process but rather scaling back the operation and only supporting those areas which are best served by the system. For example, if the system provides good coverage for the corporate archive needs but does not provide the rapid services required by current discovery programs, use the system for the archive. Then acquire newer and smaller systems that can be geographically deployed in a federated model to support the local needs of the therapeutic areas. In the past few years, the trend for automation providers has been to develop smaller more flexible systems that for the most part are almost turnkey solutions [12]. These systems are almost self-contained and can be located across the organization where the needs exists.

7.8 Position and Synchronize Inventory

The past model of sample management has been one where compounds are sent to a centralized facility and then redistributed to the end customer. This model works fine for non-core and standard services, such as screening deck preparation support or corporate archive services, but this model comes apart when faced with the needs of lead evaluation or therapeutic area chemistry and biology, areas defined as core and custom. After all, historically, chemists made the compound and delivered it just a few doors down to the biology team that processed the sample and produced the results. Preparation of samples can be improved when handled by the standard practices of the sample management team, but this is not the same as performing sample management operations distal to the chemistry or biology teams. By establishing a federated sample management satellite adjacent to the discovery area, sample management can provide services and quality management, while at the same time using the site as a collection point for new diversity that is required upstream to enrich lead generation and preserve the compound for future generations of discovery.

Inventory positioning and synchronization is the key. Position inventory where it is needed, and move it where it is required. Not all inventory is needed by all scientists at every location. So evaluate the use of the inventory to determine where opportunities exist to manage satellite inventory and where requirements for inventory synchronization must be achieved (see Figure 7.2).

In this Venn diagram:

- The **corporate collection** represents all compounds owned by the company and should be held as a centralized repository. It is a non-core standard service and its proximity to the end customer is of lesser importance than factors such as lower cost and preservation.
- The **screening collection** is a subset of the corporate archive. Not everything within the corporate archive is drug-like, so the screening collection excludes

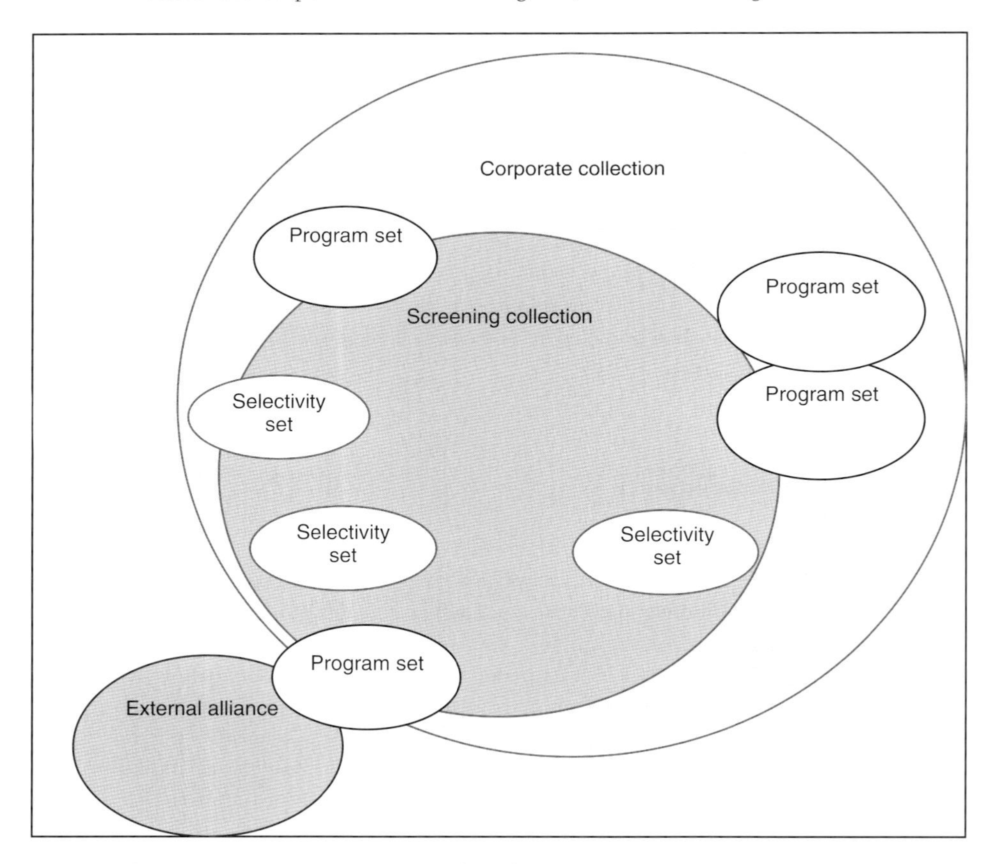

Figure 7.2 Inventory positioning and synchronization.

the non-drug-like compounds (see Chapters 2 and 15 for details of drug-like and set generation principles). The purpose of this subset is to promote lead generation and support the screening teams. A copy (or copies) of this inventory should be proximal to the screening activities to support their rapid follow-up requirements which are 'just in time,' while the preparation of the screening set is a 'build-to-inventory' activity and can be either distal or proximal, depending upon the availability of resources. Basically, the preparation of the master set, a non-core standard service, and its replication, a core standard service, need not occur at the same location.

- The **selectivity set** is usually a subset of the larger screening collection. Its preparation can be performed by 'cherry-picking' the screening inventory, with the same process as that used to process a follow-up request. The inventory is a core standard service for profiling teams, and the inventory should be moved out so that it is in close proximity to those assays.
- The **program set** starts out as recently created compounds specific to a program, making it core standard or custom, depending on how far forward the activity is deployed. The set is held under restrictions during the first part of its life cycle but then migrates into the screening collection and corporate archive, making it non-core standard. These sets present an opportunity for inventory synchronization. Since many of the activities that support lead evaluation and the therapeutic areas are not always replicated across the organization, make an effort to replicate this inventory across the organization at the point of entry, thus eliminating shipment time as part of the follow-up request critical path.
- The **external alliance** set is shown as being separate from the other sets in the diagram. This is to emphasize that these compounds are specific to the alliance and its use is covered by legal agreement. Typically these sets can only be tested in a pre-determined list of assays, and so the inventory data and assay results should be quarantined from the larger discovery community. The positioning of the inventory should be proximal to where it will be used. The standard methods of preparing and processing the set will mostly not differ from any other special set, but since its use, from the end customer's perspective, is specific to their efforts, it is a core standard inventory.

7.9 Expand the Sample Management Boundary

The idea that all activities that are defined as sample management need take place within the boundaries of the sample management organization is flawed. Many opportunities exist to position sample management capabilities outside of the boundaries of the sample management organization in direct support of discovery. In these implementations, the processes and methods of sample management are positioned in the end customer's area, making these services capable of providing dedicated support and specialization not available from the centralized sample management organization. The end result is that sample management quality can be achieved while making the customer self-sufficient, at least in part. This success is of course dependent upon managing these remote operations with the same rigor and discipline as though they were collocated with the sample management operation. When successfully implemented, these are a good choice for focused screening and discovery team local support requirements. The design of these systems can be used to minimize compound usage and at the same time capture diversity that can be fed into the screening collection. Moving services beyond the boundaries of the sample management allows for the externalization of services and customization to meet site-specific requirements. Additionally, the option of outsourcing of sample management activities has become a reality in the past decade that did not exist previously (see Chapter 18). Companies such as ASDI, Sigma Aldrich, Spectrum BioSciences, and Specs provide sample management actives that cover the range from compound procurement to reformatting of a solids collection to long-term housing and distribution of collections for archive and high-throughput screening support.

7.10 Measuring and Assessing Effectiveness and Quality

Having now established the sample management processes it is important to continuously measure the effectiveness and quality of the teams' operation and the benefit that it provides to the drug discovery supply chain. During the development phase, estimates of cycle time, cost reduction, and efficiency were made, and it is important to follow up these projections by measuring the benefits delivered. Those areas that fall short of expectations are opportunities for process improvements. Here are but a few of the measures and assessments that should be used to determine the effectiveness and quality of the operation.

- **Cycle time**: the time span between identification of the need and the receipt of the full value of that material or service [13]. These include: request to delivery and receipt to availability as relating to the end customer, and at a micro level each of the steps in the process has a cycle time that can be measured. These measurements will help identify areas that require process improvement.
- **Total cost of ownership (TCO):** this measure is particularly useful in evaluating various capital expenditure options. 'The TCO formula is really a representation of various aspects of cost and allows the supply team to estimate the order of magnitude of the cost, rather than to achieve accuracy to the fourth decimal point' [14]. Its purpose is to scope cost.

$$\mathrm{TCO} = \mathrm{A} + \text{Present Value of } (\mathrm{O} + \mathrm{T} + \mathrm{M} + \mathrm{W} + \mathrm{E} - \mathrm{S})$$

where:

A = acquisition costs
O = operating costs
T = training costs

M = maintenance costs
W = warehouse costs
E = environmental costs
S = salvage costs.

- **Program evaluation and review technique (PERT)**: this technique can be utilized for the evaluation of time, cost, and capacity, and produces a weighted average from lowest (optimistic), normal (most likely), to highest (pessimistic) values, while factoring in the probability of resource and systems outages and unavailability. It is these values that should be used to establish service level agreements, not theoretical averages which are usually based on best-case predictions. Originally developed by the U.S. Navy Special Projects Office in 1957 to support the U.S. Navy's Polaris nuclear submarine project, PERT analysis has been applied across a wide spectrum of industries to design and improve processes, including clinical settings [15]. PERT analysis has upon one occasion been applied in the sample management setting. In 2004 at Sphinx Laboratories, Eli Lilly and Company, the Sphinx Project Management team coordinated an evaluation of sample management processes that determined accurate working capacities for sample management activities including: replication, cherry picking, immediate processing, screening set solubilization, and master set creation This resulted in annual and monthly capacity values at 100 and 80% load levels to allow sample management to accurately forecast project completion, flexible allocation of resources, and the probability of operational success/failure.
- **Instrument QC**: evaluation of sample preparation precision, accuracy, and carry-over is a task that is as important as production itself. Weekly QC of liquid handling, the validation of methods at both the instrument and biological level, through the use of controls and references are important measures to insure that the quality of the product remains consistent.
- **Inventory audits**: the accuracy or inaccuracy of inventory container locations can be a major factor in the ability to meet customer turn-around service level agreements. A routine cycle count of inventory and the audit of storage areas and data will ensure that inventory is where it is supposed to be when it is required.

7.11 Conclusions

In conclusion, sample management has evolved from a simple laboratory sample preparation process to a highly sophisticated supply chain. As such it can and should be managed with the same principles applied in supply chains across a wide spectrum of industries. In doing so, the sample management leader will avoid the temptation of 'reinventing the wheel' and the pitfalls that were discovered by his or her predecessors.

References

1. Riggs, D.A. (1998) *The Executive's Guide to Supply Management Strategies*, AMACOM.
2. Stonebraker, M. and Cetintemel, U. (2005) 'One size fits all': an idea whose time has come and gone. Proceedings 21st International Conference on Data Engineering, 2005, pp. 2–11.
3. Shenhar, A.J. (2001) One size does not fit all projects: exploring classical contingency domains. *Manage. Sci.*, **47** (3), 394–414.
4. Linderman, K., Schroeder, R.G., Zaheer, S., and Choo, A.S. (2003) Six Sigma: a goal-theoretic perspective. *J. Oper. Manage.*, **21**, 193–203.
5. The Council of Supply Chain Management Professionals, *http://cscmp.org/digital/glossary/glossary.asp*. p. 114.
6. Tompkins, J.A. (1994) *The Distribution Management Handbook*, McGraw-Hill.
7. Andreae, M.R.M., Steiner, T., Hueber, M., Schopfer, U. *et al.* (2008) Closing the gap between centralized and decentralized compound management approaches. *Comb. Chem. High Throughput Screen.*, **11**, 825–833.
8. Riggs, D.A. (1998) *The Executive's Guide to Supply Management Strategies*, AMACOM, Chapter 8.
9. Gosnell, P.A., Hilton, A.D. *et al.* (2003) A library management and distribution system, society for biomolecular screening. 9th Annual Conference.
10. Tompkins, J.A. (1994) *The Distribution Management Handbook*, McGraw-Hill., Chapter 5
11. Ellis, E.D. (2003) Support and enhancement of a monolithic automation system, Society for Biomolecular Screening. 9th Annual Conference.
12. Gardner, J.H. (2002) Automated Compound Storage and Retrieval: World Markets, Trends and Opportunities, Kalorama.
13. Riggs, D.A. (1998) *The Executive's Guide to Supply Management Strategies*, AMACOM, Chapter 9.
14. Riggs, D.A. (1998) *The Executive's Guide to Supply Management Strategies*, AMACOM, Chapter 5.
15. Luttman, R.J., Laffel, G.L., and Pearson, S.D. (1995) Using PERT/CPM to design and improve clinical processes. *Qual. Manage. Health Care*, **3** (2), 1–13.

8
Solid Sample Weighing and Distribution

Michael Gray and Snehal Bhatt

8.1
The Practicalities and Technology of Weighing Solid Compounds

8.1.1
Introduction

The compound management divisions in pharmaceutical companies strive to integrate new technologies to increase efficiency and productivity in order to meet ever-increasing demands. The quality of the data derived from drug discovery screening efforts is reliant upon quality and timely sample supply. The weighing of solid compounds is a tedious and time-consuming task faced by laboratory scientists worldwide but is an essential part of the drug discovery process.

Ordinarily most operations to weigh solid compounds are performed manually due to the complexity of the process and also because of the composition of some sample material. Manually weighing milligram-to-gram quantities of varying amounts accurately and reproducibly can often represent a significant bottleneck in laboratory workflow and diverts scientists away from performing other tasks.

With the increase in the throughput of chemical synthesis and the size of compound collections generally there is an inevitable drive to automate the process of weighing solid compounds. A multitude of automated storage and retrieval systems are available for large-scale solid-compound storage, enabling compound management groups to meet pick-and-place rates required, but these are far more time efficient than manual weighing, making the weighing process the bottleneck.

Some questions posed to compound management groups concerning weighing requirements are listed below, and are all geared toward customer need and reliability:

- What is the need to change from a manual weighing system?
- Is there a bottleneck with a manual system to justify automation?
- Would increasing throughput by introducing an automated system have any benefit to the customer in downstream processes?

Management of Chemical and Biological Samples for Screening Applications, First Edition.
Edited by Mark Wigglesworth and Terry Wood.

Are automated systems reliable and flexible enough? Are automated systems that are capable of dispensing compounds representative of a typical compound collection as efficient as a human? Is the percentage of samples they are able to dispense high enough for the systems to be cost effective.

Alternatives to manual weighing are certainly available; they have their advantages but also have limitations. Some of the techniques and technologies for weighing solid compounds will be discussed and appraised in this chapter.

8.1.2
Manual Weighing

Manual weighing remains the simplest method of weighing solid compounds for drug discovery. Essentially this involves the transfer of material by a scientist from a source container to a destination container using a spatula or similar implement. Although a streamlined setup would still require IT and specialist equipment, this is probably the cheapest arrangement in terms of implementation, both internally and externally, and moving forward would require minimal support and maintenance.

A standard setup would normally comprise a powder-weighing enclosure with an air extraction or recirculation system passing the extracted air particles through a filtration system to protect the operator and the environment from exposure to potentially harmful compounds. The enclosure would house an analytical balance capable of accurately measuring milligram-to-gram quantities to two decimal places. This could cost in the region of £20k excluding any IT infrastructure, which is required if the balance is to be integrated with an inventory and tracking system. This hardware setup cost excludes the operator resource needed to weigh the compounds.

This option has a low risk of poor quality dispenses as humans can react to the type of samples they are dispensing. It also has relatively low initial investment cost; however, it has an ongoing staff resource cost. Because this method requires an operator to perform the entire task, it has limitations in throughput compared to other methods. Unless the task is operated on a shift pattern, the output is normally only performed in a normal working day. The inefficiency here is that equipment is commonly left dormant for up to two thirds of a 24 h day. Ordinarily an operator would be expected to weigh in the region of 150–300 solid compounds per day, (assuming, of course that he or she is not on holiday, in a meeting, or attending a training courses, and so on), meaning that careful consideration needs to be taken when considering throughput requirements and employee allocation. Getting this wrong could pose a potential risk in meeting customer needs and at times of high demand will result in reducing the cycle times for drug discovery. It is possible to mitigate these risks, but this usually involves paying for increased employee resources, for example, by utilizing shift patterns increasing the equipment utilization or by working in an environment where there is an excess of equipment and recruiting contingent workers as demands increase. Additionally, outsourcing of specific high-demand compound sets could be considered, as discussed in 16.

8.1.3
Automated Weighing

Automated transfer of materials in the solid state has been a challenge for many years due to the wide range of properties and textures of compounds being handled. Weighing in the milligram range further complicates matters, requiring the technology to be intelligent enough to be fast and accurate. Normally an algorithm enables the system to interpret how fast the material flows each time it is dispensed, allowing the system to store the information for the next time it weighs, making it more efficient. The attractions of automated weighing include productivity, safety, and reproducibility. Automated weighing robots can range from large fully integrated systems to small stand-alone bench top arrangements. The cost of such systems can range from tens of thousands to hundreds of thousands of pounds. While the size and degree of integration can vary, the major differences are in the methods employed to extract the compound. The typical mechanisms employ gravimetric or volumetric systems, Archimedes screw, plunging plug, extruder, controlled voltage, vibration, or a mixture of the above. Each mechanism has its place and its advantages, but really its effectiveness depends upon the composition of the material. The detail and explanation of each method is not discussed here, but the overall effectiveness of automation in general will be considered [1].

Time efficiency is the key to the appeal of this technology. Some systems can weigh a single compound in approximately 3–5 min. This is a similar rate to that of manual weighing but allows the supervising scientist to perform other tasks such as weighing those compounds which the system could not dispense or picking the next batch of compounds for an overnight run. This form of automation can run outside of normal working hours, utilizing hardware more efficiently, thus reducing the setup costs of a laboratory as fewer systems are required and reducing the running costs as fewer employees are needed to process a given number of samples. The ability to weigh orders overnight also drives down cycle times, so meeting customer demands better. This potentially produces more scope for exploration by the customer, who will have greater access to a larger number of compounds within a defined time frame.

From a safety perspective, manual weighing could pose health risks such as repetitive strain injury (RSI) and exposure to potentially reactive and toxic substances. Although the risks of manual weighing are carefully controlled, a fully enclosed cabinet housing automated weighing robotics further reduces the personal risk due to human processing errors and in turn reduces company liability.

From a quality perspective, the dispensing method, if reliable and appropriate for the material being handled, ensures that dispense after dispense is delivered in the same way, and therefore, in theory, would be expected to be more consistently accurate than a manual method. This provides confidence in any results gathered further downstream. Some systems incorporate further steps after dispensing, meaning that the next stage of a process can potentially be performed immediately in one work stream, for example, dissolution using liquid handling robotics, dispensing of solution, capping, and labeling. Again this can save significant labor within a compound management laboratory.

In early drug discovery there are large libraries of compounds which require large weighing tasks of single dispenses from potentially thousands of compound source bottles into an array of single destination vessels. This is known as *many-to-many dispensing*. Typically these compounds are only made in very small quantities (10–50 mg), so that there are issues for the design of a mechanism which can not only accurately dispense these small amounts with a very low tolerance, but which causes minimal, if not zero, wastage of material. Reliable and cost-effective automation is available, but this is perhaps more suitable for the later stages of drug discovery in areas such as clinical trials, where there is a need to dispense a small number of solids (maybe only one) into many destination vessels such as capsules. The challenge for compound management groups and the manufacturers of automated weighing devices is to design automation which will suit the variety of solid physical forms. Another main issue is the cost of the dispensing caps or heads, or the alternative mechanisms such as electrostatically charged needles. With most collections being in the range of thousands to millions, it is the initial setup cost that can cause concern. In addition, many libraries are housed in automated compound stores, which challenges the manufacturers to produce a dosing cap or head which will enable the customer to store the collection in an efficient manner. Another issue encountered concerning the mechanism of transfer is that the texture of the material across a collection can vary from a gum or free-flowing solid to a fluffy hydrostatic powder. In summary, the main concerns are cost, potential cross contamination from one sample to another if using mechanisms such as reusable tips, the need for a significant dead volume for certain methods, the ability to be able to dispense very small amounts (in the milligram range) with minimal or no loss of material, reliability, ease of use, and cost. Typically, around 80% of a standard compound collection is suitable in composition to be dispensed using automated mechanisms. These mechanisms are often based on an Archimedes screw inside the stock bottle cap (Figure 8.1).

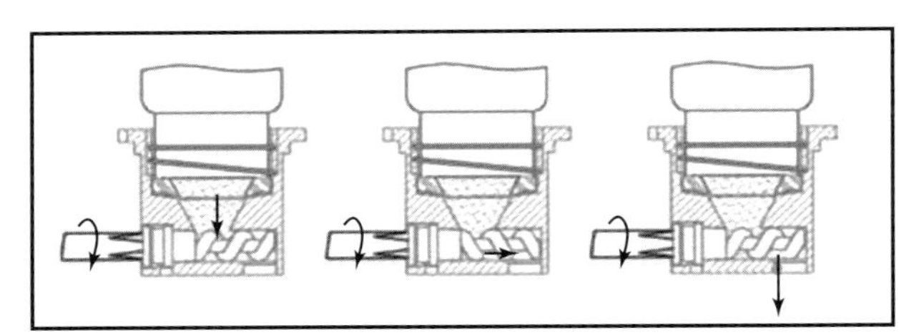

Figure 8.1 Archimedes screw cap, commonly used in early systems and still a popular mechanism. As the screw turns, solid compound moves out of the source bottle and is dispensed into the destination vessel [1].

This is a reliable and efficient way to extract solid material; however, amounts of material being synthesized these days are often small, and material left in the screw itself can cause significant loss of compound. Gums, free-flowing liquids, and hydrostatic materials would also pose a challenge for this mechanism. Other mechanisms such as electrostatic needles have the potential for cross contamination if fixed needles are used; single-use disposable needles eliminate this risk but come at a high cost. This mechanism would also have difficulties with certain materials such as gums and large particles but would be effective with hydrostatic material. These are just two mechanisms, very different in their action, but showing that each has its place and each has its limitations.

Generally, most mechanisms require a dead volume of compound in the source vial in order to be able to meet the required amount; they would not necessarily be able to extract the entire contents. Automated weighing certainly has its place and benefits, as when there is a subset of a collection in sufficient quantities, with good textural properties suitable for the mechanism being employed, and where there is a need to weigh out large orders, for example, when adding a set of compounds into a high-throughput or focused screening collection in a time-efficient fashion.

8.1.4
Volatile Solvent Transfer

As discussed above, the transfer of hundreds to many thousands of solid samples remains a challenge for compound management groups. Weighing solid compounds which are of a texture to suit automated and manual weighing has been described, but the weighing of non-crystalline samples still produces difficulties. The consistencies of some compounds produced are those of a gum or oil; others reduce to these forms having deteriorated over time. To overcome this there is a process known as volatile solvent transfer (VST), which provides an alternative to transferring these forms, and is also useful where there is a small residual amount in the stock bottle which cannot be removed through manually or automated means. This process involves dissolving a solid stock compound in a volatile solvent, transferring a portion of this solution into a destination vessel, and then evaporating the solvent leaving only solid compound in both the destination vessel and the stock bottle. Thus, automated liquid handlers can be utilized to dissolve a large batch of compounds and dispense them automatically with minimal scientist interaction. Some compound management groups already utilize this for a large number of their solid sample transfers. However, along with manual and automated weighing, this technique also has its limitations and critics.

The volatile solvent itself is not necessarily the main issue: many solvents can be used including dichloromethane (DCM), methanol (MeOH), dichloroethane (DCE), dimethlyformamide (DMF), and chloroform, to name but a few, as well as mixtures of the aforementioned. The concerns scientists would generally have about the process is that in order to completely remove the solvent there is a necessity to apply heat. This introduces the chance of overheating sensitive pharmaceutical structures, which could compromise both the sample integrity once it is dried and the viability of the data derived from screening campaigns testing these molecules. As in automated weighing, there is a range of equipment available which can eliminate solvents using evaporation, centrifugation, and heat applications in various combinations. The problem for compound management groups is what equipment and what solvents best suit their solid compound collection. Some solvents will be very aggressive and can degrade certain structures over time or cause some chemist concerns; some may precipitate out of solution or not dissolve the compound at all. However, the less aggressive solvents can be less of a risk to sample integrity and are potentially easier to evaporate at lower temperatures.

Confirming concerns over heat exposure and solvent choice, there are reports which show that an increase in temperature in VST studies using certain solvents can markedly increase the degradation of the molecule [2]. So this technique, although it resolves many of the concerns associated with automated weighing, presents different risk factors to be taken into account. An interesting development for VST would be the utilization of dimethyl sulfoxide (DMSO) as the solvent of choice. DMSO, albeit not a volatile solvent and thus difficult to evaporate, can potentially be used by some equipment available to us today. However, it is a relatively inert solvent and is commonly employed by many compound management groups as a standard for the dissolution and long-term storage of solution collections.

The use of VST is certainly efficient and serves a purpose in certain processes within the compound management groups. Implementation of VST or any of the automation choices above requires vision, perhaps persuasion, and a carefully understood process. It enables the sample manager and the chemistry team to balance the risk involved against the resulting efficiency and speed (with customer benefit at mind) to improve their work streams.

8.1.5 Sample Weighing – Summary and Conclusions

In summary, compound management groups manage vast numbers of molecules with a range of physical properties. These groups are challenged with being able to increase their throughput to meet customer demands, increasing customer flexibility and productivity, while under their own capacity, technological, and financial constraints. Instrumentation and automation have advanced, but this chapter has hopefully highlighted that there are still issues for compound management groups in all of the aforementioned areas.

The three methods summarized show the main concerns: technology cost, potential cross-contamination from one sample to another, the need for a significant dead volume with certain methods, the ability to be able to dispense very small amounts in the milligram range with minimal or no loss of material, reliability, and ease of use. The daily outputs that could ordinarily be realized by one operator using the different technologies are shown in Table 8.1 (note that a scientist could

Table 8.1 A summary of throughput and operator walk-away time by dispense method.

Dispense method	Daily throughput	Operator walk-away time
Manual weighing	150–300	Zero
Automated weighing	300–500	Estimate 5 h daily
Volatile solvent dispensing (VST)	1000+	Estimate 5 h daily

perform other activities whilst supervising automated weighing or VST, which would be impossible with the manual system).

Compound management personnel are the keepers of the drug discovery molecules and have a responsibility to maintain the integrity and safety of these molecules for future use. There are no physical characteristics of solid compounds which prevent their being dispensed: the question is which method should and could be used. The likelihood of all compounds in this stage of drug discovery in large pharmaceutical collections being dispensed using one technique (other than the manual method) is unlikely at this time. All the methods discussed, and potentially others which are being developed, could be employed, but at present they should be regarded as accompaniments and not replacements for each other.

8.2 Logistical Challenges of Transportation of Small Molecules

> *Logistics is the 'practical art of moving armies'.*
> General Antoine-Henri Jomini (1779–1869).

8.2.1 Introduction to Transportation

In a world of globalization, outsourcing, off-shoring, and external partnerships, today's pharmaceutical R&D organization is faced with increasing logistical challenges to move research samples and substances across an international network of company sites (Figure 8.2), contract research organizations (CROs), biotechnology, and academic alliances to support the discovery pipeline. As a result, the ability to efficiently manage the flow of materials across the R&D supply chain has emerged as a critical function of R&D, including even the supply chains of 'virtual' biotechs that mainly access the inventories of other companies. The span of these activities can cross several countries. In addition to the distance, this playing field is further complicated by a multitude of international, domestic, and foreign regulations and agencies. Neglecting to address distribution and logistical issues will not only yield higher costs but may have significant legal and reputational implications. To address these complex needs, the shipment coordinator must have a good

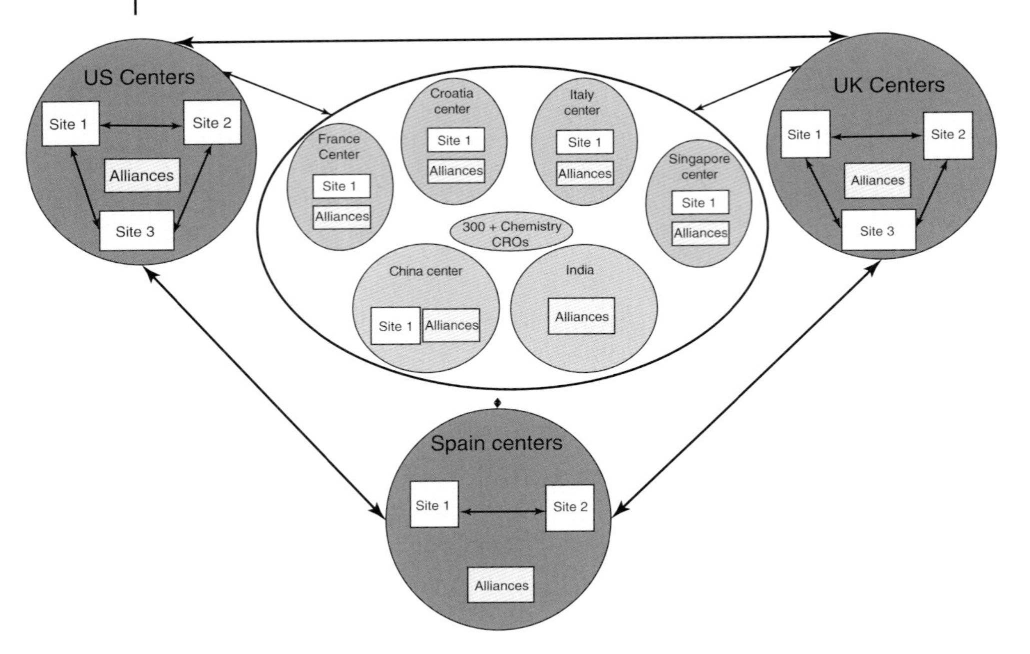

Figure 8.2 A schematic of GlaxoSmithKline's compound management global operations, exemplifying shipping interfaces across this organization as of January 2009.

understanding of what drives their supply chain dynamics as well as governmental regulations and where they come into play, in order to ensure a safe and efficient passage through this labyrinth.

The focus and reference of a shipment and its content made in this chapter only concerns small molecules, these being discovery compounds imported in limited and non-commercial quantities only, in accordance with standard industry practice. There are many other commodities within the R&D organizations that may have additional challenges and regulatory needs.

Generally speaking, the pharmaceutical industry has not focused on new ideas to handle the logistics of early-stage research compounds, though much has been done to optimize the supply chain of manufacturing and late-stage discovery and development supply chain activities. It is from these areas that innovative ideas and technologies can be adopted to improve the way compound shipping is conducted today. Although it can be difficult, developments from other supply chains can be introduced in complex, international, multi-organizational, multi-modal, and multi-systems environments of R&D organizations. By and large, the successful supply chain depends on how an organization experiments, develops, and learns to use innovative approaches such as packaging and tracking devices, temperature monitors, electronic invoices, or software solutions in real-world settings. This chapter attempts to give insight into the logistical challenges and possible solutions for the supply chain of small molecules in the pharmaceutical industry.

8.2.2
Complexity of Logistics and Compliance Challenges of Supply Chain

The small molecules contained in compound banks pose significant challenges for shipping and handling due to the smaller amounts involved (solids as well as liquids) and their unknown chemical properties, for example, temperature sensitivity and toxicity. Product life and stability of small molecules may be impacted by exposure to environmental extremes, time spent in certain temperature ranges, humidity, agitation, and so on [3]. Hence, these compounds place rigorous demands on both the shipping organization and the courier services; demands which impact product handling, storage, packaging, and mode of transport.

The illustration presented in Figure 8.3 demonstrates that, throughout the supply chain, the product shipment must go through many stages of transportation. This includes, from the initial packaging of the compounds, ground transportation, customs departure, air transportation, receiving country's customs, and back to ground transport for the final delivery to end user. Each step of the process is handled and governed by different groups and regulatory agencies and guidelines; including Department of Transportation (DOT), US Custom and Border Protection Agency (US CBP), International Air Transport Association (IATA). The goal here is to be in compliances with various regulatory agencies such as; Export administration regulations (EAR), Screening for denied parties, trade sanctions and anti-boycott regulations, Shipment Details such as; 1). Description of articles 2). Amount 3). Valuation and Shipper's Export Declaration (SED) 4). Declaration

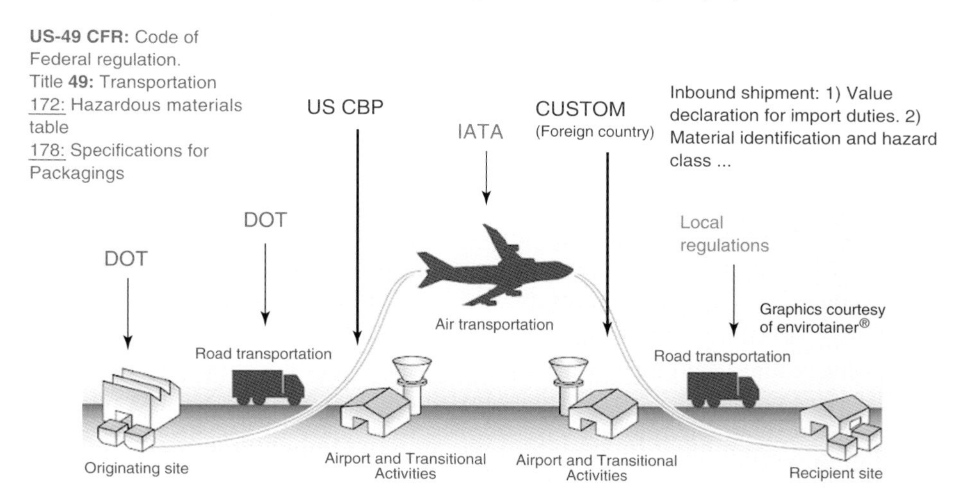

Figure 8.3 Example of logistical and compliance challenges in supply chains. Diagram modified from envirotainer [4].

of controlled substances/hazardous classification etc. If these processes are not managed properly, they can cause delays, security threats, and even loss of property.

Addressing this area is complex as it involves logistical, organizational, cultural, and legal issues as detailed below.

- **Logistical issues**
  - Scheduling of shipments to fit with your organization's requirements
  - Multiple handover points between organizations and modes of transport
  - Capabilities at distance
  - Challenges of physical environment.
- **Organizational/cultural issues**
  - Organization's ability to learn new skills or obtain specialist knowledge
  - Establishing working practice that conforms with regulations
  - Resource limitations such as the cost and constraints of tools to monitor shipments
  - Limited understanding of the latest Import/Export legislation
  - Inadequate tools and process for the valuation and identification of materials
  - No control mechanism to avoid prohibited compounds
  - Many IT tools and return on investment.
- **Legal issues**
  - Complying with regulations and understanding what regulations apply to your goods
  - Compatible and/or inconsistent regulatory environment across countries
  - Record retention and auditable processes.

These problems have been dealt with in isolation within each R&D organization and may be viewed as a market failure, perhaps due to coordination costs, incentive misalignments, cost/benefit challenges, international regulatory incompatibilities, or other mechanisms. Or it may be that organizations are relying on third-party courier organizations to ensure that the relevant standards are met [5].

During each step of the supply chain the shipper has to comply with an assortment of regulations and guidelines that are set by various regulatory agencies and span many countries. The following are some key process steps and related regulations that are mandated by agencies in the United States of America. Similar regulations are in force in other countries.

At the time of shipment preparation and during ground transportation, the shipper has to comply with regulations and guidelines of domestic transportation agencies. Transportation industries of most countries are regulated by their individual national government agencies with their own local requirements. For these reasons not all national regulations are similar. In the United States of America, transportation regulations are governed by Department of Transportation (DOT) under Title 49 of Code of Federal Regulations (CFR) (In the EU the relevant regulations are: the European agreement concerning the International Carriage of Dangerous Goods by Road (ADR) and the International Carriage of Dangerous Goods by Rail (RID).

Before passing them to the provider of air transportation, the package and its contents must be approved for export under local custom Export/Import regulations. As well as customs, shipments are subjects to compliance with many other regulatory agencies. In the US these regulations are governed by the following agencies:

1) CBP: Custom and Boarder Protection
2) FDA: Food and Drug Administration
3) EPA: US Environmental Protection Agency
4) USDA: US Department of Agriculture
5) CDC: Centers of Disease Controls and Prevention
6) DEA: Drug Enforcement Administration.

Once the content of the shipment has been cleared by customs, it can be passed to the airline. Here the package is subject to International Air Transport Association (IATA) guidelines, a trade association which represents the international airline industry.

8.2.3 Regulations and Procedures for Shipping Hazardous Materials, Hazardous Material in Small, Limited Quantity, with Dry Ice

For shipping purposes *screening compounds* are generally defined as 'a carbon based molecule, amongst many millions of similar entities, with largely unknown biological activities, that has been selected against a diverse set of chemical filters to test against a range of biological targets, typically synthesized or acquired in quantities of 1–20 mg.' Characteristically, these compounds are dissolved in 100% DMSO and distributed internally among other research sites. The distribution amounts of these materials vary from 50 nL to 1 μL. For these reasons most of the small molecule shipments are qualified under 'small quantity regulations' outlined in DOT 173.4.

Domestic and international shipping and packing guidelines vary slightly in scope and limitations, but both include special dispensations for smaller quantities of dangerous goods. The two sets of small-quantity regulations are very similar in scope and content but have a number of limitations that must be adhered to. It is important to consult the original texts of both the DOT and IATA regulations before setting up small-molecule shipments in your organization.

8.2.3.1 Domestic Regulations

In the United States, the shipment of dangerous goods (referred to as *hazardous materials*) is covered in DOT Title 49 CFR (parts 100–185 [6]. An exception to the regulations is made for dangerous goods in restricted quantities termed '*small quantity regulations*' outlined in DOT 173.4. These small quantities are considered exempt from regular DOT hazardous goods requirements. Most of the small molecules, solids as well as solubilized in DMSO, can be packed and shipped utilizing these small quantity regulations.

8.2.3.2 General Shipping Procedures and Associated Regulations

1) Small quantities may be sent through air or road transportation with any of the major courier companies that follow DOT 49 CFR 173.4 small quantity regulations. The maximum quantity of dangerous goods per inner container (glass/plastic vials, microtiter plates, etc.) cannot exceed 30 mL for acceptable liquids.
   a. Small quantity exceptions according to DOT 49 CFR 173.4:
      I. 30 mL for liquids, 30 g for solids, of hazardous materials, (please refer additional requirements for Division 6.1, Packing Group I, Hazard Zone A or B and radioactive materials).
      II. Materials of trade exceptions (173.6)
      III. Meet Definition in 171.8, 0.5 kg, or 0.5 L for PGI material or 30 kg or 30 L for other material
      IV. If the material is listed in the Hazmat table and it does not fit into any available exceptions then the material must be shipped in compliance with this procedure.
2) This inner container must be protected with removable caps or seals and must be placed within a securely sealed secondary package, for example, a plastic bag sealed with wire, tape, or self-sealed bags. Each inner container must be placed within a securely sealed secondary package. It is recommended that sufficient cushioning and absorbent material (that will not react chemically with the dangerous goods) is used to surround each inner container. This material must be capable of absorbing the entire contents of the inner container.
3) The secondary packages must be securely packed in a strong outer package (box) (such as a dry ice cooler with an outer cardboard box) which complies with DOT mandated drop and compressive load tests without breakage or leakage from any internal container.
   a. Packaging (49 CFR 173)
      I. UN stamped hazardous materials packaging systems are approved by the United Nations Transportation Board for the transport of hazardous materials and are obligatory for hazardous materials shipping.
      II. UN packaging systems are designed and constructed to prevent a release of hazardous material during transportation.
      III. To obtain UN stamped hazardous material packaging systems you may contact one of the following vendors: ULINE (*www.uline.com*), Labelmaster (*www.labelmaster.com*), or Lab Safety Supply (*www.labsafety.com*) [7].
      IV. Packaging system requirements are specific to the type of hazardous material offered for shipping.
4) The gross weight of the package must not exceed 29 kg (64 pounds).
5) Markings (49 CFR 172.300) and Labels (49 CFR 172.400).
   a. Proper markings and labeling on shipment containers are very important for shipping a hazardous material. Note that packages with improper or incomplete information are typically not getting delivered to their destination.

This could also result in you being subject to fines or imprisonment by the US DOT.

I. Labels and markings must be attached to or printed on the package. Labels must be durable and printed in English.

II. Any package containing a hazardous material must have the following as per requirements listed in the hazmat table 49 CFR subpart B, 172.101:

i. Name and address of shipper

ii. Name and address of receiver

iii. Proper shipping name – column 2 hazmat table

iv. Numeric hazard class and division – column 3 hazmat table

v. UN/NA ID number (obscured and away from other markings) – column 4 hazmat table

vi. Packing group – column 5

vii. Choose proper label(s) from – column 6 in hazmat table (packages may have more than one label)

viii. Additional info if required (limited quantity, RQ).

6) Shipping papers (49 CFR 172.200)

a. The accurate completion of appropriate shipping papers is required for all hazardous materials shipments and must contain the following:

I. Appropriate DOT shipping description including proper shipping name, hazard class, UN/NA ID number, and packing group – hazmat table

II. Total quantity

III. Emergency response information (24 h telephone number)

IV. Shipper name and address.

7) Shipments with dry ice (carbon dioxide; CO_2 solid): As an industry standard practice, all small-molecules solutions are being shipped using dry ice. In general, with three times the pound-for-pound cooling capacity of ordinary wet ice, it's easy to use and is very cost-effective. The product stays clean and dry and avoids any mess that may occur from broken gel packs. The greatest advantage of shipping with dry ice is that it has no liquid state – it sublimes (changes directly from a solid to a gas).

a. Dry ice, UN 1845, hazard class 9, is classified by DOT and IATA as a 'miscellaneous' hazard, which means it is a material that presents a hazard during transportation. Dry ice is characterized as an explosion hazard (releases large volumes of carbon dioxide gas), asphyxiation hazard (carbon dioxide can create an oxygen deficient atmosphere), and contact hazard (as a cryogen, dry ice causes severe frostbite if left in contact with skin). It is absolutely essential that the person, who performs dry ice shipping and receiving activities, follow and understand the regulatory requirements, including package-compliant protocols and mandatory safety training [8].

b. A hazardous material declaration form must be filled out for any dry ice shipment whether it contains a hazardous material (dangerous good) or not. This is important to ensure all dry ice shipments are categorized and labeled properly.

c. Packages refrigerated with dry ice are normally shipped by air in order to reach their destinations quickly. Most companies insist on shipping on Monday, Tuesday, or Wednesday due to the requirement of keeping samples frozen. Another reason for not making shipments on Thursday or Friday is that the shipment may arrive at the destination during the weekend where customs officials or receiving personnel may not be available.

d. Labeling: The outermost container must be labeled with a hazard class 9 miscellaneous 'dry ice' label, UN 1845, and net weight of dry ice in kilograms. This label satisfies the US DOT and International Air Transport Association (IATA) marking and labeling requirements when accurately completed. The dry ice label must be outside the outer container on a vertical side of the box (not the top or bottom). The maximum allowable net quantity of dry ice allowed per package is 200 kg.

8) Air Bill: The carrier air bill must be filled out properly and must be signed by the shipper. If the shipment contains a hazardous material, then the Shipper's Declaration of Dangerous Goods must be signed and included with the shipment. Note: a copy of the air bill and Shipper's Declaration of Dangerous Goods must be kept on file for the appropriate period of time as required by DOT.

8.2.3.3 International Regulations

International shipments of dangerous goods are covered in Section 2.7 of the IATA regulations[3]. As above, restricted quantity regulations exist for international shipping, contained in IATA Section 2.7.1 and referred to as '*dangerous goods in excepted quantities*.' Dangerous goods in excepted quantities, in contrast to DOT regulations, are considered dangerous goods under IATA regulations but are exempt from significant portions of the dangerous goods regulations applicable to larger quantities.

1) All international shipments must be sent using a courier service that follows IATA regulations. Some major couriers that provide this type of services are FedEx, UPS, DHL, BioCair, and World Courier.
2) All packaging groups from class 3 of IATA dangerous goods are acceptable for chemical compounds. However, packaging group III is commonly used for small-molecule compounds. Dangerous goods are divided into classes on the basis of the specific chemical characteristics producing the risk. Packing groups are used for the purpose of determining the degree of protective packaging required for dangerous goods during transportation.
   a. **Group I:** Great danger and most protective packaging required. Some combinations of different classes of dangerous goods on the same vehicle or in the same container are forbidden if one of the goods is group I [9].
   b. **Group II:** Medium danger.

c. **Group III:** least danger among regulated goods, and least protective packaging within the transportation requirement.

3) As referred in Section 8.2.3.1, each inner container may not contain more than 30 mL. All other regulations in regards to construction material, liquid fill, and closure security regulations also apply.
4) Each inner container must be placed within a securely sealed secondary package.
5) Sufficient cushioning and absorbent material (that will not react chemically with the dangerous goods) must surround each inner receptacle and be capable of absorbing the entire contents of the receptacle.
6) The same package drop and compressive load test regulations as domestic regulations apply.
7) IATA regulations state that each inner receptacle must be placed within a securely sealed secondary packaging the total contents of which may not exceed 500 mL for packing group II liquids and 1 L for packing group III liquids.
8) Labeling – each package must be labeled with the label as illustrated in Figure 8.4, having minimum dimensions of 100 mm × 100 mm (4 in. × 4 in.). This label must be filled in and signed by the packer. The 'Nature and Quantity of Goods' section of the air waybill must be completed with the words 'dangerous goods in excepted quantities.'

It is very important for any organization to be in total compliance of 'Import/Export' regulations. Performing with compliance is a 'good business, it's the right thing to do.' It is an essential part of company's policies of Ethics and Compliance. One must make every honest effort to abide by the laws of the country where one's company does business. Good compliance maintains an uninterrupted global supply flow with integrity and respect. It prevents any mishaps during transportation of materials such as possible regulatory hold, detentions, or seizures, and so on. Moreover, better compliance minimizes delays and any potential penalties during the cross-border transit and prevents risks of possible regulatory audits. Increasing compliance not only provides a secure supply flow, but also drives operational and cost efficiencies. A steady and timely delivery of material increases overall productivity and organizational effectiveness. Any knowing or unknowing violation of laws will be detected sooner or later and will have significant consequences.

For example, the U.S. DOT enacts and enforces all hazardous materials (hazmat) transportation laws in the United States. Compliance with DOT regulations is a requirement for any person who offers a hazardous material for transportation. The civil penalties for violation of DOT regulations can be up to $32 500 per day, per incident. Criminal penalties can be as much as $250 000 per day, per incident with up to five years in jail for individuals. For organizations fines of $500 000 per day, per incident are possible [7].

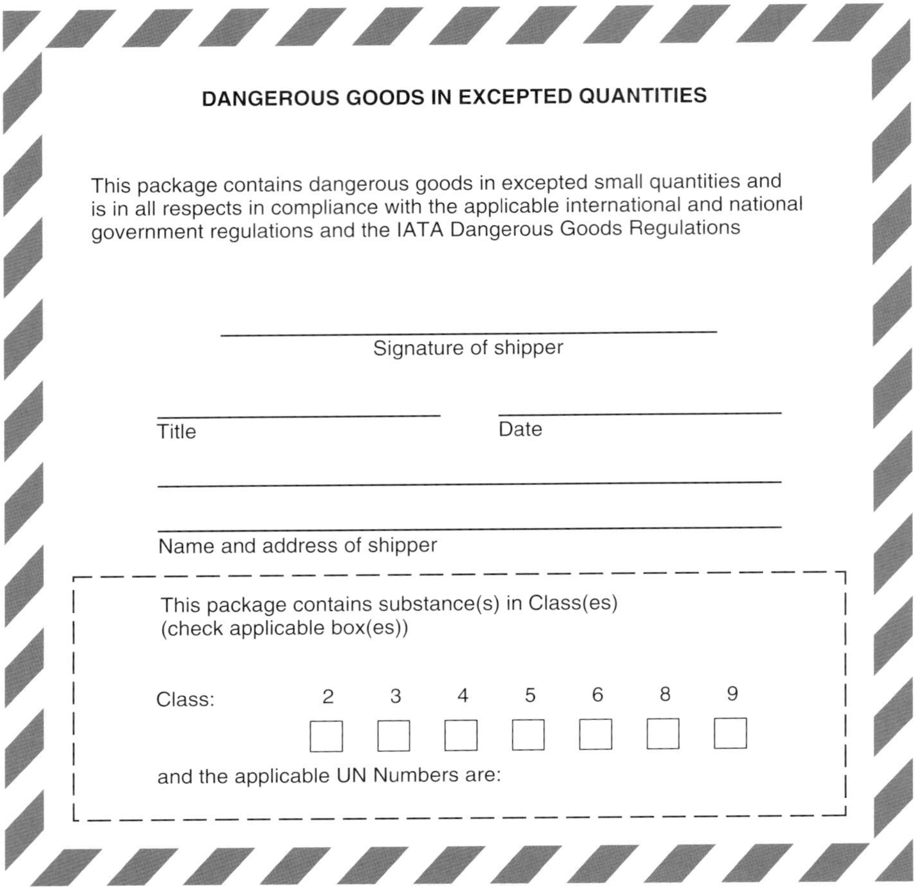

DANGEROUS GOODS IN EXCEPTED QUANTITIES

This package contains dangerous goods in excepted small quantities and is in all respects in compliance with the applicable international and national government regulations and the IATA Dangerous Goods Regulations

Signature of shipper

Title

Date

Name and address of shipper

This package contains substance(s) in Class(es)
(check applicable box(es))

Class: 2 3 4 5 6 8 9

and the applicable UN Numbers are:

Figure 8.4 IATA regulations labeling for dangerous goods in excepted quantities [6].

8.2.4 Cold Supply Chain Challenges

Large pharma and biotech companies maintain their collections in two forms: solid and solution. Typical collection size varies from a few thousands to many millions of small-molecule compounds in one or more stores located at various geographic locations. Characteristically, compound management groups offer a range of services, including compound storage, routing of compounds for biological assays, management of compound requests (both internal and external), and shipping to world-wide locations. As the virtual global inventory of compounds is available to everyone in the organization, compound movement between the internal sites, alliances, and CROs is becoming a daily event in the global operation for all companies.

Normally, solution compounds in DMSO are the first preference for the early lead discovery. In many global companies, the compound collection and high-throughput screening (HTS) scientists are not co-located. This requires transportation of millions of samples in assay-ready plates to support various HTS processes. These solution compounds are usually shipped in volumes of microliters or less. Keeping such quantities of sample from degradation and water uptake can be challenging and will always require transportation at $-20\,^{\circ}\mathrm{C}$ in order to maintain chemical integrity. These compounds have to stay frozen prior to packaging as well as post-packaging and during the entire transportation process.

Cold-chain systems rely heavily on protective insulated containers such as dry ice coolers for the shipment of temperature-sensitive supplies and products. As shipping technologies advance there are innovative ideas such as highly insulated, reusable, and temperature-monitored containers and even containers that contain radio frequency (RF) devices [10] for the tracking of the shipments.

Consequently, the protective package design and control is a significant determinant of cold-chain performance, but what happens when supporting systems breakdown or the unexpected comes into play? This could be something as ordinary as a delayed flight or a truck breakdown. Adding a 24 h delay to a cold-chain shipment may reduce the effectiveness of the compounds and in some case ruin an entire shipment if measures are not taken to manage a situation. In situations like these there is usually a 'plan B,' but timing is critical and timely information enables the most effective response. Addressing this area is difficult and requires careful logistical support and communication.

To deal with such instances, it is crucial to have a strong relationship with your transportation service provider. The service provider can provide logistical support before and during the journey of the shipment by using the right kind of cooler and if necessary refilling the refrigerant (dry ice). It is advisable to find a vendor or courier who provides a 'door-to-door' service during the entire transportation process, from providing ground and air transportation to representing you at the customs of the departure and arrival countries. It is also vital to establish an effective communication with the service provider, such that you know of any issues en route and so that you can advise both them and your customers when problems occur. To build a strong partnership it is very important to establish a reliable and uninterrupted supply flow for your samples. It is imperative that the company set up a service level agreement with clear expectations and accountabilities for your company; the shipper, and the service provider. Making a mistake in the handover of documents or samples will not excuse either party if this should result in shipments that contravene regulations.

8.2.5
Collaboration with Subject Experts

For any research organization it is difficult to train bench scientists to be knowledgeable of the numerous regulatory and compliance requirements. To handle the convoluted business of supply chain and increasing enforcement of cross-border regulations, the organization must also build a strong alliance with essential internal functional groups: the 'experts.' To establish an uninterrupted flow of supply chain, all subject matter experts must work together to form a synergy. Although a necessity in small organizations it is rare within large companies that one person or functional group will have all the expertise needed, so cumulative knowledge is essential to form a successful support structure. Such cross-functional experts can be from environmental, health and safety, dangerous goods and safety (DGS), logistical group for transport regulation and custom, finance personnel for the revenue and valuation of material, internal/external service providers, and foremost support from the IT group and legal departments. Therefore, specialist personnel or departments are required to ensure a smooth flow of research samples to and from the discovery engine.

8.2.5.1 Process Development and Standardization

In order to meet this ever-growing business need, all of the components of supply chain processes must be streamlined and standardized. Such standardization and well-defined processes will allow the organization to perform cross-border trade in a manner that will abide by the various governmental laws and regulations for transportation, customs, and revenue agencies. Once the entire shipping process has been developed and documented, it is important to have regular revisions, and necessary amendments should be made and communicated to all involved personnel. Furthermore, to execute shipping process with consistency, accuracy, and a complete compliance, the organization should establish appropriate internal controls for the key process elements. A list of some of the key process steps that the shipper must understand is given below:

1) Proper identification and classification of materials
2) Legitimate valuation
3) Effective communication and information sharing
4) Record retention
5) Management of control substances
6) Export Administration Regulations (EAR) such as Denied Party List, Trade Sanctions, and Anti-Boycott Regulations.

1) **Material identification:** Uniformity and meaningful material identification are two of the key factors to meet regulations and build trust within the custom authority. As any large research institute may transport from hundreds to millions of compounds each day, it is not practical to list them individually. In addition to the large number, most of the compounds are proprietary to the company and intellectual property (IP) rights must be maintained. This is one of the most difficult areas and is often where company and regulatory agencies (such as various custom agencies, FDA, etc.) encounter most problems. The company wants to preserve the identity of its unique compounds, while regulatory agencies want to be sure that these chemical and biological materials are in full compliance (for disease control, control

substance, drug enforcement, etc.). Historically, the regulatory environment has changed with alterations in regulation within a country or when moving into new geographical areas, such as the recent expansion of biotechnology centers in India and China. Hence, it is difficult to find the right balance between enforcement of security and IP preservation of the company's chemical and biological assets. For this particular reason, shipments with research compounds are sometimes being delayed at customs. The majority of research compounds are novel, and no chemical, biological, or any other hazard data may be available. Such compounds should be classified according to the Corporate Environment, Health and Safety protocols. It is an industry standard practice to identify the research compounds generically as 'Pharmaceutical compounds' or 'Pharmaceutical research compounds.' However, if the compound has a known hazard and has a material-specific material safety data sheet (MSDS), this must be used for suitable classification. In many developing countries, such as India and China, the use of the generic names may not be sufficient, and shipment will require additional sample detail. Authorities in these countries look for a full disclosure of chemical names or structures, which can be difficult to reconcile with protection of the shipping companies IP, although, as experience and relationships of both customs agencies and shipper have developed, it has been observed that providing the name of the core structure or chemo-type can fulfill their requirements. However, this remains a difficult area to operate in, and many companies have often experienced shipments being detained at customs if adequate compound information is not provided. Another issue with importing compounds, particularly to China, is that very hefty import duties amounting to 30% of the total value of the shipment are currently being imposed. Thus, in some cases, the cost benefit of moving samples to these areas for generation of screening data is quickly eroded, so beware of this hidden cost.

2) **Valuation:** Any materials moved across borders must have a value for customs purpose (VCP). In general, R&D moves an array of materials including bulk active pharmaceutical ingredients (APIs) used in the manufacture of drugs), APIs in final dose form, discovery compounds, custom reagents, biological samples, live animals, diagnostic kits, equipment/instruments, and so on. Failure to appropriately disclose the VCP for any material moved cross border can result in a huge liability to customs authorities globally. There must be a consistent process within R&D to place a value for any material arriving at the customs. The value must be based on real facts and data. In many organizations, the valuations of all R&D materials are formulated meticulously and intensively based on real material values and associated service costs. The methodology used to formulate the valuation must be supported by auditable information contained in the group's finance and inventory system. The World Trade Organization (WTO) has determined a set of rules for the determination of a custom value when importing goods. These rules have been formalized in article VII of the General Agreement on Tariffs and Trade of 1994 and implementation thereof. The WTO valuation rules have been adopted and

implemented by all members of the WTO. Although there are some variations in the application and implementation of these rules, the WTO has established a uniform method for determining the value of imported materials.

3) **Communication:** Establish a good communication channel between cross-functional groups, intra-site, third party service providers, and CROs. This can be achieved through setting up various interest groups, forums, or working parties, ensuring that shipments are booked in and out within your defined process. At various stages each support group collects/logs necessary information in paper or electronic formats; ensuring that you, as a compound management service, have access to this tracking is also important.
4) **Record retention:** Maintain a good record of shipments as company policy and as a governmental requirement. All records of movement of goods in cross-border trade including all documents relating to importing and exporting any commodity must be retained for at least five years in the United Kingdom, seven years in the United States. To err on the side of caution, many companies adopt an internal mandate that goes beyond the governmental requirements.
5) **Control substances:** The compound management leader must be familiar with how to manage controlled drugs. Controlled substances or forbidden compounds must not be shipped with regular research compounds. It is vital that the shipper has a process and tool to identify such compounds at the time of shipping. Mistakenly shipping such materials can result in a hefty fine and in the case of a reccurring event the responsible company may lose its export license.
6) **Export administration regulations (EAR):** Chemical samples of compound management do not required any export permit under EAR; however, every compound must fulfill the requirement of Denied Party List, Trade Sanctions, and Anti-boycott regulations (for example, compounds cannot, at the time of publication be sent to Iran, etc.).

8.2.6 Software Solutions

Besides simplification and standardization of shipping processes, consistent execution of the six elements above requires at least a well-organized and agreed process. For any company shipping large quantities of samples this should also include a well-designed software solution (IT application). This should be a system which allows each functional group to record their pertinent information, has access to other groups' information and is able to indicate appropriate actions within the same system. Such an IT system can provide a common platform with essential fields where important records can be captured and electronically streamed between the requestor (bench scientist), the classifier (shipper), and the final recipient. Therefore, such a single system would provide remote association between the cross-functional groups with secure and reliable exchange of information and create a speedy and seamless shipping operation. This IT application can also harmonize the entire shipping process and provides control mechanisms where everyone follows the same process every time. Besides business continuity, the system can be designed to provide automated compliance by mandatory requirement of key information including:

- **Built-in regulatory/DEA filters:** Preventing accidental transportation of prohibited material.
- **GxP material identification:** FDA Regulations of material identification for various 'Good Practice' compliance requirements (i.e., GLPs: Good Laboratory Practices, GMPs: Good Manufacturing Practices, etc.).
- **Material identification:** Proper hazard classification.
- **Built-in material valuation tool:** Accurate shipment valuation, reduce/eliminate calculation errors, prevent over/under payment of customs charges, and reduce/eliminate customs delays.
- **Built-in tool to aid transport classifications:** Enables the internal controls within an organization to ensure compliance with thousands of transportation regulations for various hazardous materials.
- **Built-in reporting tool:** Fulfills vital record retention requirement. Track, report, and data preservation. Pertain, validate, and create auditable process in case of any regulatory inquiries.
- **Features with automated email notification:** Electronic information sharing can provide necessary and timely communication for all involved parties.
- The shipping system can be a web-based system where, besides the internal use, the transport service providers and the third party customers can also have access. The system can also capture the carrier's tracking number, which allows real-time tracking of the shipment. In case of emergency, the responders can search the system as needed to retrieve detailed shipment information to aid in accident/incident response.

8.2.7 Conclusion

Successful globalization and externalization of pharmaceutical R&D functions require the flawless synchronization of organizational interfaces, affiliations with subject experts, inter-organization systems, and the exchange of material and information. Additionally, transportation of temperature-sensitive research samples in small volume by land and air cargo demands seamless cold-chain control to ensure product quality and integrity.

It is also important to remember that dangerous goods regulations are not written to specifically address the shipment of early drug discovery samples. There are various groups currently working with the various regulatory and transport organizations to address shortcomings of the existing regulations for research sample shipments, but such efforts are not necessarily co-ordinated into a collective voice. Hence it is difficult to envisage a unilateral agreement. However, any organization can implement an uninterrupted and compliant sample supply chain through rationalized shipping processes and with regulatory compliance controls.

References

1. Comley, J. (2009) Automation of solid/powder dispensing. Much needed, but cautiously used! *Drug Discov. World*, Summer, 39–51.
2. Matson, S.L. *et al.* (2009) Best practices in compound management for preserving compound integrity and accurately providing samples for assays. *J. Biomol. Screen.*, **14** (5), 476–484.
3. Isom, E. (2006) *Requirements for Outsourced Cold Chain Logistics and Storage in BioPharma Development*, BioProcess International.
4. Graphics from the Envirotainer, Cold Chain Management Services, *http://www.envirotainer.com.* (accessed 11 November 2011).
5. Higgins, A., Mangan, A., Kerrigan, A., Laffan, S., and Klein, S. (2009) Activity, ICT, and material infrastructure in complex multi-organisational settings: an assessment of innovation potential for pharmaceutical cold chain transport and handling. BLED 2009 Proceedings, p. 28.
6. The U.S. Department of Transportation (2011) The Hazmat Table 49 CFR Subpart B,172.101, PHH50-0119-1110, *http://www.dot.gov, http://www.phmsa.dot.gov/hazmat/library.* (accessed 10 November 2011).
7. Bentley, A. (2007) *Shipping and Handling of Natural History Wet Specimens Stored in Fluids as "Dangerous Goods" – Hazardous Materials*, Natural History Museum & Biodiversity Research Center, University of Kansas, National Park Service, Conserve O Gram Number 11/13, *http://kguerrero.net/documents/exporting_specimens-dangerous_goods_us.pdf.*
8. Regulations and Procedures for Shipping Hazardous Materials, *http://www.nyu.edu/mail.* (accessed 11 November 2011).
9. Bishara, R.H. (2006) The application of electronic records and data analysis for good cold chain management practices. *Am. Pharm. Outsourcing*, **7**, 1–5.
10. (2009) Guideline for Shipping Items 1 on Dry Ice that Are Not Dangerous Goods, Rev. 2, 1–5. *http://safety.dri.edu/Forms/shippingnonhaz.pdf.* (accessed 11 November 2011).

9
Managing a Global Biological Resource of Cells and Cellular Derivatives

Frank P. Simione and Raymond H. Cypess

9.1
Introduction

Managing a biological resource of living cells and cellular derivatives that effectively supports global scientific research, development, and quality assessment requires a complex and focused operational infrastructure [1]. The first step is to determine the needs of the scientific community and ensure that the available materials meet those changing needs. A robust program requires processes for acquiring, authenticating, preserving, producing, developing, and distributing the needed resources. The materials themselves must be well managed, and a system is also required for documentation and data management.

The complexities of global biological resource management include ensuring equitable and compliant access and distribution. This necessitates taking steps to protect material ownership rights and intellectual property, as well as comprehensively understanding all requirements and regulations for handling and transfer of biological materials. Use of materials derived from humans and animals requires attention to good practices, guidelines, and regulations covering research on human subjects and animals. Finally, a robust process is necessary for assuring the continued quality and standardization of the materials managed by the resource.

Culture collections, currently termed biological resource centers, have evolved to support these increasing biological resource needs. The World Federation of Culture Collections (WFCC) has nearly 600 member collections, most of which are collections of microorganisms that range in size from private collections to large collections like the American Type Culture Collection (ATCC). Collections like the Deutsche Sammlung von Mikroorganismen und Zellkulturen GmbH (DSMZ) and the European Collection of Cell Cultures (ECACC) provide cell culture resources to scientists around the world, primarily with government financial support. In addition, commercial entities such as Sigma–Aldrich and Invitrogen have started to offer traditional and specialized cell resources for which there is greater demand than for many of the resources supported by the biological resource centers.

Management of Chemical and Biological Samples for Screening Applications, First Edition.
Edited by Mark Wigglesworth and Terry Wood.

As a global tools and reagents provider, ATCC has been delivering biological reference standards and related materials to the global scientific community for the past 87 years. The resources ATCC has amassed and currently holds have been developed and contributed by scientists representing the vanguard of life sciences research in academia, government, and industry, including those who have been and currently are members of the ATCC scientific team. Key competencies developed and enhanced by ATCC during its long history include the ability to identify, acquire, authenticate, protect, and equitably distribute biological resources that are useful to the scientific community.

Assuring that the critical tools and reagents supporting state-of-the art research are available and accessible, without over burdening the process with materials that are of little or no use, requires good management of these processes. ATCC has developed and enhanced a non-profit business model that ensures that state-of-the-art and high-quality biological resources are available and accessible when needed. This is accomplished by utilizing a proprietary customized management system that assures the resources entrusted to it are not compromised or jeopardized in any way without reliance on external support. The elements of this management system as it applies to the ATCC holdings of cells and cellular derivatives, important to the development of pharmaceuticals, biologicals, and other biomedical products, are delineated in the following sections.

9.2 Diversity of Collections

Culture collections and biological resource centers vary in the diversity of the materials they offer to the scientific community. Small collections that are developed for use by an individual investigator or group of researchers often consist of large numbers of items representing a specific group of cells or organisms. Commercial organizations that offer cells and reagents for scientific research often limit their offerings to those items that are in demand and will generate revenue. Large biological resource centers support a wider range of scientific research by offering a broader diversity of resources, usually by relying on support from academia or government.

The collection of materials ATCC has acquired and maintains consists of a diversity of biological resources, tools, and reagents. The biological resources include microorganisms and viruses, as wells as human, animal, and plant cell lines (Table 9.1). The idea for a central collection of microorganisms originated with the Society of American Bacteriologists (SAB), now the American Society for Microbiology (ASM), which created the collection of bacteria that eventually became the first ATCC collection of biological resources. Microbes continued to be the primary source of collection materials until the culture of human, animal, and plant cells in the laboratory began in earnest.

Related materials such as nucleic acids, clones, proteins, and other derived materials were later added.

Table 9.1 ATCC collections.

Collection	Number of items
Cell lines	4500
Primary cells	15
Bacteria	18 000
Fungi	49 000
Protists	3500
Animal and plant viruses	2500

The collection of cell lines at ATCC was initiated in 1962 and currently contains cell lines from approximately 150 species. The types of cells include established normal and tumor lines, immortalized cells, hyridomas, stem cells, and primary cells. These are further subcategorized into tissue types, such as neurobiology cell lines and genetic variant fibroblasts, and are complemented by derived resources such as nucleic acids and proteins.

Each collection of cells is given a designation for cataloging and for expediting user searches (Table 9.2).

Designations are assigned to the collections based on how the cells were originally deposited into the ATCC. Certified cell lines (CCL) are fully authenticated lines, while cell repository lines (CRL) are minimally tested for microbial contamination and verification of species of origin, and both are part of the ATCC general collection. The tumor immunology bank (TIB), human tumor bank (HTB), and the hybridoma bank (HB) designations represent lines that were transferred to ATCC from government-supported collections. The most recent additions to ATCC's collection of cells are the stem cell resource center (SCRC) and the primary cell solutions (PCS). Table 9.3 lists the current ATCC holdings of cells lines and cellular derivatives.

In addition to the general collection of cells and those deposited through government funded programs, ATCC also manages cell lines that are part of special collections or were received as patent deposits, and houses a collection

Table 9.2 ATCC cell line designations.

Designation	Type of cell line
CCL	Certified cell line
CRL	Cell repository line
TIB	Tumor immunology bank
HTB	Human tumor cell bank
HB	Hybridoma bank
SCRC	Stem cell resource center
PCS	Primary cell solutions

Table 9.3 Cells and cellular derivatives.

Cell type/use/derivative	Designation	Number of items
Hybridomas	CRL, HB, TIB	1200
Tumor cell lines	CCL, CRL, HTB	1200
Tumor/normal matched cell line pairs	CCL, CRL, HTB	50
Stem cells	SCRC, CRL, HTB	100
Primary cells	PCS	15
Tools and models	CRL, HB, TIB	250
Neurobiology cell lines	CRL, HTB, HB	60
Genetic variant fibroblasts	CCL, CRL,	100
Genes and bioactive compounds cell lines	CCL, CRL, HTB, TIB, HB	3000
Immortalized cells telomerase reverse trascriptase (h = human) (hTERT)	CRL	6
DNA	CCL..D, CRL..D, etc.	60

of human and animal cells developed at the Naval Biosciences Laboratory (NBL) in Oakland, California, and transferred to ATCC in 1982. The NBL lines are not accessioned into the ATCC general collection, nor are they fully authenticated; they are available on a limited basis by special request.

The collections contain cells from a number of human and animal tissue sources including, but not limited to, bladder, bone marrow, brain, breast, eye, heart, intestine, kidney, liver, lung, blood (e.g., lymphocytes and macrophages), ovary, prostate, and skin. Although most of the cells are derived from human, mouse, and rat sources, a number of cell lines are available from other mammalian and non-mammalian species. More than 1000 hybridomas are available from ATCC listed in the catalog by the antigenic determinant recognized by their expressed monoclonal antibodies.

Tumor lines are available from a variety of species, including tumor cells from primary sites or from sources in which the primary or metastatic status is unknown. These lines represent diseases including, but not limited to, carcinomas from a number of sources, adenocarcinoma, astrocytoma, fibrosarcoma, glioblastoma, hepatoma, insulinoma, leukemia, lymphoma, lymphosarcoma, melanoma, neuroblastoma, osteosarcoma, retinoblastoma, and sarcoma. In addition, a number of tumor/normal matched cell line pairs from the same patient are available representing several of the metastatic diseases. Approximately 950 cancer cell lines are part of the ATCC cell culture collection.

In recent years, ATCC has added stem cells and primary cells to its diverse collection in response to the needs of the scientific community. Research involving stem cells has increased in recent years because of the potential for therapeutic use, and primary cells provide a research tool that is more closely related to cells *in vivo*. The stem cell collection includes embryonic and somatic stem cells, embryonal carcinoma cells, as well as mouse and human feeder cell lines. The primary cells available from ATCC are accompanied by basal growth media, growth kits, and reagents, all available from ATCC under the Primary Cell Solutions™ brand as a system of matched components designed to maximize growth and functionality, and maintain normal morphology.

Many of ATCC's cell lines are used as phenotypic or process models, tools for production or expression, and in other applications. These include, but are not limited to, transfection, expression, signal transduction, feeder cells, hybridoma fusion partners, factor assay systems, and differentiation. Others uses include retroviral packaging lines, recombinant retroviral expression, T cell receptors, and IL-3 signaling. Neurobiology cell lines are available from humans and other species, and the collection of skin fibroblasts consists of lines derived from individuals with genetic disorders and other disease states. In addition, the collections of cells are further cataloged by their expression of, or response to, specific genes, gene products, and bioactive compounds.

Purified genomic DNA is available from ATCC from normal and tumor human cell lines, including pairs from the same individual, which allows a comparison of normal cells, such as Epstein Barr virus (EBV)-transformed B-lymphocytes and tumor cells. The holdings include DNA from human breast, lung, liver, and prostate cancer lines, as well as non-human lung and kidney lines, and these are cataloged by adding the letter D to the ATCC number of the cell line from which they are derived (e.g., CCL-240D).

As a full service cell biology resource, ATCC supplements the cells and derivatives holdings by also providing culture media, serum, and other reagents to support the use of the biological materials. High-performance media and sera are formulated and tested to ensure maximum response from ATCC cells in culture. These include media such as Dulbecco's Modified Eagle Medium (DMEM), Eagle's Minimal Essential Medium (EMEM), F-12K, Iscove's Modified Dulbecco's Medium (IMDM), Leibovitz's L-15, McCoy's 5A, and Roswell Park Memorial Institute medium (RPMI)-1640, as well as fetal bovine serum (FBS), including FBS qualified for use with mouse and human embryonic stem cells, and calf bovine serum and horse serum.

Cell culture reagents available from ATCC support the growth and use of ATCC cell lines. These include L-glutamine supplements for support of mammalian and insect cell growth in culture, minimal essential medium (MEM) non-essential amino acids provided as a supplement to enhance cell growth, penicillin-streptomycin-glutamine solution to assist in minimizing the risk of bacterial infection during cell growth, and Hank's balanced salt solution for washing cells or use as a diluent or an inorganic base for media preparation. The cell culture reagent offerings are complemented by the availability of cell culture grade water that meets water for injection (WFI) quality standards.

Kits and panels are also available from ATCC that provide cells and reagents in a set ready for use. These reflect ATCC's continued commitment to ensuring convenience and quality in scientific research, and represent ATCC's efforts to continue to add value to the tools and reagents that it offers. The ATCC® Breast Cancer Cell Panel consists of 45 ATCC cancer cell lines provided as a set for convenience in research on genomic abnormalities. A stem cell line analysis kit, the ATCC® enzyme-labeled fluorescence (ELF) Phosphatase Detection Kit, provides an assay for determining the status of stem cell differentiation, and two kits, MTT and XTT, are available for evaluating cell proliferation.

ATCC's product and service offerings are supplemented by advice and assistance to those using cell lines in their research and development activities, such as how to properly reconstitute frozen cells and cultivate and maintain them for optimal response. These published guidelines include information on subculturing cell line monolayers, cell line verification testing, cryogenic storage, and the effects of passage number on cells in culture, and they are part of ATCC's efforts to enhance its service to researchers by providing information on working with the biological reference materials (ATCC Technical Documents – *www.atcc.org/culturesandproducts/technicalsupport/resourcesforcellbiology*).

9.3
Sourcing and Acquisition

The original purpose for creating ATCC as a collection of bacteria was to provide a central resource for scientists to contribute strains to be shared with their colleagues. This model worked not only for bacteria, but later for other microorganisms, viruses, cell lines, and derived materials like nucleic acids until value and ownership became critical parts of biological materials management. After 1980, scientists and their institutions began to recognize the value of their holdings and became more reluctant to provide materials to be made available to other scientists without use restrictions. This has led to the establishment of practices for ensuring protection of ownership and intellectual property rights when biological materials are shared with others. ATCC has put in place tools such as material deposit agreements (MDAs), material acquisition agreements (MAAs) and material transfer agreements (MTAs) to enable this process.

Another challenge in sourcing and acquiring biological materials is the need to prioritize based on both usefulness to the scientific community and the resources available to support the management and distribution of the materials. This requires knowledge of both advancements in science as well as the technological resources available in the scientific community that support these advancements. ATCC directly engages the scientific community through its team of scientists and marketing professionals to ensure that the most useful materials are acquired. In some areas, ATCC acts in partnership with the Federal government to develop collections for specific uses, and its project management teams then focus on acquiring the specific resources to support these initiatives.

That portion of ATCC's activities that are not covered by government funding for specific acquisitions requires a business-like approach to ensure that the investment of resources results in a return measured by subsequent use of the materials. Because of limited resources, ATCC assures that the materials it acquires are supporting the broadest usage in its tripartite global user base, that is, industry, academia, and government. This requires targeting specific resources available in the scientific community and providing support and options to prospective depositors. ATCC provides assistance in transport and compliance when necessary, and offers depositors and their institutions options such as gifting the materials to ATCC or supplying them under a bailment arrangement whereby we hold the cells on behalf of the depositor without owning them.

Sourcing of cell lines and related materials requires not only continuous vigilance of developments in scientific advancement but awareness of where new resources originate and how they are used. Assessing the needs of users enables ATCC scientists and marketing professionals to locate resources that will fully support those needs. Part of the process is establishing a dialogue and relationship with those who have developed or possess cells or reagents that may be of broader use. When an agreement is reached to make the material more widely available the processes for acquisitioning and accessioning begin.

Acquisition of new resources for global distribution has become a complex process involving good management and control of the transfer and use process. When potential depositors are approached they must be assured that their materials will be shared equitably and that the use of the materials is consistent with protection of their ownership and intellectual proper rights. The deposit and transfer agreements that ATCC has developed are designed to ensure that depositors understand the conditions for sharing of their materials, and users understand the use conditions and restrictions. While use restrictions may at times seem cumbersome, use is not restricted when the proper agreements have been executed between the user, the depositor, and ATCC.

New lines received at ATCC for addition to the collections are assigned a catalog designation based on the type of material and its source. Accessioning cell lines into the ATCC collection includes preparing preserved token lots from the submitted material, seed lots from the token material, and finally distribution lots from the seed material [2]. Quality checks are performed at each of these levels and the goal is to assure that cells made available to the end-users are as close in passage as possible to the original deposit. The token lots are prepared immediately upon receipt of the depositor's material to assure that the cryopreserved cells are as close in passage as possible to the original. Seed or master banks are prepared to provide starting material for production of distribution lots. When a cell line has been through the entire process, including full authentication, it is then listed in the catalog and made available to the user community.

9.4 Authentication and Characterization

The need for well-qualified cells and other reagents is universally understood as a strong underpinning for assuring effective, reproducible, and timely research and development. The recognition that cells can become contaminated and mis-identified dates back to the 1950s just after the successful establishment of cells in culture, when it was quickly recognized through chromosome analysis that interspecies cross-contamination had occurred [3].

New and more sophisticated assays developed in recent years have revealed that many cell lines currently used in research and development may not be what they have been claimed to be [4–6]. Additionally, it is also becoming apparent that over-subculturing cells can lead to changes that affect experimental outcomes [7–9]. These changes are reported to affect not only cell morphology and development, but gene expression as well. Cells can also undergo mutation during continuous subculturing, and faster growing cells can eventually outnumber the slower growing cells in a heterogeneous population of cells.

The use of cells exchanged among colleagues can lead to unknown and unwelcome characteristics if the cells are not fully authenticated. The costs to research laboratories in both money and time can be prohibitive when continuous verification of authenticity and quality is performed. Procuring cells from a source that maintains low-passage materials and continually verifies the quality of the starting materials is the best solution for assuring expedient and cost-effective research results.

Assuring users of cells and cellular derivatives that the materials are what they are supposed to be and will do what they are reported to do requires a robust system of authentication and verification. Also, users are constantly looking for additional information about the materials they use, and this is generated from ATCC in-house evaluation, data obtained from users of the material, or through collaborative activities.

Table 9.4 lists assays performed by ATCC for authentication, characterization and verification of the cell lines and cellular derivatives. Performance of many of these tests requires both sophisticated equipment as well as knowledgeable and skilled scientists.

Table 9.4 Authentication and characterization assays.

Assay	Purpose
Viability	Recovery following preservation
Cellular morphology	Changes due to aging
Microbial contamination testing	Bacteria, fungi, mycoplasma
Mycoplasma (cultivation; hoechst; polymerase chain reaction (PCR))	Mycoplasma contamination
Endogenous virus testing	Viral contamination
Single tandem repeat (STR)	Species verification (intraspecies)
Isoenzyme analysis	Interspecies
Cox 1 gene	Interspecies
Cytochrome oxidase (CO1) DNA barcode	Interspecies
Immunophenotyping flow cytometry	Cell surface antigens
Cytogenetics (karyotyping, FISH)	Interspecies
Immunocytochemistry	Antigen expression
PicoGreen and propidium iodide	DNA quantitation

9.4.1
Viability

Assessment of viability on cryopreserved cells is performed to ensure that an adequate population of cells is recovered for establishing them in culture. This can be accomplished by several means including dye exclusion and counting of viable cells.

9.4.2
Cellular Morphology

Cellular morphology is a critical indication of health and normal growth, and can vary depending on factors such as the differentiation state of the cells. The morphology of cells can change when conditions such as plating density or varying media and sera combinations are used. During the accessioning and bioproduction processes cells grown at both low and high densities are checked morphologically to ensure changes have not occurred.

9.4.3
Microbial Contamination

Testing of cell lines for contaminants such as bacteria, fungi, mycoplasma, and viruses is an important part of the full authentication process at ATCC, and includes using the BacT/ALERT® BD automated microbial detection system, backed-up by standard batteries of selected growth media and temperature regimens [10].

9.4.4
Mycoplasma Detection

Rigorous testing is performed on all ATCC cell lines to ensure they are free of mycoplasma contamination. Mycoplasma contamination can affect cell growth and function, including transfection, and gene expression. Cultivation trials and Hoechst staining are performed in conformity with the Food and Drug Administration (FDA) points to consider protocols. Screening of cells for mycoplasma contamination is also performed routinely using a PCR mycoplasma detection system developed by ATCC scientists [11–13].

9.4.5
Virus Testing

Testing cell lines for adventitious viruses can be carried out in both *in vitro* and *in vivo* assays. Indicator cells that are susceptible to potential contaminating viruses are used to monitor for the presence of the virus, and observation of the cell population for cytopathic effects provides an indication that a virus may be present. Testing cell lines for the presence of reverse transcriptase is used to determine the presence of retroviruses in cell types traditionally known to harbor retroviruses based on their origin. Embryonated eggs and mice can be inoculated with cell lysates to test for the presence of murine and other endogenous viruses. ATCC employs *in vitro* test methods on selected cell lines to determine the presence or absence of adventitious viruses [14].

9.4.6 Short Tandem Repeat (STR) Profiling

Multiplex PCR is used to amplify polymorphic markers in the human genome in order to generate short tandem repeat (STR) profiles of human cell lines in the ATCC collection. This assay is used to verify the identity of the cell lines and is performed at all levels of the production process. The results for each assay are compared to the baseline STR profile from the STR analysis of the first, or token, lot of cells prepared from the original material received from the depositor. The STR database for ATCC cell lines is available to users of ATCC cells upon request to assist in their assurance of the authenticity of the cells during their use [15].

9.4.7 Isoenzyme Analysis

Isoenzyme analysis is used to verify information provided by the depositor regarding the species of origin. This analysis consists of differentiating the enzymatic components based on their electrophoretic properties. Isoenzyme analysis is part of ATCC's accessioning protocol, and each distribution lot is assayed to verify the species of origin [16, 17].

9.4.8 Cox 1 and CO1

The Cox 1 gene and cytochrome c oxidase subunit 1 (CO1) assays are used to verify the species of origin of selected cell lines. Other assays include Cox 2, cytochrome B, and NADH dehydrogenase.

9.4.9 Karyotyping

To ensure that a stable genotype is maintained by cells in culture, karyotyping is used to monitor any changes in chromosome morphology and composition. Karyotyping is routinely performed as part of the authentication process for many of ATCC's traditional cell line holdings, as well as all embryonic stem cell lines. Fluorescent *in situ* hybridization (FISH) is also used to detect the presence or absence of specific DNA and RNA sequences [18].

9.4.10
Immunophenotyping and Immunochemistry

Immunophenotyping is performed on cells using flow cytometry to assay for cell surface antigens, and immunochemistry assays on selected cell lines are performed to determine the presence of marker antigens.

9.4.11
Pico Green and PI

The quality control of DNA preparations involves quantitative analysis using the Pico Green® reagent to more accurately determine the DNA concentration. Other methods include propidium iodide (PI) as well as the traditional 260 nm absorbance determination.

To improve the standardization and efficiency of performing authentication assays, ATCC created a core services unit that works closely with a robust quality control unit, to perform genomic and other molecular assays. This provides less costly and more expedient testing, coupled with consistency in authentication testing protocols and practice throughout ATCC.

The importance of continued vigilance and application of new technologies to the authentication processes is reflected in the findings at ATCC that some of its cell lines were actually unknowingly misidentified. When DNA profiling via STR analysis was implemented at ATCC it was discovered that six cell lines in the collection were not as originally reported by the depositor when they were acquired many years earlier.

9.5
Cryopreservation, Storage, and Production

The ability to preserve living biological materials at low temperatures has been critical to ATCC's ability to maintain a large array of microorganisms, cells, and related materials for more than 85 years. Microorganisms were originally freeze-dried to ensure their stability and assist in the ease of distribution; however, with the advent of liquid nitrogen storage in the early 1960s, ATCC was in a position to both more effectively stabilize the microorganisms and adopt a means for low-temperature storage of more complex cells, including human, animal, and plant cells. This capability has allowed the development of collections of living cells and helped ensure their stability during long-term storage [19, 20].

Cell lines are preserved at ATCC using protocols that have traditionally proven to be effective in ensuring optimal recovery of the preserved population following freezing and storage [21, 22]. These protocols include ensuring first that the cells are in a healthy state, and providing cryoprotective agents to assist in protecting them during the freezing process. ATCC uses dimethylsulfoxide (DMSO) as a cryoprotective agent for most of the cells, and preparation for freezing includes acclimation to the cryoprotectant prior to starting the freezing process. Controlled cooling is achieved using programmable freezing units, and the cooling rates have been developed following years of experimentation and proven success.

After the freezing process is complete the cells are transferred to all-vapor storage liquid nitrogen units where they are maintained at temperatures below −140 °C to ensure they remain below the critical temperature for ice crystal reformation. The liquid nitrogen units are specifically designed to maintain the desired temperature in the inventory space at all times. Additionally, stocking and retrieval operations are conducted in a manner that ensures that the cells remain below the critical temperature at all times until prepared for shipment on dry ice. Although the temperature of dry ice is only −79 °C, ATCC experience and experimental testing has established that most cell lines will remain viable at this temperature during transport and for short periods of time prior to immediate use upon arrival. More sensitive cells are packaged and shipped in liquid nitrogen carriers that maintain the temperature of the cells well below the −130 °C critical temperature for biological materials.

The freezers are housed in ATCC's biorepository, which is equipped with a sophisticated alarm system to alert trained operators to any potential temperature excursions. This includes redundant monitoring at the freezer, in the ATCC facility-wide system, and via an off-site independent monitoring service. When an alarm is received, operators on 24 h call are notified and respond immediately. Standby freezers are maintained in operating mode to provide a transfer option if a freezer completely malfunctions. The facility is equipped with a 6000 gal bulk liquid nitrogen stand tank, and the electrical power is backed up by an emergency generator. The facility is secure, with access limited to authorized personnel only, and is monitored via intruder alarms, motion detectors, and surveillance cameras.

Because of increased production requirements and to minimize handling of cell cultures by laboratory personnel, ATCC invested in robotic systems for production of the cell lines. Two TAP SelecT™ robotic units have been online for more than three years in a clean-room environment. These units have enhanced the capacity for bioproduction of cell lines to meet increasing demand for these resources. Minimizing human handling of the cells during production has reduced the risk of contamination during production processes.

9.6 Data Management

Good data and information management are critical elements of a well-managed biological materials handling program. Components of these include information on the types of materials housed, their inventory location, and scientific information about them both from external sources and internal laboratory evaluation, as well as information documenting transfer and use. The different information management needs require complex and often expensive resources for cataloging, documenting, and manipulating the generated data. ATCC's enterprise resource planning (ERP) system consists of a commercially available system (QAD, Inc.) which has been extensively modified by ATCC to meet the requirements of its production, inventory, and distribution activities.

Information on ATCC cells and cellular derivatives initially comes from the depositor, and reliance is placed on both detailed and accurate information. Deposit forms have been created to try to cover much of the information that is needed for assuring some robustness and accuracy in the type of material and its performance characteristics, or usefulness to scientific applications. Additional information is often obtained via publications about the deposited material, including both historical and ongoing published studies. This effort keeps the user of the materials up-to-date on novel characteristics and use information, but requires constant vigilance and updating by ATCC staff.

Scientific data associated with each item in the collection is obtained from published literature, both historical and contemporary, and through laboratory testing at ATCC or via collaborative arrangements. The greatest challenge in documenting and cataloging scientific information generated on the cells and their derivatives is standardization of the data. This requires sophisticated mechanisms for transferring and translating laboratory data into usable and standardized databases.

To ensure that the scientific data collected on the cell lines is current, ATCC continues to invest in the addition of resources for generating scientific data. In recent years, ATCC has added the tools to enable genomic and proteomic analysis of cells and cell products, including PCR analysis, genome sequencing, and protein analysis. Scientific data generated on the cells is tabulated in internal databases, and then organized and cataloged for access by users of the cell lines. The useful data is then added to the product profile for each cell line in the ERP system, and the information is provided via product sheets and certificates of analysis.

9.7 Quality and Standards

Quality and standardization are the hallmarks of a good biological resource program, and ATCC has traditionally assured that its resources perform as expected with repeated use in different laboratories. Assuring quality requires not only paying attention to the laboratory characterization of the cells but ensuring that the cells are maintained in a manner that allows full traceability to the original source. When cells are exchanged between laboratories outside of a resource like ATCC, assurance of quality, standardization, and traceability are lost.

To provide overall management and control, ATCC has a corporate-wide quality management system (QMS) based on quality requirements as defined by the US FDA and International Standards Organization (ISO). This program consists of a quality assurance unit that documents and records control and management processes, and a separate quality control unit that performs quality testing on all cells in the collection. A corporate quality manual outlines the roles, responsibilities, and standards of practice at ATCC in compliance with accepted quality practices. Written procedures are in place for production and quality testing of each ATCC product, and document control is managed through an electronic document management system. The program includes internal and external quality audits, as well as a Corrective Actions and Preventive Actions (CAPA) program.

ATCC is ISO 9001:2001 certified and ISO Guide 34 and ISO 17025 accredited. As a certified reference material (CRM) manufacturer, ATCC's accreditation is unique in that the processes used to produce the cell products are fully covered. This allows ATCC to apply the standard to all products, including those to be received in the future. ISO 17025 is an accreditation for laboratory testing and all of ATCC's testing processes are covered. Production of CRMs necessitates adherence to the requirements for traceability and homogeneity found in ISO Guide 34, and ATCC has adapted these requirements to production of CRMs where these are required by the end-users.

In 2007 ATCC became the first biological resource to be accredited by the American National Standards Institute ANSI as a Standards Development Organization (SDO). This designation recognizes ATCC as a developer of written standards for biological materials and their use, and involves the development of a consensus process that includes participation by all stakeholders, including users. To ensure participation by all interested parties, ATCC developed the Consensus Standards Partnership (CSP), where membership is open to all stakeholders and other interested parties.

Using this process, ATCC developed and published ANSI Standard ASN-0001-2009: Standardization of *in vitro* Assays to Determine Anthrax Toxin Activities available through the ANSI Standards Store. A second ANSI standard, ASN-0002: Authentication of Human Cell Lines: Standardization of STR Profiling, is nearing publication. This standard was developed by a working group composed of experts in cell culture authentication and characterization.

In response to a need for better standardization of biological materials used in proficiency testing, ATCC established a Proficiency Testing Program to allow manufacturers and providers of proficiency testing materials to obtain well-qualified and traceable biological components. Proficiency testing is a mechanism by which diagnostic laboratories are evaluated in order to maintain their accreditation as a testing laboratory. ATCC initiated the Proficiency Testing Program in response to a need for assuring that the biological materials provided for these assessments are accurate and safe. Users of this program enter into agreements with ATCC to obtain materials that have been prepared and qualified to their specifications, ensuring that the materials are represented accurately, and provide a safe environment for users of the proficiency testing panels. Currently there are 13 organizations participating in ATCC's Proficiency Testing Program.

The ATCC Standard Resource was launched in 2006 to allow users of recognized standards to determine which ATCC cultures conform with these standards and hence, which are applicable for their use. The resource provides a user-friendly tool for obtaining information on the standards and the biological materials required for use with the standards, and also facilitates the acquisition of the required materials.

9.8 Order Fulfillment and Distribution

As a global resource for diverse biological materials ATCC has developed a sophisticated and robust system for managing access to its holdings and delivering them to scientists around the world. Order fulfillment is the mechanism by which potential users are qualified and provided with access to the materials, and distribution is the process by which materials are transported to the user. Each of these is heavily overlaid with quality and compliance requirements to ensure accuracy, product protection, and adherence to all regulations. A robust ERP system coupled with sophisticated enhancements added by ATCC is used to manage the handling of the information and requirements in the processing of requests and delivery of materials.

All potential users of ATCC materials must register both their institution as the responsible entity, and themselves as the ultimate end-user before any material can be provided. Institutional registration ensures that an entity is ultimately responsible for compliance with both safety and regulatory requirements, as well as all use restrictions with regard to ownership and intellectual property protection. An institution can have multiple end users; however, when an end user changes institutions they are re-assigned to their new entity's account. This process ensures accurate documentation and accounting of all activities with regard to transfer and use of these materials.

The ability to deliver critical and labile biological materials anywhere in the world is a hallmark of ATCC's distribution system. Currently ATCC has 12 partners located in Europe, Asia, North America, Latin America, and the Middle East that assist in the delivery of the cells and related materials to users in all areas of the world. In addition, ATCC staff work closely with expeditors and freight forwarders to ensure the timely, safe, and secure movement of biological materials in transit. Compliance specialists maintain a daily vigilance in order to be aware of changes in regulations and requirements to ensure that importing, exporting, and domestic transfer of biological materials can be made efficiently and in a timely manner.

Because cells and other biological materials are sensitive to environmental changes, it is imperative that they are protected from deterioration during transfer. ATCC has a diversity of transport capabilities designed to handle its diverse holdings, allowing transfers ranging from ambient to liquid nitrogen temperatures, as well as multiple temperatures in between. Specially designed packaging provides protection for products against physical damage as well as temperature excursions, and temperature monitoring is used for some transfers to verify that temperature excursions did not occur.

9.9
Offsite Biorepository Management

Many of the materials currently in the ATCC collections are not available elsewhere and would be lost to science if they were not secured by ATCC. To ensure that these resources are not lost, ATCC has established a backup biorepository in which all holdings are duplicated. This backup storage protects against loss sustained by a physical disaster at the main ATCC facility, and is further supported by a business continuity plan that allows for the gradual replacement of the entire ATCC collection if total loss was experienced.

Initially, ATCC relied on other facilities to provide this backup, including a government facility and later a private company. However ATCC has invested in a backup facility owned and operated solely by ATCC that is equipped with all of the support requirements available at the corporate facility, including backup emergency power, remote monitoring, alarm notification, and security. Development of this redundant facility is part of ATCC's continuing commitment to invest in assuring that the valuable biological resources entrusted to it continue to be available for scientific research and development.

Good management of biological materials is a core competency of ATCC honed through 87 years of continued experience. Leveraging that competency into providing support for other bioresources is now part of ATCC's services offerings. This includes both design and development of facilities, as well as full management of a biological resource program. ATCC's broad experience includes design and construction of biorepositories and management of biological materials for other organizations both at their location and using ATCC resources. These services are a logical outgrowth of any good biological resource center, and are recognized by government and other organizations as a resource for use in developing their programs.

Examples include design and development of a turn-key respository for the United States Army's Critical Reagents Program and management of the Centers for Disease Control and Prevention (CDC) biorespository in the Atlanta, Georgia area for the past 14 years. ATCC has also developed and managed biological resource collections for Federal government agencies, including the National Institute of Allergy and Infectious Diseases (NIAID) collection of resources for malaria research, that is, the Malaria Research and Reference Reagent Resource Center (MR4) managed by ATCC for the past 12 years, the Biodefense and Emerging Infections Research Resources Repository (BEI Resources) since 2003, and the Microbiology and Infectious Diseases Biological Resource Repository (MID-BRR) starting in 2010, as well as the CDC Influenza Reagent Repository (IRR) since 2008. While these resources are focused on infectious diseases, each of them requires the use of cell lines to support the development and production of the microbial and viral holdings.

Since 1998 ATCC has also been supporting the National Cancer Institute (NCI) Divisions of Cancer Epidemiology and Genetics efforts to develop and authenticate cancer cell lines and genetic materials. This program involves the procurement of

DNA, RNA, transformed lymphocytes, and lymphoblastoid cell lines in support of genetic linkage studies. It involves the processing of human biospecimens to establish the cell lines, and extraction and purification of human DNA and RNA.

For more than 30 years ATCC has been providing services in the safekeeping of biological materials for other organizations and in support of biotechnology patenting. The safe deposit service includes secure maintenance of cGMP-compliant microbial and cell banks for pharmaceutical and biotechnology companies. In 1981 ATCC was designated as one of the first International Depositary Authorities (IDAs) by the World Intellectual Property Organization (WIPO) for the deposit of biological materials in support of biotechnology patent applications. Today ATCC is the largest of 40 IDAs located around the world in terms of the diversity of biological materials accommodated for patent deposit, and the number of deposits received.

9.10 Regulatory and Legal Compliance

Regulatory and legal compliance is an important aspect of good management of biological materials, and ranges from assuring safety and compliance for high-risk materials to proper management of human and animal derivatives. As a consequence, ATCC has developed a robust program that provides continued assurance that all materials housed at ATCC are handled and transferred in full compliance with all requirements and regulations. This requires constant vigilance over all activities associated with the transfer and use of the materials as well as the constantly changing regulatory requirements.

The process developed and used by ATCC consists of three phases: (i) evaluation of the product to be added to the collection, (ii) assessment of the institution and end-user, and (iii) processing of requests and transfers. For each item added to the ATCC collections a comprehensive review is conducted by the specific collection scientist in collaboration with ATCC compliance specialists. The review is based on information known about the material, including biosafety level and required permits, and licenses. When this review is completed the information is entered into the ERP system and the product released for distribution.

Each institution and end-user must establish an account with ATCC by providing a profile of their organization and their ability to handle material obtained from ATCC. Accounts are tiered to accommodate the needs and capabilities of each institution so that those requiring access to low-risk materials do not need to meet the more stringent requirements for high-risk materials. Other assurances include the execution of any special documents between ATCC and the requesting institution such as MTAs and customer acceptance of responsibility (CAR). After an account application is fully vetted, the institution and end-user are entered into the ERP system, and requests for materials can then be accommodated.

When requests for materials are received, the ERP system matches up the requestor's account information to the information associated with the material requested. The system is designed to alert order processing and compliance staff if discrepancies occur between the two databases, as well as any special requirements that must be met before releasing the material for distribution. In addition, the system screens all names, the institution as well as end-user, against a list of names compiled from government watch lists such as the department of commerce entities list of export violators. If a match is found, further vetting of the institution or end-user is required to assure the match is not legitimate. The screening lists are updated on a daily basis by the compliance specialists.

Many of the cells in ATCC's collection are derived from humans, and steps are taken to assure that the processes used in obtaining the cells have met the guidelines and requirements for use of human-derived materials. ATCC's Institutional Review Board (IRB) provides oversight and guidance on acquisition of human-derived cells, and reviews and approves specific acquisitions when required. For legacy cell lines the acquisition and consent processes are not well delineated; however, all newer acquisitions are vetted through this process.

9.11 Ownership and Intellectual Property Management

When ATCC's cell line collection was established in the 1960s the practice of providing materials to a central resource for use by the scientific community was full and open. However with biotechnology advances and commercial use of biological materials, depositors began to recognize value in their materials and became reluctant to share them without some protection against indiscriminate commercial use. To assure that these materials continue to be made available in support of research and development activities in the general scientific community and to protect the rights of the depositors of the materials, ATCC established mechanisms for equitable access. This necessitated creation of MAAs and MTAs, as well as mechanisms for allowing use beyond the basic research use limitation on all ATCC materials.

To support this activity, ATCC created a licensing and intellectual property program fully staffed with technical and legal experts who review deposit conditions and assure maximum protection of the rights of the depositors. Depositing cell lines with ATCC starts with negotiating an agreement on the conditions for deposit, and depositors and their institutions are provided the opportunity to submit materials under several conditions. If the material is gifted to ATCC, it can then be made available for research use only by ATCC under all the terms and conditions established by ATCC for use of the materials for research only and not for commercial application.

When a depositor or their institution choose a bailment option, that is, providing the material under conditions which assure ownership and control by the depositor, the conditions for use are negotiated with the depositor. The conditions are documented in the MTA, and ATCC negotiates the best options for protection of the owner's rights while ensuring that the material is available for broad research use. If conditions for access and use are too restrictive, ATCC provides the depositor with options that allow their control under a business relationship with ATCC.

To ensure that recipients are meeting the requirements for use of the material, an MTA is established with the user's institution that outlines the conditions for use. This agreement establishes the terms for research use only, and provides options for commercial use only after negotiation with ATCC, and ultimately with the depositor. The MTA also requires assurance from the user that the material will be used in a safe manner in the laboratory, and is not used in humans or animals without review and approval by ATCC, and limits use to the recipient's laboratory only to avoid any potential violation of the agreement. Transfer of materials received from ATCC to third parties is not allowed unless prior approval is obtained from ATCC under a subsequent transfer agreement.

9.12 Collaborations

Because manufacturers of quality kits and reagents use ATCC cell lines in their product development and optimization processes, ATCC forms partnerships with other life science companies to ensure that cell lines available from ATCC are performing as needed. These partnerships also provide opportunities to enhance the use of ATCC cell lines for specific purposes. Our partners include academic institutions and government agencies as well as private sector organizations.

ATCC collaborated with Roche Applied Science to develop a database of ATCC cell lines transfected with Roche's FuGENE® 6 and FuGENE® HD Transfection Reagents. The database includes links to protocols and information on the transfection of the designated ATCC cell lines. The cell lines developed for use with the FuGENE® transfection reagents and protocols are available on the Roche web site (*https://www.roche-applied-science.com/sis/transfection/index.jsp?id=tfn_010100*), and include COS-1 cells (ATCC® CRL-1650™), HeLa (ATCC® CCL-2™), and CHO-K1 (ATCC® CCL-61™). Using these lines and the designated expression plasmids provides a good starting point for transfection studies; however, these are suggested protocols, and it is incumbent upon the user to determine the optimal conditions for their use.

Amaxa's Nucleofactor® is used to transfer various substrates, such as DNA, siRNA, and peptides, into cell lines, allowing manipulation of cells not previously amenable to gene transfer. In collaboration with ATCC, Amaxa has developed optimized protocols for high transfer efficiencies using ATCC cell lines, with good cell survival and minimal metabolic impact. Cell lines, such as MCF-7 (ATCC® HTB-22™), were used to provide optimal transfection and viability as well as uniformity of reporter gene expression in the 96-well format. The ATCC cell lines and protocols can be found on the Lonza web site (*http://www.lonzabio.com/cell-line-list.html*).

ATCC scientists have collaborated in research and development projects [23] with research groups around the world, including Johns Hopkins University, King's

College, Hadassah Research Institute, The Valley Hospital in New Jersey, and Innovative Biosensors, Inc. Other collaborations include business partnerships with organizations such as Geron, through which hTERT immortalized cell lines are provided. Partnerships with organizations around the world have been formed to assist ATCC in distributing its cell lines globally, and to provide additional means of obtaining new materials of use to the scientific community. These partners are located in Europe, Asia, Australia, Israel, Canada, and Mexico, and provide more efficient and less costly access to ATCC materials by scientists in the global community.

Arrangements have been made with several organizations for ATCC to manage special collections of materials for use by the scientific community, such as the Johns Hopkins Special Collection, developed in collaboration with Johns Hopkins University, which offers a number of cell lines for research use, as well as clones, vectors, and vector kits. This collection was developed to provide a mechanism for wider use of novel biological materials developed by Johns Hopkins University scientists while assuring that the rights of the university are protected. The ATCC Mantle Cell Lymphoma Bank was developed in collaboration with the Lymphoma Research Foundation (LRF) to ensure the availability of well-characterized cell lines for use in lymphoma research.

9.13 Conclusion

Management of any collection of biological resources for use in scientific research presents many challenges. The primary goal is to ensure that the materials are traceable for continuity of research, are of high quality and meet performance standards, and are equitably accessible. These challenges require building a robust system that includes all of the operating parameters needed to support continued maintenance and availability of the resources under conditions that ensure meeting the public scientific trust. Careful preservation of the biological materials is necessary to ensure longevity and traceability, and an overall quality system is required at all levels of the operation. Assurance of compliance in the distribution and use of the materials guarantees continued availability and a focus on ownership and intellectual property issues that maintains equitable access and use.

ATCC has built a biological materials management program that meets all of the requirements for good biological resource handling using a proprietary business system built on ATCC's long history. Financial concerns continue to plague the traditional culture collections, and commercial sources of materials focus only on those that make good business sense. The key to ATCC's successful business model is to ensure that the scientific community is provided with all resources needed, not just those that support profitability, while at the same time providing for the continued security of the entrusted resources for use long into the future.

References

1. Cypess, R.H. (ed.) (2003) Biological Resource Centers: Their Impact on the Scientific Community and the Global Economy, American Type Culture Collection, Manassas.
2. Hay, R.J. (1988) The seed stock concept and quality control for cell lines. *Anal. Biochem.*, **17** (2), 225–237.
3. Hughes, P., Marshall, D., Reid, Y., Parkes, H., and Gelber, C. (2007) The costs of using unauthenticated, over-passaged cell lines: how much more data do we need? *Biotechniques*, **43**, 575–586.
4. Chatterjee, R. (2007) Cases of mistaken identity. *Science*, **315**, 928–931.
5. American Type Culture Collection Standards Development Organization Working Group ASN-0002 (2010) Cell line misidentification: the beginning of the end. *Nat. Rev. Cancer*, 10. doi:10.1038/nrc2852.
6. Liscovitch, M. and Ravid, D. (2007) A case study in misidentification of cancer cell lines: MCF-7/AdrR cells (re-designated NCI/ADR-RES) are derived from OVCAR-8 human ovarian carcinoma cells. *Cancer Lett.*, **245**, 350–352.
7. Esquenet, M., Swinnen, J.V., Heyns, W., and Verhoeven, G. (1997) LNCap prostatic adenocarcinoma cells derived from high and low passage numbers display divergent responses not only to androgens but also to retinoids. *J. Steroid Biochem. Mol. Biol.*, **62**, 391–399.
8. Briske-Anderson, M.J., Finley, J.W., and Newman, S.M. (1997) Influence of culture time and passage number on morphological and physiological development of Caco-2 cells. *Proc. Soc. Exp. Biol. Med.*, **214** (3), 248–257.
9. Behrens, I. and Kissel, T. (2003) Do cell culture conditions influence the carrier-mediated transport of peptides in Caco-2 cell monolayers? *Eur. J. Pharm. Sci.*, **19** (5), 433–442.
10. Cour, I., Maxwell, G., and Hay, R.J. (1979) Tests for bacterial and fungal contaminants in cell cultures as applied at the ATCC. *TCA Man.*, **5**, 1157–1160.
11. Chen, T.R. (1977) *In situ* demonstration of mycoplasma contamination in cell cultures by fluorescent Hoechst 33258 stain. *Exp. Cell Res.*, **104**, 255–262.
12. DelGiudice, R.A. and Hopps, H.E. (1978) Microbiological methods and fluorescent microscopy for the direct demonstration of mycoplasma infection of cell cultures in *Mycoplasma Infection of Cell Cultures* (eds G.J. McGarrity, D.G. Murphy, and W.W. Nichols), Plenum Press, New York, pp. 157–169.
13. Macy, M.L. (1980) Tests for mycoplasmal contamination of cultured cells as applied at ATCC. *TCA Man.*, **5**, 1151–1155.
14. Hay, R.J., Kern, J., and Caputo, J. (1979) Testing for the presence of viruses in cultured cell lines. *TCA Man.*, **5**, 1127–1130.
15. Masters, J.R., Thompson, J.A., Daily-Burns, B., Reid, Y.A., Dirks, W.G., Packer, P., Troji, L.H., Ohno, T., Tanabe, H., Arlett, C.F., Kelland, L.R., Harrison, M., Virmani, A., Ward, T., Ayres, K.L., and Debenham, P.G. (2001) Short tandem repeat profiling provides an international reference standard for human cell lines. *Proc. Natl. Acad. Sci.*, **98** (14), 8012–8017.
16. Macy, M.L. (1978) Identification of cell lines species by isoenzyme analsysis. *TCA Man.*, **4**, 833–836.
17. O'Brien, S.J., Shannon, J.E., and Gail, M.H. (1980) Molecular approach to the identification and individualization of human and animal cells in culture: isozyme and allozyme genetic signatures. *In Vitro*, **16**, 119–135.
18. Lavappa, K.S. (1978) Trypsin-Giemsa banding procedure for chromosome preparations from cultured mammalian cells. *TCA Man.*, **4**, 761–764.
19. Simione, F.P. (1992) Key issues relating to the genetic stability and preservation of cells and cell banks. *J. Parenter. Sci. Technol.*, **46**, 226–232.
20. Simione, F.P. (1999) Cryopreservation: storage and documentation systems

in *Biotechnology: Quality Assurance and Validation*, Drug Manufacturing Technology Series, Vol. 4 (eds K.E. Avis, C.M. Wagner, and V.L. Wu), Interpharm Press, Buffalo Grove, IL, pp. 7–31.

21. Hay, R.J., Cleland, M.M., Durkin, S., and Reid, Y.S. (2000) Cell line preservation and authentication, in *Animal Cell Culture* (ed. J.R.W. Masters), John Wiley & Sons, Inc., and Oxford University Press, New York, 69–103.

22. Hay, R.J. (1978) Preservation of cell culture stocks in liquid nitrogen. *TCA Man.*, **4**, 787–790.

23. Selvan, S.R., Cornforth, A.N., Rao, N.P., Reid, Y.A., Schiltz, P.M., Liao, R.P., Price, D.T., Heinemann, F.S., and Dillman, R.O. (2005) Establishment and characterization of a human prostate carcinoma cell line, HH870. *Prostate*, **63** (1), 91–103.

10
Development of Automation in Sample Management

Gregory J. Wendel

10.1
Introduction

Sample management (SM) today, whether applied to small-molecule compounds or biological specimens, is central to discovery and research activities at organizations ranging from the small academic laboratory or start-up biotech all the way to the large, global pharmaceutical company. Given the high degree of sophistication which is typically applied, it is easy to forget the humble beginnings of SM just a few decades ago. In this chapter we will review some of that history to provide perspective for how far the field has evolved [1]. We will then survey the automation solutions that are currently available for both storage and sample handling needs to the SM laboratories as well as appropriate accessories. Lastly, a case study is presented to illustrate the topics discussed in this chapter.

10.2
Historical Background

As recently as the late 1980s, screening of compounds was localized at the level of the individual chemist or program. With the growth of mainstream assays such as ELISA it became practical to think of primary screening as a campaign to test a multitude of compounds against an assay. The numbers of compounds assayed in a screen grew from the tens of thousands to hundreds of thousands and beyond [2], in an attempt to find chemical templates for new drug targets. Researchers increasingly called upon laboratory automation companies such as Zymark, The Automation Partnership (TAP), and RTS Life Science to develop larger and larger scale automated systems to execute these screens. But a high-throughput screening (HTS) cannot operate without a supply of compounds to test against. As screeners reached out to groups within their organizations for compounds to put through their screens they found that the available compounds were often problematical in numerous ways. Compounds were stored as dry powders or films requiring dissolution and formatting to be accessible to the automated systems. There

Management of Chemical and Biological Samples for Screening Applications, First Edition.
Edited by Mark Wigglesworth and Terry Wood.

was limited or no control over the type of container that was used for storage. Details of the quantity and history of the compounds with respect to storage conditions, previous freeze/thaw or dissolution events, and purity were incomplete or nonexistent. Each chemist would store his compounds in a manner that made sense to him, but may not be consistent with others within the group, making it hard to share compounds across projects. Considerable effort was therefore expended preparing compounds for each HTS campaign, and this was repeated for every experiment.

By the early 1990s it became apparent within larger organizations that there were substantial benefits in collecting the compounds from all of the chemists into a centralized store. Often this process started with *ad hoc* SM groups forming within the screening group, borrowing time on the HTS automation equipment to perform compound formatting and distribution tasks. It was not until dedicated SM groups were ultimately formed that the execution of these tasks became seen as vitally important to the success of the screens. The formation of these groups went a long way to ensuring that all of the compounds from across the organization would be available to all projects within the organization and in standardized format; the concentration and where possible the purity of the compounds would be ascertained. All transactions relating to the compounds could now be recorded in a central database, creating an accurate, up-to-date inventory, with all of the requisite records for a complete compound history.

Where centralized stores certainly had organizational advantages, this way of working was not established without challenges. Support of HTS involves making available sets of large numbers of compounds in predefined plate formats for screening, utilizing formalized automation platforms. This would be followed by responding to cherry-pick requests and dilution curve generation for hit validation/confirmation screens. These two support functions require two fundamentally different capabilities. The provision of large sets of compounds for primary screening entails plate replication on a 96- or, more commonly, a 384-channel liquid handler capable of aspirating from all wells of a source plate in one step and then dispensing to one or more daughter plates. Storing the compounds in source plates with sufficient volume to make multiple daughter sets facilitates this plate creation. When it comes to cherry-picking, however, this format of compound storage is problematical. Multiple plates will need to be accessed to find the compounds of interest, exposing the rest of the compounds on those plates to additional plate manipulations, freeze/thaw cycles, and uneven well depletion across the plate.

The introduction of automated tube stores addressed these complications by making compounds individually accessible. Storing each compound in its own individual bar-coded tube offers the advantage that compounds are removed from the protective environment of the store only when they are requested. Further, the tubes can be formatted into racks that can be accessed by a 96- or even 384-channel head for improved operational efficiency. However, this format placed additional demands upon the automation, requiring it to be able to physically manipulate the tubes, read the barcodes, and remove and replace caps. It also necessitated the movement of legacy compound collections from their plate-based storage into tubes, a process which can be costly and time-consuming. Further, making copies of a large screening deck is difficult and time-consuming from a solely tube-based store, which is why many SM organizations maintain both plate-based master plates for screening deck replication and tube-based storage for cherry-picking and medicinal chemistry support.

SM has continued to evolve to support the advances of HTS into lead optimization. The requirements put upon the SM group for lead optimization support entail significant differences from the support of HTS. The medicinal chemistry groups supporting lead optimization produce new compounds on a weekly or even daily basis. These compounds need to be added to the collection and then made available for additional studies with very quick turn-round times. These follow-on studies involve a wide variety of assays which often do not utilize standardized plate formats, formats which may differ significantly from those used in HTS. The processes within SM need to be flexible to respond to weekly compound submissions and requests from the MedChem group. High levels of flexibility and adaptability are required to provide the compounds in unique plate layouts, volumes, and concentrations. Different automation capabilities are required to execute against these demands. Mechanisms for the rapid registration, quantification, dissolution, and ultimately storage of submitted compounds tend to operate on single containers as opposed to the multiwell plate format of a screening deck. Instruments with individual cannulae can access vials and tubes as well as microplates, enabling tube-to-plate transfers as well as accessing any well on a plate.

10.3 Automation of Sample Management Today

The SM laboratory of today must be able to meet the demands of its many customers within the organization, and must do so in a timely manner. To be successful, the group must be able to provide a wide array of services. These services can be considered in a number of phases starting with receipt of the sample(s) and preparation for storage. Once the samples are in the proper format and entered into the database they move on to safe storage. Eventually, in response to a request from a user, or some other demand, the samples will be retrieved from storage, processed in some way, and distributed to one or more final destinations.

It is this last group of activities which can present some of the biggest challenges for the sample manager, and which will have the most significant impact on automation hardware choices. At its simplest, the sample processing may be just making replicates of compound plates for use in screening, but this is rarely the case. Most often there are requests to cherry-pick specific compounds from those in the store, potentially to be delivered in a dose response plate. In this case, if the compounds are stored in tubes, the storage system can support retrieval of specific tubes containing the compound, removing the burden from an external resource, or if the compounds are in plates then the liquid handler will need to access specific wells in a plate.

One of the most impactful decisions to be made in setting up automated systems for sample handling is the choice between designing and implementing an integrated automation system or installing a collection of smaller, focused workstations. Fully integrated systems contain all of the instruments and accessories that are necessary to perform the functions to fulfill all of the requests put upon SM. The integrated system will include sample storage and retrieval, liquid handling, bulk reagent addition, plate and tube manipulation, plate sealing, centrifugation, labeling, and any other operations that will be required, plus a robotic manipulator to move plates, tubes, and consumables between the instruments. It will include a software system that enables the user to create a sequence of operations and orchestrate the movements of the robot, and all of the instruments to perform those functions. It will also interact with the corporate database to track all of the information. A well-thought-out integrated system can be a boon to the SM team by removing the burden of repeated manual operations and enabling the team to work on other tasks that truly require a human touch. Once installed and validated, the system can be set up to run unattended through the night and even weekends, producing large quantities of plates.

The workstation model of laboratory automation is the polar opposite of a fully integrated model. In this system, each instrument is independent from the next, and the plates, tubes, and consumables are manually moved between them. The orchestration of events can be coordinated through a local information management system (LIMS) and events tracked, but it is still in principle a manual system. It is very difficult to achieve significant throughput in a workstation model, errors will inevitably be made, and the staff will become fatigued. However, the workstation model excels in enabling great flexibility in the processes and is typically significantly less expensive to implement than a fully automated system [3]. When it becomes necessary to offer a wide range of services and formats, such as in support of a medicinal chemistry project, then workstations can be the most efficient model for achieving that goal. New layouts and distribution schemes can be created in a matter of hours on a workstation liquid handler and the attendant peripheral operations, as opposed to the days or even weeks that it would take to create and validate on an integrated system.

Both the integrated and workstation models have their place in the modern SM laboratory. In an optimum situation, both would be present in the same laboratory, creating an ideal high-capacity, flexible laboratory. Budget and space constraints may make this impractical in many situations. The collection size, expected growth, screening, and MedChem requirements must all be taken into account when making the design choice. For smaller organizations such as academic laboratories, and small biopharmaceutical companies the workstation model is the best place to start. It offers the best flexibility to meet the changing demands of the organization, while being considerably less expensive to implement.

Fortunately for the sample manager there is an abundant array of available choices when it comes to automating his processes. These choices cover not only the needs of the large multi-national pharmaceutical company, which may have several million compounds in its collection, but are also scaled to address the needs of the small organization which may have fewer than 100 000 samples to manage. The laboratory automation industry has evolved from the early days when devices originally designed to serve another purpose were pressed into service in SM. The choices available today are often designed with the unique needs of SM in mind and therefore provide robust solutions.

10.4 System Building Blocks

The first step in the selection of automation solutions is to perform an exercise of gathering all of the requirements for the system and creating a functional requirements specification. This is an involved process, which is critical to the overall success of an automation project and is beyond the scope of this chapter. What is important here is to understand that this process will generate a set of detailed specifications that can be used as guidelines in the selection of automation instrumentation. The required instrumentation will generally fall into one of five categories:

10.4.1 Storage Systems

The starting point for all of SM is to decide where and how the samples are to be stored. This is affected by the format of the storage vessel, whether it is tube or plate based, the temperature that must be maintained, how the samples are to be accessed internally, and how they enter and leave the storage device.

10.4.2 Liquid Handling

The majority of the functions within SM require the use of a liquid handler at some level. The processing of samples which arrive in a liquid format is typically easily accomplished. Samples which are not already in a compatible container, for example, samples provided in a deep-well microplate, which need to be in tubes, can be processed on an automated liquid handler. Virtually all of the liquid handlers reviewed below have options for 96-channel with disposable tips for transfer from microplate to tube-rack, or the reverse.

Many of the available liquid handlers also offer individual channels which can be used for accessing individual tubes or vials and individual wells of microplates if necessary. One of the major enhancements in the implementation of these channels, that significantly expanded the versatility and abilities of the SM groups, was the addition of variable pitch to these channels and individual control over the depth in the tube/well and volume aspirated/dispensed. These enhancements separate modern liquid handlers from their more constrained predecessors which had been designed with the needs of the screeners in mind. These capabilities opened up the automated processing of cherry-picks and compound submissions in tubes and vials.

Once a sample is retrieved from storage it will either be passed directly off to a process external to SM such as screening, or some transformation to the sample is performed within SM. Those transformations that involve taking an aliquot of the sample require the use of a liquid handler to perform the transfer of the aliquot from the storage container to the secondary container as well as other steps such as dilutions, serial dilutions, creation of screening plates, creation of mixtures, or other functions. A typical screening plate contains many microliters of compound. When used in an assay, an aliquot of the compound is transferred out of the screening plate and into the assay plate in a liquid handling step. Since the volume transferred is small, screening plates can typically be used for many iterations across several assays.

An increasingly common activity in SM laboratories is the creation of Assay-Ready Plates for screening. Assay-Ready Plates differ from screening plates in that they are plates into which a small aliquot of compound, sufficient for a single assay, is transferred by SM. During the screening run the other components of the assay are added directly into this plate, making this the assay plate. By their very nature, Assay-Ready Plates are single-use plates, resulting in the need for multiple production runs in the place of a single run for the creation of screening plates. Assay-Ready Plates are beneficial to the screening laboratory since they eliminate the need for a liquid handling step, with its attendant disposable tip exchanges, washing, and so on, resulting in significant throughput gains. The creation of Assay-Ready Plates has been made possible by the introduction of acoustic dispensers, a topic that is discussed in greater detail in Chapter 11.

10.4.3 Accessories

Other functions that are generally required include the sealing and unsealing of microplates, lid handling, capping and decapping of tubes, sorting/reordering of tubes, reading of barcodes, application of labels both human readable and with barcodes, centrifugation, mixing, bulk reagent addition, and thawing/incubation.

10.4.4 Plate Handling, Integration

The flow of sample containers between each of the devices within the overall system must be performed either by a human operator or, more typically, by a robotic arm. This flow involves not only the physical movement of the container, but also the coordination of the actions of each device and the flow of data to and from the devices to the central informatics database. In larger systems this coordination is handled by a system integration package.

10.4.5 Data Management

One of the critical functions that SM performs is to maintain integrity of the samples. One element of this requires a data management system that can track all operations that are performed on the sample containers, hence creating a history of sample creation and the processes it has undergone. This data management system must interact with the system integration package, but also in some cases with individual instruments.

Below we will survey available automation systems and review each from the point of view of these needs. This survey is not intended to be an exhaustive listing of all vendors and instruments but a representative overview.

10.5 Storage Systems

Many solutions are currently available for the storage of samples, ranging from the manually accessed shelf in a freezer up to room-sized custom-built automated stores.

A key factor to consider when specifying a store is how the sample collection may grow in the future. Many of the storage solutions available in today's marketplace are modular. A modular product offers the advantage of a reduced initial capital investment while still allowing for future expansion. It is essential that flexibility and modularity are designed into both the hardware and software of such systems. For larger collections with high throughput requirements and integrated external processes a modular product may not provide the optimal solution. A higher initial investment is often required to implement such systems. Here, recognizing potential system development and expansion requirements are key success factors. Partnering with a capable and experienced supplier ensures that complex systems can be adapted to accommodate changing business needs and will continue to deliver value to the organization.

10.5.1 Features

Choosing among these options when contemplating the purchase of a storage solution requires examining many factors which will ultimately result in a well-informed choice.

10.5.1.1 Size

One of the first questions to be asked is how large a collection will be stored. Below a certain number of samples the cost and complexity of an automated store cannot be justified. That point is specific to each individual organization but is likely to be in the range of 25 000–50 000 tubes or 1000–2000 microplates, either of which will fill two to four shelves of a typical laboratory freezer. Beyond four shelves it becomes impractical to search for a set of specific containers and return them to their starting points on a routine basis. Automated stores are available with storage sizes starting at 30 000 tubes or 800 microplates. Larger, custom-built automated stores are available which will handle any number of sample containers even up to installations storing several millions of samples.

10.5.1.2 Format

Hand in hand with the question of store size are the questions of what format will be used for storage, along with the resolution of access. Will the system store microplates, deep-well plates, tubes, vials, or some combination of these? Especially relevant in the case of tubes and vials is the question of how will they be retrieved, will it be at the level of a rack to access a specific tube/vial or is picking required to retrieve just the specified tube?

10.5.1.3 Temperature

Do the samples need to be held at −80 °C, which is typical for biological specimens, or at −20 °C (for small-molecule compounds), or room temperature? Generally as the storage temperature decreases, the complexity and cost of the system increases, sometimes dramatically. Lower temperatures will require larger, more robust refrigeration components, as there is more heat to dissipate, and the design and maintenance of the internal mechanisms become more complicated. In addition to the initial purchase price there is the cost of increased maintenance and electricity usage as the operating temperature is lowered.

10.5.1.4 Environment

Do the samples need to be kept dry? By its very nature, low-temperature storage is dry, but what are the concerns if the samples are being held closer to or even at room temperature? Is an inert environment required? Many of the available systems can be purged with an inert gas, although this can be costly in terms of gas usage.

10.5.1.5 Internal Manipulation

A key function which can be performed is cherry-picking of tubes, assuming that the samples are stored in tubes. This presents the opportunity for the storage system to take on some of the burden typically borne by the liquid handler for the processing of cherry-pick requests. Having the storage system cherry-pick tubes from storage and present them to the operator or integrated system as a completed 96-tube rack has the advantage that all of the liquid handling steps can be performed with a 96-channel head. This is a significant throughput advantage over using the individual channels to cherry pick samples from complete plates.

10.5.1.6 Robotic Interface

The purpose of an automated store is not only to hold samples in the desired conditions, but also to make them available to the operator on demand, which requires some sort of input/output mechanism. The question to be asked is whether or not the storage system will be part of a larger, integrated system. If it is, then the I/O mechanism must present the container, in its output function, in a robot-friendly manner. Even if it is not part of an integrated system it still needs a mechanism to retrieve the specific container and return it to the user.

10.5.2
Example Hardware

The **Active Sample Manager (ASM) System** from Hamilton Storage Technologies (*www.hamilton-storage.com*) is an automated SM system for HTS of vials, tubes, and microplates at a temperature ranging from ambient to −20 °C. Utilizing a modular design, the system can be expanded with one to five store modules for increased capacity and tube processing speed, as each module has an independent tube picker. Applications include HTS screening as well as inventory and order fulfillment for compounds or biologicals. The ASM System includes integration-friendly hardware and software. The fully automatic hand-off arm allows for the integration of Hamilton MICROLAB® STAR robotic liquid handling workstations and other third-party devices. Its small footprint does not require infrastructure or facilities changes, and allows it to be rolled into a laboratory through a standard door for rapid implementation. The application protocol interface (API) has an open architecture for easy integration with LIMS or other inventory management software. Each module of the ASM can store up to 2160 microplates or 207 360 0.5 mL tubes or other consumables including 1.4 mL tubes and 1 dr vials. Tube processing rates are 300–1000 h^{-1} and it can deliver 120 plates or tube racks per hour.

The **ASMST System** version of the ASM is a high-throughput microplate handling system that uses a unique removable lid (SealTite®) to automatically seal and unseal microplates. This system removes and seals the same plate with the same lid for the life of the plate. The temperature range on the ASMST is from ambient to −20 °C. The ASMST utilizes the same modular design as the ASM and has the same features for liquid handling and software integration. The system can be expanded with one to five store modules for increased capacity. Each ASMST Store holds up to 2160 microplates.

The **Sample Access Manager (SAM) System** is another automated SM system from Hamilton Storage Technologies for storage of tubes and plates. It is available in two platforms, the −20 °C platform has a temperature range from +20 to −40 °C, and the −80 °C platform has a temperate range from −55 to −80 °C. The SAM has an integrated −20 °C tube picking compartment which can read both 1-D and 2-D bar-coded tubes. The capacity of the SAM is smaller than that of the ASM; it can store up to 803 microplates or 34 080 0.5 mL tubes or a mixture of consumables including 1.4 mL tubes and 1 dr vials. Tube and plate processing rates are lower than those of the ASM. However, its compact size makes it ideal for individual laboratories.

The RTS Life Science '**ARange**' (*www.rts-group.com*) of automated storage and retrieval systems currently includes three products named A3 SmaRTStore™,

A4 Sample-Store™, and A5 Compact-Store™. The three products each provide a range of environmental storage conditions at temperatures down to −20 °C with low humidity and inert options. All RTS products utilize industrial quality robotics to ensure the highest levels of reliability and availability.

A3 SmaRTStore™ is a flexible, compact, and space-efficient automated sample storage system adopted by increasing numbers of organizations, ranging from academia through biobanking to large pharmas. With the highest density storage in its class, A3 SmaRTStore™ delivers reliable, flexible, and high-integrity automated SM for up to 500 000 tubes.

A4 Sample-Store™ is a mid-size automated storage and retrieval system typically for 350 000–1.5 million tubes, although expansion beyond this is easily achieved. Designed to offer the best combination of storage density, performance, sample integrity, and system longevity, the system enables reliable, unattended operation, enabling overnight runs of thousands of tubes to be picked without error. A4 is scalable and can be readily extended/expanded post installation without requiring samples to be removed from the store. A range of I/O solutions ranging from tray-based, through carousel stack-based, to fully integrated options are available.

A5 Compact-Store™ is a large-scale solution providing space-efficient storage, typically for collections larger than 1 million tubes. Single or multiple picking stations offer potential retrieval throughputs in excess of 100 000 tubes per day. A range of output buffer options are also available and the system can be readily integrated with automated liquid handling stations.

All of the ARange products are capable of storing and retrieving a range of labware types including tubes, plates, and vials. The A3 and A4 stores offer integral cherry-picking, ensuring that samples are kept at the store temperature while picking takes place. RTS storage systems are equipped with the RTS d-Sprint™ software for robotic control and SIS sample tracking and ordering database. These feature-rich applications provide a depth of functionality which normally ensures that client's operational requirements are fully catered for 'out-of-the-box' without the need for special development. RTS stores are easily integrated into LIMS or corporate inventory systems.

Matrical Bioscience's **MiniStore** storage systems (*www.matrical.com*) are optimized for quick retrieval of individual and/or bulk samples for processing, automating the process of storing and tracking thousands or even millions of samples, all in one secure space-saving location. The MiniStore offers flexible sample storage temperatures from ambient to −20 °C, an ideal fit for satellite laboratories and dedicated libraries. The expandable, modular design of the MiniStore enables systems to fit within existing or new laboratory space constraints and sample capacity requirements. MiniStore systems can be enlarged after installation by expanding the cold room and adding additional storage modules. This design ensures that future sample demands/growth can be supported. Depending upon the configuration, the MiniStore will hold between 86 000 and 5 700 000 individual 0.5 mL tubes and between 1500 and 98 600 standard microplates.

MiniStore storage systems accommodate a variety of sample types and formats, including 96 tubes, 384 tubes, dram vials, scintillation vials, and microplates. Systems are equipped to fit current media standards with the ability to store and manage multiple media types within the same system. End-of-arm tool changers cherry pick, store, and manage multiple media types. Sample types can be added to the system at any time with the addition of tool grippers. The database knows what sample type is going to be stored or retrieved, and the robot picks the tool needed to perform the task. Matrical's Vitesse™ operating software keeps a log file of operations performed in the system. Order logs are maintained to show what samples have been put into the system, where they were stored, and any additional information that the user may have about them.

The **comPOUND** system from TTP LabTech (*www.ttplabtech.com*) comprises a high-density sample storage unit and an additional suite of specialized delivery and processing modules to enable integration into any compound management (CM) or screening system. The comPOUND offers the ability to store up to 100 000 1.4 mL samples in each self-contained module measuring just 1.2 m × 1.65 m × 2.4 m. There is also the option of conversion to double-density – holding up to 200 000 × 0.5 mL, or a mixture of both types. Sample integrity is maintained using a hermetically sealed chamber providing an inert environment set to a user-defined temperature ranging from −20 °C to ambient. Every sample is tracked in and out of the store. By using pneumatics for high-speed cherry-picking, comPOUND minimizes the number of moving parts in the store's controlled environment, increasing reliability. This also means that samples can be delivered directly to the laboratory, while the store is housed elsewhere. The comPOUND module can identify and deliver a sample in 5 s. When multiple modules are linked to form a larger sample library, samples can be accessed from all modules simultaneously to maximize throughput. This means that processing speeds actually increase as the library grows.

10.6 Liquid Handler

The liquid handler in an SM laboratory shares center stage with the storage system. The liquid handler is where all of the transformations and distributions of the samples are performed. It is often the workhorse of the system, kept busy most of the day performing a variety of tasks. Given this central role, care must be taken in selecting the right instrument for the laboratory. One of the most important considerations is to scale the liquid handler to the needs of the laboratory.

10.6.1 Features

The applicable features which must be evaluated when selecting a liquid handling platform include:

10.6.1.1 Deck Size

The deck of a liquid handler is where all of the consumables, samples, and destination plates are positioned for access by the system components. The size of the deck has a direct impact on the complexity of operations that are to be performed, and how much walk-away time the operator is given. For example, with a 9-position deck, there is not enough space to make a 384-well compound plate with dilution from 4 × 96-well source plates and four tip racks unless the tips and/or source plates are stacked, but multiple copies of a source plate can be stamped with just one source and one set of tips. With a larger deck, it is possible to load multiple source plate, destination plate, and tip rack sets that can be processed sequentially, enabling the operator to attend to other tasks, or more complicated workflows can be accommodated. However, while uses can always be found for extra deck space, an excessively large deck will needlessly impinge upon precious budgets and laboratory space without adding any extra value in the laboratory.

10.6.1.2 Head Format

The typical consumable used in unit operations in drug discovery screening is the SBS microplate. The density of wells on the microplate varies from 96 to 384 to 1536, but the spacing is always some multiple or fraction of 9 mm. This means that for straight forward plate replication, even if there is a change in well density, either a 96-channel head or a-384 channel can be used. Certainly, if every transfer is going to be 384-well plate to 384-well plate, then a 384-channel head should be selected. But, a 384-well head can access a 96-well plate using tips from a tip rack that is formatted for 96 tips, and a 96-channel head can replicate a 384-well plate in four steps instead of one. When deciding upon the head format option the predominant plate format and transfer operation should be given extra weight.

10.6.1.3 Head Volume Range

As the number of channels in the head is increased, the volume range decreases. Heads with 96 channels can typically pipette up to 200 μL but have decreasing accuracy below 5 μL, although low volume performance of a 96-channel head can be improved by using smaller tips. On the other hand, 384-channel heads are limited to the 25–50 μL range but can deliver acceptable performance at fractions of a microliter.

10.6.1.4 Individual Channels

To be able to access vials or large tubes or to perform reformatting of plate layouts involving remapping of well positions, it is necessary to use channels which can be activated independently of each other. These channels are each mounted on an actuator that can be separately activated, has its own syringe or valving to achieve volume displacemen, and in many instruments have variable-span capability to simultaneously access containers that are not 9 mm apart. In a plate remapping method, such as a cherry-pick, these will be used to aspirate from wells scattered across one or more plates and then dispense into a single column in the destination plate.

10.6.1.5 Gripper

A gripper integrated into the liquid handler can be very beneficial in many methods. It can be used to temporarily remove lids from plates to give access to the head or cannula; it can move consumables around the deck and in some cases even reach outside of the instrument to access external devices.

10.6.1.6 Tip Loading

Unless the head and/or independent channels are equipped with fixed cannulae, disposable tips will need to be available to the head during the operation of a method. Some instruments can attach tips from their storage location whereas others require a dedicated tip-loading station, taking up a deck location. Additionally, some instruments have options for integrated tip servers which will make more tips available upon demand; these are typically not a requirement in stand-alone applications.

10.6.1.7 Barcode Reader

Every consumable on the liquid handler deck, with the exception of tips, will most likely have a barcode which is used for data tracking. Having a barcode reader integrated into the liquid handler allows it to be smarter, enabling it to verify that that is the proper source; destination plates are then loaded. A more customized liquid handler method might also be able to use the barcode value to look up values such as pipetting volumes, delay times, or step iterations in a database or worklist file, allowing for specialized processing of each plate.

10.6.1.8 Tube/Vial Gripping

Some instruments have the ability to grip tubes, allowing them to reformat tube racks, weigh tubes/vials with an optional balance, pick up, and possibly even decap vials or otherwise manipulate these consumables.

10.6.1.9 Integration Options

If the liquid handler is to be part of a larger automation system, then it must provide a suitable interface. The deck must be open enough so that a robotic arm or other plate transport device can pick and place consumables and plates, and the software must have interfacing options that will allow the integration software to control its operations to some degree.

10.6.1.10 On-Deck Accessories

In addition to the standard components offered by the suppliers of liquid handling equipment, many accessories can be added to the deck of most systems for specialized operations. Microplate shakers such as the Thermo Scientific Variomag will mix the well contents after liquid additions. Temperature-controlled locators from vendors such as MeCour, Inheco, and Torrey Pines Scientific will maintain the temperature of sensitive compounds and reagents. Vacuum filtration stations, refillable reservoirs, and other specialized devices can be obtained through third-party suppliers such as Hypertask, Acme Automation, V&P Scientific, the instrument vendor, or the system integrator.

10.6.2 Example Hardware

Hamilton's **STAR Line** (*www.hamiltonrobotics.com*) of automated pipetting workstations offers performance for sample preparation needs, being designed to serve as a completely stand-alone pipetting system or be integrated to multiple on- and off-deck Hamilton or third party devices. The proprietary Compressed O-Ring Expansion (CO-RE) technology facilitates highly robust tip attachment, enabling a positional precision of 0.1 mm in all axes, critical for 384-well plates. The STAR utilizes proven air-displacement pipetting technology, offering pipetting accuracy and precision across a wide dynamic range (i.e., 2% CV at 10 μL volumes). This technology allows for both capacitance-based as well as pressure-based liquid level detection, the latter being critical for nonconductive reagents and volatiles. In addition, air-displacement pipetting affords the use of Hamilton's proprietary Total Aspiration and Dispense Monitoring (TADM) software, which monitors and records all pipetting steps, providing process control. Trace files are retained so that the end user can go back and review pipetting steps if some downstream discrepancy is found. Actual real-time feedback can be provided while the pipetting is occurring, to allow for on-the-fly error handling/error recovery.

The STAR Line offers 96 multi-channel pipetting heads with 1–1000 μL volume dynamic range and 384-head options with 1–50 μL volume ranges. The 96-channel heads have the ability to pick up single and multiple rows and columns, quadrants, or a single tip. Also available are 2–16 independent 1 mL liquid channels with a 1–1000 μL dynamic range or 2–8 independent 5 mL liquid channels with a 500–5000 μL dynamic range.

Hamilton's **VENUS** instrument control software is used for the creation of automated methods or scripts, supporting robotic and pipetting steps across a broad spectrum of applications. The software allows users to easily program the routines using step-by-step instructions. For tube/vial and plate handling, Hamilton offers several carrier options to position these onto the pipetting deck and gripper options to manipulate these objects. The STAR Line workstations can be easily integrated with in-house LIMS systems, and the programming can be customized when necessary.

The STAR Line's three available deck sizes provide options to meet the needs of any SM laboratory. The autoload option facilitates automated loading of all samples, reagents, buffers, and labware (plates and tubes) carriers onto the deck with simultaneous scanning of 1-D barcode labels for complete sample and reagent tracking.

The **NIMBUS** is Hamilton's compact automated pipetting workstation, offering speed, flexibility, precise tip positioning, superior pipetting performance, and affordability, all in a space-efficient footprint.

Available as a four-independent (1 or 5 mL channels) system for flexibility or a 96-multi-channel (1 mL) head system for speed, both NIMBUS platforms incorporate Hamilton's novel CO-RE technology, which facilitates tip and tool pick-up via a robust lock-and-key style attachment. Three different deck configurations are available: 12-position (3 × 4) SBS microplate positions, 11-position (3 × 3) + 2 positions on a height-adjustable subdeck (for integrating tall labware and devices), and 8-position (2 × 4) + 3 × 32-tube 'Shift-n-Scan'* modules. The Shift-n-Scan is a tube rack/1-D barcode scanning system which holds up to 96 tubes in a small space for pipetting, with simultaneous barcode scanning.

Tecan's current **Freedom EVO®** series of liquid handlers (*www.tecan.com*) incorporate the ability, introduced to the market by Tecan in 1994, to spread eight pipetting channels any equidistance from 9 to 42 mm, allowing simultaneous access to eight test tubes or vials followed by access into eight wells in the column of a 96- or 384-well microplate. This mechanical advantage is coupled with a dedicated independent syringe for each tip and the ability to detect the liquid level of aqueous or DMSO solutions in the tubes/wells. This design allows tubes to have a varying level of liquid in each vessel, thus eliminating the danger of a tip plunging to the bottom of the tube and the liquid overflowing the vessel. Also, individual tip tracking of the liquid level during aspiration and staying a definable distance from the meniscus minimizes the amount of sample that comes in contact with the outside of the probe. These channels use either disposable (single-use) tips or stainless steel washable tips with various chemically resistant and biologically appropriate coatings and lengths.

The Freedom EVO®, with deck sizes available from 75 to 200 cm, can incorporate up to three arms on a system. Available different liquid handling arms from 2- to 16-channel liquid handling arms can pipette discrete individual samples for ultimate flexibility, with liquid volumes ranging from 100 nL to 5 mL. This range can be extended with DynamicFill™ Technology to 50 mL and higher. With optional lasers that can read the absolute position of the pipetting tips, the Freedom EVO® can aspirate and dispense into and out of not only 384 well plates, but even 1536 well plates. With the safety option Pressure-Monitored Pipetting (PMP™), pipetting errors are detected by comparing recorded and real-time-simulated pipetting pressure signals. The instrument can also incorporate either the MultiChannel Arm™ 96 or MultiChannel Arm™ 384 for high-density pipetting.

The customer can select multiple options for moving tubes, plates, and even racks around on the work surface and, to a certain degree, off-deck within the reach of the robotic manipulator arm (RoMa) arm. The entire system is controlled by EVOware® a user interface and scheduling software package which is available in many configurations designed to support a variety of workflows. The power of EVOware coupled with the availability of software drivers for third-party devices such as barcode readers (1D and 2D), vortexers, mixers, incubators, readers, and many other devices plus the off-deck reach of the RoMa all make it possible to configure the Freedom EVO as a small system capable of much more than just liquid handling.

The Agilent **Bravo Automated Liquid Handling** Platform (*www.chem.agilent.com*) is a fast and versatile small-footprint liquid-handling system. Two Bravo platform models are available: standard and SRT. The standard model fits in most standard laminar flow hoods. The SRT model, at 3 in. (7.62 cm) shorter, provides extra clearance for hoods with short internal height. The Bravo pipetting deck is compact and consists of nine microplate positions that can be configured to specific assay needs. The Bravo Platform uses high-accuracy pipette heads for dispensing from 100 nL to 200 μL in 96-, 384-, and 1536-well microplates with either disposable or fixed tips. Pipette heads can be changed in minutes yielding enhanced flexibility, and numerous platepad options are available to enable a wide range of assays. The unique open design permits access from all sides for simple system integration as well as standalone use, and removable positions in all models permit through-deck accessories or labware waste output. Powered by VWorks Automation Control software with an event driven scheduler, easy-to-use interface, and error recovery technology, complete control is easy for all grades of operator.

10.7 Accessories

Many of the workflows which are typical in an SM laboratory are facilitated by the use of specialized instruments. Each of these is quite small; they tend to be dedicated to a single function, and many of these functions can be performed manually. Despite their modest nature, they have a dramatic impact on the efficiency in the laboratory, and are absolutely vital to the practicality of performing workflows in an integrated system.

10.7.1 Common Devices

The most common devices include:

10.7.1.1 Plate Seal/Unseal

During storage microplates and deep-well plates are frequently sealed to preserve the integrity of the contents. These seals are either applied with an adhesive or heat-bonded to the top of the microplate. Adhesive seals run the risk of exposing the well contents to the adhesive. In using heat sealing the top of the plate is temporarily exposed to heat, which may lead to plate warping after repeated sealing.

The **Agilent PlateLoc** Thermal Microplate Sealer (*www.chem.agilent.com*) is a thermal sealer that exhibits speed, a small footprint, ease of use, and dependability. It automatically accommodates many deep-well, assay, PCR, and compound storage microplates. The PlateLoc Sealer can be part of a robotic integration, featuring an extended-travel plate stage, RS-232 serial port, and ActiveX control. In order to minimize system downtime when replenishing consumables, the instrument features an easy-to-access, top-loading seal roll support.

Thermo Fisher Scientific (*www.thermoscientific.com*) offers a series of heat sealers including the Thermo Scientific **ALPS 25** manual heat sealer, which is a hand-operated sealer designed for low-throughput environments. This compact and portable instrument can seal a wide range of plates of differing heights when supported by plate carriers. The heating element of the sealer is pre-set for optimal sealing conditions to produce permanent or peelable seals when used in conjunction with Thermo Scientific sealing foil or film sheets.

Also available is the Thermo Scientific **ALPS 3000** compact heat sealer created for robotic integration in high-throughput environments; it also allows manual, benchtop control. This instrument features RS-232 connectivity options, sealing rates of up to 10 plates per min, and top-loaded seals for easy access and changing. A variety of available heat seals meet any laboratory application, including PCR, colorimetric, fluorescence, long-term storage, low-temperature storage, piercing, and resealing.

To access the contents of the wells the seal must be broken or removed. Pre-slit seals are available with are easy for tips to pierce, but these must be re-sealed after use. Mechanical piercing units which punch out the seals are available for use on foil seals. After access, the plate can be resealed for further storage, and then pierced again. This sealing/piercing cycle can be repeated numerous times, but eventually the seals become too thick for further access. An alternative approach is to remove seals by hand, a very tedious process when many plates are involved, or they can be mechanically peeled by a device such as the **XPeel** from Nexus BioSystems (*www.nexusbio.com*). The XPeel automatically removes seals from a broad range of plate types, all at a single touch of a button. At the core of this patented process is the seal removal mechanism. Instead of utilizing a mechanical approach, which is prone to malfunction leading to unreliable results, the seals are removed by XTape, a proprietary adhesive medium which results in a more consistent peel. This returns the plate to its original condition, ready for another seal.

10.7.1.2 **Plate Label**

New plates pre-labeled with a user's barcode sequence can be obtained from plate manufactures, removing that burden from the SM team. However, there may still be instances where a label with a barcode and/or other information needs to be applied to a plate. This is true in both walk-up instrumentation, and as part of an integrated system. Several options are available for barcode labeling instruments:

The Agilent **Microplate Labeler** (*www.chem.agilent.com*) is a benchtop device which works equally well in stand-alone or integrated systems. Using a 600-dpi thermal transfer printer, it prints up to six fields (barcodes and human readable fields) per label. The instrument's software can access a variety of data sources, including comma- or tab-delimited files and spreadsheets, and can be integrated with an ODBC-compliant LIMS.

10.7.1.3 **Tube Sorting**

A laboratory that does not have a large, integrated sample store with integrated tube picking may still have a need to process tubes. For example, if tube storage is in racks, kept in a freezer, a cherry-pick request will require the manual picking of specific tubes out of multiple racks, a task which is prone to error. Several devices exist which can be used perform this task. Using a list as a request input, they can pick the tubes out of racks, read the barcodes on the bottom of the tubes for identification, and, with the use of an optional balance, they can also weigh the tube.

One such system is the **XL20 Tube Handling Instrument** from *BioMicroLab* (*www.biomicrolab.com*). This benchtop robot is available in both 9 and 20 position models, for the processing of 96 tube rack samples in volume sizes from 0.5 to 1.4 mL tubes. It performs automated tube sorting, 2D scanning, and weighing using either a text based file input or can be integrated into the LIMS via the ActiveX Tool Kit.

The **Xtp-1152** Tube Picker/Tube Sorter from FluidX (*www.fluidx.co.uk*) is another example of a benchtop instrument which can pick and sort all common types of microtube including 0.5, 0.65, 0.75, and 1.4 mL tubes. While it is capable of tube picking, 2D reading, and tube weighing, it can also perform some liquid-handling steps using either a single fixed or disposable tip system for reagent addition and sample transfer functions between tubes. An optional decapping/capping module enables liquid-handling steps to be completed with a wide variety of screw-cap tubes and cryovials.

10.7.1.4 Centrifuge

There are many reasons for needing to centrifuge a microplate. Pipetting steps can introduce air bubbles into wells, small droplets can be present on the underside of seals, or precipitates may form which need to be driven to the bottom of the wells/tubes. For a workstation system many stand-alone laboratory centrifuges can be used with the addition of swing buckets made for microplates. Companies such as Drucker, Eppendorf, and Thermo Scientific Sorvall offer a wide range of benchtop and floor model centrifuges which cover the needs from small laboratories up through larger institutions.

For an integrated system, an automated centrifuge is required. This must allow for the robot to introduce and remove the microplate. An example of an automation-friendly centrifuge is the Agilent Microplate Centrifuge and optional Centrifuge Loader from Agilent Technologies (*www.chem.agilent.com*). This small centrifuge is ideal for high- or medium-throughput applications such as PCR purification, cell harvesting, and air bubble removal in high-density microplates, with customizable settings for acceleration and deceleration. The Microplate Centrifuge and optional Centrifuge Loader can be accessed by most laboratory microplate handlers/robots.

10.7.1.5 Mixing

When multiple liquids are added to a microplate well, or if a well containing solids such as cells, has been static for some time, there is a need to mix the contents of the well. On an automated liquid handler this can be accomplished by running a mix cycle with the liquid handling head consisting of multiple aspirate/dispense steps. Alternatively, a microplate shaker such as the Thermo Scientific Variomag can be added to the deck.

Acoustic technologies have also been used to enhance mixing. Microsonic Systems (*www.microsonics.com*) offers the **HENDRIX SM100**, which excites Micro-Electrical-Mechanical Systems (MEMS)-based transducers with RF power to produce broad beams of ultrasonic waves. These waves have a very high level of lateral ultrasonic thrust, causing a mixing vortex within the microplate well and delivering rapid and controllable mixing of the sample – especially in low-volume assays.

Products from Covaris (*www.covarisinc.com*) use Adaptive Focused Acoustics™ (AFA) to send acoustic energy wave packets from a dish-shaped transducer that focuses on a small, localized area. AFA is a form of mechanical energy which causes bubbles to form from the naturally occurring dissolved gases and vapors of biological specimens and chemical fluids. The collapse of these bubbles creates a controlled, gentle mixing environment. When driven at high energy intensity, tissue disruption and can be accomplished in the well. AFA is an isothermal, non-contact sample preparation method that is suitable for a variety of applications including creating nano-suspensions and dissolution/micronization.

10.7.1.6 **Bulk Reagent Addition**

When there is a need to fill an entire plate, or some portion of it, with a bulk reagent for sample dilution/distribution or assay assembly, the bulk reagent dispenser (BRD) is most often used as the dispensing tool. BRD systems are either based upon a pressurized liquid with some sort of solenoid valve, a syringe pump, or a peristaltic pump, and most often dispense on a column-by-column basis. Most are able to address the 96- and 384-well plates, with some being able to address 1536-well plates. Flexibility to dispense to selected columns and/or wells, and in some cases specific wells, is increasingly available. All of the BRDs reviewed below can be used either in a walk-up mode common to the small SM laboratory or can be integrated as part of a larger automated system.

The **BioRAPTR FRD** Microfluidic Workstation from Beckman Coulter (*www.beckmancoulter.com*) is a high-precision, high-speed, and non-contact BRD. For maximum flexibility and control, the BioRAPTR workstation can dispense 100 nL to 60 μL of reagents, including cellular solutions, into any well of a 96-, 384-, 1536-, or 3456-well plate. The BioRAPTR workstation can independently dispense up to eight reagents per dispense head in one pass. Each reagent has a short, dedicated fluid path that minimizes dead volume and reduces contamination. The user can easily switch between interchangeable dispense heads and input the dispense tables through Microsoft Excel. The Synquad systems from Digilab (*www.digilabglobal.com*) is another solenoid-based non-contact dispenser. It provides the user with the ability to accurately transfer target substrate in picoliter to nanoliter aliquots.

IDEX Health & Science (*www.idex-hs.com*) supplies the Innovadyne **Nanodrop Express** system, which is a 16-channel syringe/solenoid-based BRD for low-volume, high-precision pipetting, with a volume range of 50 nL to 500 μL per channel. The system aspirates and dispenses a broad range of liquids, including DMSO, and features the Nanobuilder software system. This comprehensive package enables a wide range of applications and data manipulation for processes such as the creation of dose response curves and complex plate maps.

BioTek (*www.biotek.com*) offers the **MultiFlo™** Microplate Dispenser with a choice of either peristaltic pump and/or microprocessor-controlled syringe drive channels. One MultiFlo can accommodate two of each for a total of four reagents that can be dispensed in parallel, decreasing cycle time, deck space, and plate movement. 1–3000 μL can be dispensed into 6- to 1536-well plate formats. Another peristaltic-based BRD is the Thermo Scientific **Multidrop** Combi reagent dispenser (*www.thermoscientific.com*). This BRD utilizes disposable dispensing cassettes, which are optimal in the dispensing of live cells with reproducible results, without the fear of cross contamination, in addition to their suitability for straightforward reagent distribution.

10.7.1.7 **Tube Inspection**

In many organizations it is standard practice to subject samples to a purity check before loading into an automated store; however, measurement of the sample volume, or inspection for precipitates, either does not occur or is only undertaken on a random sample basis and/or subjectively (for example, as a result of manual inspection). This approach can ultimately lead to problems further along the drug discovery/development process.

The RTS **Tube Auditor**™ (*www.rts-group.com*) is a high-speed, non-contact volume measurement and precipitate detection instrument for SBS format microtubes that helps to solve this problem. Available as a stand-alone unit or integrated into a system, Tube Auditor offers significant benefits over manual and other semi-automated solutions, facilitating rapid and accurate determination of sample quantity and quality. Knowing the volume of sample in the source tube, and knowing whether there is any precipitate prior to plating, is key to avoiding screening empty wells or samples at the wrong concentration. Tube Auditor enables the user to ascertain:

- If there is any sample in the source tube
- How much volume of sample is in the tube
- If any of the sample has precipitated out of solution.

10.8 Plate Handling, Integration

When considering an automated system, a choice must be made between a fully integrated, turn-key system and a more open, custom integration. The turn-key solutions offer the advantage of proven robustness and ease of use, which may be appropriate for workflows which are predictable and unlikely to change. Several examples are:

- The large scale **POD**™ automation platform and smaller scale **Access**™ laboratory workstations from Labcyte (*www.labcyte.com*) add automated plate handling to Echo® liquid handlers. Each system can be configured with devices and accessories to address key requirements for assay miniaturization. Both platforms are designed to reduce the steps required to design and schedule automation routines in contrast to typical integrated systems. Labcyte automation control software uses Echo® software applications to schedule automation events and take full advantage of the low-volume, any-well to any-well power of acoustic transfer; creating routines optimized for efficient use of the Echo platform. With each Echo application it only takes a few clicks to map dose-response curves of any number of points, in vertical and horizontal patterns, or map hits from a pick list to complex destination plate patterns. Labcyte makes it easy for any researcher, regardless of their experience with automation, to walk-up and quickly schedule automation routines.
- The **Agilent BioCel** (*www.chem.agilent.com*) System automates any microplate-based protocol. The new Direct Drive Robot brings increases in speed and precision of the system. With modular cells, options for various enclosures and environmental control BioCel Systems are tailored to individual needs. Operated by Agilent VWorks scheduling software, BioCel Systems will deliver walk-away time and throughput.
- More open integrated systems offer the ability to be reconfigured as the needs of the laboratory change. These can be more difficult to install initially but do allow the customer to integrate specific instruments and create protocols to precisely meet their needs. Prime examples of this open architecture are the **MicroStar** and **NanoCell** systems from HighRes Biosolutions (*www.highresbio.com*), two system platforms that have extensive capabilities for customization to meet the specific needs of the laboratory. Both of these systems feature the MicroDock system, where instruments are mounted on carts that then 'dock' with the system pod. This makes it possible to add/remove inventory and devices from the system with the touch of a foot pedal, supply power, communications, and gases automatically without the need to handle cabling or tubing, reconfigure the system daily (adding new dispensers, pipettors, and readers) for different assays, and move devices and inventory easily from system to system. This system also makes it easy to add a new device quickly and economically to take advantage of evolving technologies.The **MicroStar** is available in one, two, or even three-robot integrations (featuring Stäubli robots), operates with the Cellario™ scheduling software (with full error recovery and over 80 device drivers) and is available in 6-sided, 9-sided, or 12-sided configurations. The smaller **NanoCell** offers all the flexibility and modularity of the MicroStar but does so over a very small footprint. Coupling an industrial-grade robot with microdocks and microcarts, NanoCell is ideal for a wide range of applications including compound delivery and cherry-picking applications.

10.9 Case Study: Evolution of a Compound Management Group

As an illustration of the evolution of automation in CM over the past few decades, it would be instructive to examine the development of the CM group at the Broad Institute in Cambridge, Massachusetts (*www.broadinstitute.org*). Over the course of three years this CM group underwent an evolution from a collection of loosely coupled manual processes with marginally organized storage systems and data flows into a fully integrated organization equipped to support both HTS with compound plates and cherry-picks as well as an emerging medicinal chemistry group. The thought processes involved in the decision making in the system and process design along with the specific choices made will serve to illustrate the evolution in the SM field over time.

10.9.1 Background

Eli and Edythe L. Broad, in partnership with Harvard University, its hospitals, and MIT, founded the Broad Institute in 2003 as a 10-year 'experiment' in philanthropy and in collaborative scientific organization with $100 million (later doubled to $200 million). The goal of the founders was to test if this model of science combined with 'venture philanthropy' could accelerate the transformation of medicine. Only four years after its launch, the experiment was declared an unqualified success by the Broads, Harvard, and MIT. In recognition of this success, the Broads announced an endowment of $400 million in addition to their previous gifts. In 2009 the Institute formally separated from Harvard and MIT to become a permanent non-profit organization, with Harvard and MIT continuing to play a major role in its governance.

Research at the Broad is structured into programs, one of which is the Chemical Biology Program. This program, which collaborates with many facets of genome biology at Broad, comprises researchers in the community who study and alter the physiology of cells and organisms with small molecules. A central goal of chemical biology is to harness the power of synthetic organic chemistry to discover and to elucidate molecular pathways fundamental in cellular, developmental, and disease biology. The creation and use of small molecules to probe the genome is a fertile area of research that facilitates the translation of biological insights into powerful new medicines. The Chemical Biology Platform comprises scientists from a wide range of disciplines (chemistry, biology, computational science, software, and automation engineering), innovating and working cooperatively toward these goals. The platform team works in an extraordinary research environment that has high-throughput research capabilities in organic synthesis and small-molecule screening. Informatics and computational analysis teams integrate these capabilities to enable collaborating researchers to design new experiments and, not enabling members of the global research community to benefit from the resulting discoveries.

10.9.2 Starting Condition

When the Broad Institute was formed in 2004, Harvard University's Institute for Chemistry and Cell Biology (ICCB) transitioned into the Chemical Biology Program. Some of the assets and groups within ICCB were moved into the Chemical Biology Platform including its small-molecule collection and the group dedicated to supporting the compound needs of the researchers. This group, which had started screening activities in 1998, was the foundation of the CM Group at the Broad Institute.

While part of ICCB and in the early days at the Broad, the compound collection was driven by the needs of individual research groups. Compounds were purchased from commercial vendors for a specific project and then added to the general collection. There was minimal thought given to library design, resulting in overlap in some areas. In addition to these purchased collections, individual compounds from both internal research groups and external collaborators were submitted for inclusion in screens. These compounds were delivered in a variety of containers and formats. The structure and quantity of information supplied with these compounds was also of variable quality and consistency. Another source of compounds was the burgeoning Diversity-Oriented Synthesis (DOS) group, which was synthesizing sets of structurally related compounds around common cores. The informatic activities related to compound submissions had evolved over time to support these diverse source streams and had subsequently become a tangled morass of file transfers and data transformations with a two-week processing time.

The gathering of these diverse compounds into screening plates was also driven by the screeners and not by an overarching logical design. Screens were small and directed at specific subsets of compounds, resulting in screening compound source plates being made specifically for the screen, often with unique layouts; some were even made with dilution curves built into the plate. Plate names were selected by the requestor, resulting in arcane plate names that made sense to no-one else and without barcodes for tracking. All of this led to significant overlap between sets of screening plates and needless depletion of the compound source.

Screening plates and source compound deep-well plates were kept in several manually accessed −20 °C freezers. This storage situation was problematical in a number of ways. Firstly, retrieval of specific plates was a time-consuming manual process which periodically failed to find all of the requested plates. This situation was exacerbated by errors and omissions made when plates were returned to storage. A related problem was the open nature of these freezers. It was not uncommon for screeners to 'help themselves' to plates, resulting in further fragmentation of the plate sets and degradation of the CM group's ability to find specific plates. The most significant problem created by this type of storage was the impact on compounds from cherry-pick requests. To access a specific compound, all of the compounds on that deep-well plate were thawed and then refrozen, a cycle that generally should be avoided as much as possible. Further, depletion of only some of the wells on a source plate made the entire plate unsuitable as a precursor to new screening plates.

The physical manipulation of the compound was performed with both manual and automated solutions. For plate replication and dilution, a variety of instruments, such as a Robbins Scientific Tango outfitted with 96 tips, were used for parallel pipetting operations. These instruments have the advantage of reproducibility and rapid processing, but they were not linked into a database, so the volume transfers were manually tracked. Operations such as cherry-picking and/or serial dilutions were performed either with a hand pipettor or a Tecan Genesis with an eight-channel pipettor. Both manually executing a cherry-pick and manually creating the list for the automated platform ran the risk that mistakes could be made, delivering the wrong compound to the requestor. As a consequence of all of these manual processes, the delivery time for a cherry-pick with a serial dilution was typically four to six weeks.

10.9.3
Roadmap to Evolution

As the Chemical Biology Platform at the Broad began to scale up for increased screening it was clear that the CM group would need to increase its capabilities in tandem. An aggressive development program, which would have been familiar to a compound manager of the 1990s [2, 4], was initiated in 2007. This development entailed steps to:

- Document the integrity of the current holdings
- Design, procure, and implement automated solutions
- Standardize the workflows.

Each of these steps is discussed in detail below.

10.9.4
Current Holdings Integrity

As of 2007 the collection consisted of approximately 90 000 compounds. These had been acquired from a variety of sources over the preceding years, were stored in several different formats, and, given that they had gone through numerous freeze/thaw cycles with no intermittent quality control checks, were of unknown volume and purity. Right at the beginning of the evolution the decision had been made that all of the Master stocks would be held in individually addressable tubes; thus the first step was to physically move everything from the 96-well deep-well blocks into Matrix tubes. This work was performed on a Hamilton StarLet liquid handler with a 96 disposable tip head. During the transfer an aliquot was taken for quality control (QC) by LC–MS, allowing the team to establish the quality of all of the compounds.

The results of the QC analysis showed that 81% of the compounds were above the purity threshold of 75%. The remaining 19% of the compounds either showed significant degradation or, more commonly, there was no evidence of the compound in the sample. This latter condition can be attributed to compound precipitation, complete compound degradation, mis-identification of the compound in the database, or inadequate LC–MS resolution.

10.9.5
Automated Solutions

The process of designing and procuring the automated solutions, which was started before the work of transferring the compounds to individual tubes, began with gathering of the requirements. By looking at the current state of the CM group and the screening center and then examining the anticipated growth in screening demands and interest in supporting the compound needs of other groups within the Broad Institute, the following high-level requirements were developed:

- All compounds would be kept in individually addressable containers
- All compounds would be stored at −20 °C
- Anticipated capacity was 300 000 Mother DMSO stocks at 10 mM in 0.5 mL tubes, which are the working liquid stocks that are accessed for plate production or cherry-picks and are routinely replaced. Additionally, 300 000 Master stocks, which are dry films or powders in 1.4 mL tubes or 4 mL vials were anticipated within five years. Master stocks are the prime source of compounds from which all other samples are derived.

  On a weekly basis the system would be expected to produce
- Multiple sets of 250- 1536-well Assay-Ready Plates from Master plates
- 250 new 384-well screening plates from Mother tubes
- 10 cherry-pick, dose-response plates averaging less than 100 compounds.

  The following actions would also be required:
- Process 1000 Master tubes from the DOS group into new Mother stocks and screening plates
- Screening plates would be in 384-well plates with multiple sets available to serve the three automated and one walk-up screening systems.
- Dry compounds either from an internal medicinal chemistry group or commercial supplier would need to be processed through intake and registration, and made available for testing within two days.
- All events and transfers would be captured in real time in the institutional database.

Combining these requirements with an examination of the existing commercially available instruments, budgets, and timelines led to decisions on the direction to take in the development of the laboratory infrastructure. Perhaps the most important decision made was that the laboratory would be designed as a combination of a tightly integrated system to enable mass production and walk-away time and workstations for focused, low-throughput operations. This yielded a laboratory that was at once highly efficient and flexible, as has been achieved by other automation systems [3, 5–10].

10.9.5.1 **Storage Format**

The desired storage format was to maintain 200 μL at 10 mM in DMSO Mother stocks in individually addressable containers. Plastic 0.5 mL tubes from Matrix with the solid SepraSeals were selected for this purpose. These tubes are supplied as 96-tube racks with standard SBS footprint in an 8-by-12 array with a 9 mm spacing, all of the tubes having a unique 2D barcode on the base. The larger dry compound stocks were to be kept in 1.4 mL glass Matrix-style tubes from TradeWinds with solid SepraSeals, also with 2D barcodes on the base.

10.9.5.2 **Storage Systems**

The requirements to store everything at −20 °C and to have the tubes individually available meant that an automated store was needed that could accommodate racks of tubes, to keep them in an environmentally stable, secure environment, and then retrieve specific tubes on command. This translated into a storage system with a tube picker to select tubes from storage racks with tube pick rates of 300 tubes per hour or better. Additionally, the storage system must have entry points for both manual insertion and retrieval of racks of tubes and an automation interface for robotic insertion/retrieval to allow it to be part of a larger integrated system.

To meet these requirements, a pair of Universal LabStores from Nexus Biosystems were selected. These systems have an I/O module that allows an operator to insert trays holding six SBS racks of tubes for input, a tube picker that can reformat the tubes from the racks into a higher density tray, doubling the storage density, shelves for storage of the trays for random access, and an automation interface that presents completed racks of cherry-picked tubes for robotic access. The first storage system was configured to process 1.4 mL tubes, used for Master storage. The second system was configured to process the 0.5 mL Mother tubes. Each system has a capacity of approximately 300 000 tubes. The systems also incorporate 2D barcode reading to identify the tubes being presented at the I/O. This scanner is also used to double-check the tube IDs after picking. The flexible nature of the shelf storage also allowed storage of microplates and 4 mL Vials in unused space as the compound collection was growing.

10.9.5.3 **Liquid Handling**

The requirements for plate production, including screening plates and cherry-pick plates, drove the decision on the liquid handlers. It was clear that not all of the required work could be done on one system, as the throughput demands were too high and the required functionalities did not all exist on one instrument. Further, the requirements indicated that both an integrated instrument(s) and a workstation instrument were needed.

Two primary liquid handlers were selected to meet these requirements. The first was a pair of Hamilton Starlets. One of these was purchased with a 384-channel disposable tip head. It was primarily intended to be used for plate-to-plate replication including reagent addition from a refillable reservoir. The availability of tips racks with 96 tips also allowed this system to perform reformatting between 96- and 384-well plates. The second Starlet was purchased with a 96-channel disposable tip head. These longer tips were well suited to pipetting into and out of the deeper Matrix tubes for plate production and cherry-pick and dilution series processing.

The second liquid handler selected was the Echo 555 from LabCyte. This unique system uses no tips to affect its liquid transfers, and has the ability to transfer as little as 2.5 nL of sample. The Echo was ideally suited to the production of 384- and 1536-well Assay-Ready Plates with no risk of cross-contamination from tips, no need for intermediate dilution plates, and minimal sample consumption [11].

All three of these liquid handlers were installed on Microcarts from HighRes Biosolutions to make them available to the MicroStar integrated system described below.

10.9.5.4 Accessories

Several accessories were required to fulfill the operational requirements. Both source and destination plates needed to be sealed to avoid contamination and solvent loss. For automated operation the PlateLoc thermal plate sealer from Agilent was selected for its ease of use and high reliability in automated systems. This was combined with the XPeel automated plate seal removal system from Nexus to provide the integrated system with the ability to remove seals as necessary.

All of the plates produced in the system required a label so that they could be identified both by a human operator and by the downstream automated systems. The Agilent Microplate Labeler was selected to provide this capability. This labeler was easy to integrate on the automated system, was known to be reliable, and would take label information from the institutional database.

The decision to use the Matrix tubes for compound storage hinged upon the availability of an automated system for removal of the solid SepraSeal caps from the tubes. Because of the relationship with Nexus Biosystems for the two LabStores, the decision was made to use their XCap tube decapping instrument. This instrument can simultaneously remove and replace all 96 caps from a rack of tubes and either dispose of the caps to waste or hold for recapping. The removal of the caps is a difficult mechanical process which the XCap performs well.

The addition of compounds and reagents to wells of a 384-well plate with an automated liquid handler can result in bubbles being introduced into the wells. These bubbles may cause problems in downstream liquid handling steps. For this reason, an Agilent Microplate Centrifuge was added to the system. All destination plates are spun down after liquid addition and seal application.

10.9.5.5 System Integration

The MicroStar pod automation from HighRes Biosolutions was selected for system integration. The MicroStar features a central Stäubli robot that is used for microplate and tube rack manipulation and nine docking stations for instruments. Individual instruments, or combinations of multiple small instruments, are mounted on wheeled carts that can be easily docked onto the pod for system configuration. The Cellario software coordinates the activities of the installed instruments and schedules them for the most efficient operation. Cellario also interacts with the institutional database to receive orders and provide event processing information back to the database.

In the system installed at the Broad, three of the available docking stations were occupied by the Nexus LabStore, leaving six available for instruments. Depending upon the needs of the process that was being run, different carts would be docked to provide functionality. The docking system is designed so that the reconfiguration was performed in less than an hour, including loading of consumables. This quick system changeover resulted in the system being occupied with productive work 75% of the time, including overnight and at weekends (Table 10.1).

10.9.5.6 Integrated vs Walk-Up

As noted earlier, the choice between an integrated system and a collection of workstations is pivotal to the creation of an efficient, responsive CM facility. Both have their place in the laboratory, and commitment to one or the other may limit throughput or expansion capabilities. The selection of the MicroDock system from HighRes Biosolutions for the integrated system was made precisely because it supports operations in either the integrated or workstation mode. By installing all three of the liquid-handling instruments as well as most of the peripheral accessory instruments on Microcarts they were all available to the MicroStar integrated system and also could be used in a workstation mode off-line. Two MicroDocks were installed in the laboratory, not associated with a system, to provide power and services to the carts for use off-line. This setup allowed the laboratory to quickly reconfigure for optimal efficiency at any time. There was no need to have duplicate instruments (i.e. one set fixed on a system and the other available in workstation mode), resulting in considerable cost and operational savings.

Table 10.1 Examples of the system configuration for two common processes run through this system.

Dock	Plate production from tubes	Assay-ready plate production
1–3	Nexus LabStore	Nexus LabStore
4	Hamilton StarLet, 96-head	Plate carousel with destination plates
5	Plate carousel with destination plates	Plate carousel with destination plates
6	Nexus XPeel Nexus XCap	Nexus XPeel Nexus XCap
7	Liconic incubator (tube defrost and storage)	Liconic incubator
8	V11 Centrifuge	LabCyte Echo
9	V11 PlateLoc sealer V11 Labeler	V11 PlateLoc Sealer V11 Labeler

10.9.6 Workflow Standardization

Selecting the proper hardware and systems to perform the functions required by the customers of the CM group was certainly critical to the success of the modernization of that group. However, the benefits from all of that effort and expense would not have been realized without making fundamental changes to the workflows and processes within the group. The effort that was put into this redesign process equaled that put into the design and implementation of the instrumentation and systems for the laboratory.

The group had grown slowly over the years from its roots in ICCB, and its processes had been developed, not through a thoughtful process, but rather through organic growth in response to demands at the time. While this that ensured current demands were being met, it did result in inefficient processes. As the procedures became ever more complex, it became difficult to respond to even the simplest of standard requests. Further complicating the situation, the culture was such that the CM group was expected to do anything that a researcher requested. The result of this was that cherry-pick requests could take four to six weeks to fulfill and compound submission was a two-week process.

The methodology for this redesign effort started with delineation of the activities that the CM group would be supporting in the future and an alignment of the existing workflows with those activities. This effort revealed that there were existing workflows that did not contribute to the future goals of the CM group and which were disruptive to performance. A prime example was the acceptance of compounds, often in numbers of less than 10, to the screening deck from outside institutions. This broad range of submitters did not use a common container type, and the structure information supplied was in a variety of formats. The result was that an inordinate amount of effort was expended to physically format these compounds to be consistent with systems, and to format the information to be database-compatible for compounds that were of questionable value to the overall library design. This submission pathway was closed down, along with other processes that did not support the future direction of the group. Once the clutter was removed, the group was free to focus on performing its core functions while simultaneously designing the improved workflows. These fell into three areas: supplying screening plates, responding to cherry-pick requests, and compound registration.

10.9.6.1 Screening Plates

Having just gone through the effort to transfer the entire collection from deep-well plates into tubes and performing QC on all of the compounds, it was an ideal time to redesign the contents of the screening plates [12]. First, duplicates and compounds of no interest were removed from the screening deck. Coupled with this was the grouping of the compounds by area of activity, settling on a common screening concentration, and a logical plate map-naming convention. For plating purposes the compounds within the activity groups were ranked by maximum diversity so that each screening plate was as diverse as possible. Informer sets of a small number of compounds were also selected, which provided a sampling across all the compound sets to be used in preliminary screens and assay development. Rules were established on the lifetime of the plates and use of plates from the screening deck to prevent over-sampling of specific plates. As new compounds were added to the compound collection they were added to the groups, maintaining the logic of the design. All of these efforts led to clarification of the screening deck and a predictable cycle for creation and replacement of the plates.

10.9.6.2 **Cherry-Picks**

The workflow requiring the greatest degree of modification was the cherry-pick process. Providing specific compounds for confirmatory testing after a screen is a central function of CM groups. It is critical to be able to supply the selected compounds in proper dose curves in a timely manner. The legacy cherry-pick process at the Broad had grown over several years, was complex, and contained many ambiguities. The process started with a researcher sending a list of compounds, in a format of his choosing, to a member of the CM team and specifying, in the text of the e-mail or a hallway conversation, the required dilution series and plate layout. The CM team was expected to interpret this information and supply the requested plates. In practice, what often happened was that a series of clarifying messages and conversations would ensue to gather the required information. The team member would then go through a manual process of creating a plate map, entering it in the database, pulling the compounds from a freezer, and then make the plate. Time lags were excessive in this process and the error rate was unacceptably high.

To create an improved, robust system, a number of steps were taken. First, the historical records were examined to look for the most commonly requested plate layouts and dose curves, and the feasibility of creating these was validated against the capabilities of the new automated systems. In consultation with key requestors, a small set of standard plate layouts were designed that suited their assay needs and which were efficient for the CM group to produce. These were formalized in new guidelines for conducting retests.

In parallel with the specification of a small set of standard layouts, a request template was established and distributed to the requestors for their use. This template had data entry points for all of the information required to create the retest plates; it included a list of supported layouts and assisted the requestor in making sure that they had provided the CM group with all of the pertinent details. This formalized template greatly reduced the amount of back and forth communication that had been previously required to gather the information; it provided the CM team member with one place for consolidating the information, and ensured that all CM team members were working in the same manner.

To rectify the problematic communication cycles between the requestor and the CM team, a formalized order request system was established with a single e-mail address based upon a software bug-tracking system. All communication between the CM team and the requestor was through this system, with the result that all of the communication and related information was in a single location that was visible to both the requestor and the CM team. As steps were taken to fulfill the request, the tracking system was updated, keeping the requestor constantly aware of status.

The final step in creating a robust cherry-pick processing workflow was to create the tools in the database to support the steps of the workflow. These tools were designed to read the template file and then create the required plate-maps and pick lists for the automated systems. During processing of the request, the CM team member would record the completion of steps in the database interface so that information regarding compound stocks and newly created plates was current and in real time.

All of this redesign had the desired positive effects on execution of cherry-pick requests. Missed or delayed requests were eliminated as were errors in the execution and recording of the information in the database. Requestors had a more consistent and reliable supply of retest plates with a reduction in request fulfillment time from four to six weeks to five days. The stress on the CM team was greatly reduced by being able to work in more consistent ways, making the team significantly more efficient. This efficiency increase had the unexpected consequence of actually increasing ability to honor special requests.

10.9.6.3 Compound Registration

The legacy compound registration process had grown in much the same fashion as the legacy cherry-pick process, with similarly detrimental effects. The compound registration process had developed to the point where it required the creation and transfer of seven different files through three software layers involving four steps of human intervention. In total it required nearly 2 h from a member of the CM team, spread across as much as 10 days to register a set of compounds.

The path undertaken to simplify this process was similar to that taken on the cherry-pick process. It started with the suspension of external compound submissions, which had been a major contributor to difficulties in registration. Next, the submission process was divided between those from internal chemists and those from external compound purchases.

The task in compound registration requiring the most significant human involvement is curation of the structure information for storage in the database. This requires a sound understanding of the chemistry involved in the synthesis of the compounds. In the case of compound submissions from internal chemists, the latter have the specialized knowledge required, especially as it applies to their compounds, so they are uniquely suited to performing this task. Therefore, they were supplied with the software tools necessary for them to self-register the compounds directly from their electronic notebooks. They were also supplied with the specific vials used for storage and, given their access to the analytical balances, they were able to enter the delivered mass directly into the database. The result of these changes was that the internally generated compounds were registered with just 30 min of the chemists time, and the only task remaining for the CM team member was to physically record the supplied sample into storage.

Compounds entering the registration process by way of purchases from external vendors continue to go through essentially the same legacy process with a few notable improvements. The first was to centralize purchases through the CM team. This enabled the team to create a request fulfillment system for tracking and request templates, much as was done with the cherry-pick process to standardize the information gathering. Next, by working through a selected commercial compound accumulator the structure data was supplied in a much more consistent manner, reducing the curation load. Finally, by batching the processing of commercial compounds, including transfer to the specified vials into specific periods during the week, the work became more predictable and efficient.

10.10 Results

The evolution of the CM group at the Broad Institute spanned nearly three years. Where possible the steps involved in this evolution were concurrent. In the requirement gathering and design phase the initial focus was on deciding upon the storage format and hardware plus the liquid-handling platforms, so that these could be ordered as rapidly as possible. This resulted in the new liquid handlers arriving early in the process, enabling the work to reformat the existing collection to begin. Having accomplished that milestone, the group could support making screening plates and responding to cherry-pick requests as the more involved task of designing, building, and validating the integration platform progressed. Workflows evolved constantly, taking advantage of new hardware and software systems as they became available. The benefit of this approach was that the CM group was able to keep up with the increasing demands of the screening group, with only temporary interruptions in service. By the middle of 2010, the CM group had achieved its major goals, and was supporting the platform's needs with consistent quality and results.

The effectiveness of the process used in this evolution and the flexibility of the integrated systems from High-Resolution Biosolutions and workflows was validated in 2009 when it was decided within the screening group to transition to the use of Assay-Ready Plates for as many assays as possible, many of them in 1536-well format. The initial designs for the CM group did not foresee Assay-Ready Plates as a demand that the systems would have to meet. However, screening was having trouble meeting its throughput targets, and the use of Assay-Ready Plates was seen as a means to significantly increase throughput. Processes were rapidly developed in CM to process requests for Assay-Ready Plates with the necessary informatic support. A LabCyte Echo 555 was already mounted on a HighRes micro-cart for use on the systems, making it a trivial exercise to add it to the CM system, reconfigure the other storage carousel and accessories micro-carts, and program the scheduling software to process source compound plates through the Echo into Assay-Ready Plates (see Table 10.1). This new workflow was running within a matter of weeks and then continued to improve in robustness and speed over the following two months. It quickly became a standard protocol that ran unattended on evenings and weekends without disrupting the other tasks within the CM group and enabled the group to consistently stay ahead of the demands for Assay-Ready Plates from Screening.

References

1. Archer, J.R. (2004) History, evolution, and trends in compound management for high throughput screening. *Assay Drug Dev. Technol.*, **2**, 675.
2. Houston, J.G., Banks, M.N., Binnie, A., Brenner, S., O'Connell, J., and Petrillo, E.W. (2008) Case study: impact of technology investment on lead discovery at Bristol-Myers Squibb, 1998–2006. *Drug Discov. Today*, **13**, 44.
3. Quintero, C. and Kariv, I. (2009) Design and implementation of an automated compound management system in support of lead optimization. *J. Biomol. Screen.*, **14**, 499.
4. Rutherford, M.L. and Stinger, T. (2001) Recent trends in laboratory automation in the pharmaceutical industry. *Curr. Opin. Drug Discov. Dev.*, **4**, 343.
5. Schopfer, U., Engeloch, C., Stanek, J., Girod, M., Schuffenhauer, A., Jacoby, E., and Acklin, P. (2005) The Novartis compound archive – from concept to reality. *Comb. Chem. High Throughput Screen.*, **8**, 513.
6. Schopfer, U., Hohn, F., Hueber, M., Girod, M., Engeloch, C., Popov, M., and Muckenschnabel, I. (2007) Screening library evolution through automation of solution preparation. *J. Biomol. Screen.*, **12**, 724.
7. Kuzniar, E. (1999) Inventory management and reagent supply for automated chemistry. *Comb. Chem. High Throughput Screen.*, **2**, 239.
8. Keighley, W.W. and Wood, T.P. (2002) Compound library management. An overview of an automated system. *Methods Mol. Biol.*, **190**, 129.
9. Fillers, W.S. (2004) Modular tube/plate-based sample management: a business model optimized for scalable storage and processing. *Assay Drug Dev. Technol.*, **2**, 691.
10. Harris, C.O. and Schweiker, S.L. (2001) Optimizing production of serially diluted compounds and distribution to multiple targets. *J. Autom. Methods Manage. Chem.*, **23**, 179.
11. Zaragoza-Sundqvist, M., Eriksson, H., Rohman, M., and Greasley, P.J. (2009) High-quality cost-effective compound management support for HTS. *J. Biomol. Screen.*, **14**, 509.
12. Chan, J.A. and Hueso-Rodriguez, J.A. (2002) Compound library management. *Methods Mol. Biol.*, **190**, 117.

11
Applications of Acoustic Technology

Eric Tang, Colin Bath, and Sue Holland-Crimmin

11.1
Introduction

Compound management (CM) has witnessed numerous technological developments over the last two decades. Advances, primarily aimed at improving the efficiency and quality of the samples curated by CM, are clearly evident. In particular, customers operating one key process, high-throughput screening (HTS), have played a pivotal role in driving improved quality and efficiency requirements [1] to support expanding compound collections. The paradigm shift from testing a few compounds a week to testing hundreds of thousands per day has resulted in large-scale investment in advanced automation, sample tracking, and informatics. The output from this investment can be appreciated from the large banks of solubilized compounds, automated vial storage, miniaturization, delivery of 96-, 384-, 1536-, and 3456-well plates, identification with 2D barcodes, optimized environmental storage conditions, and predefined compound shelf life.

One of the key challenges in supporting the evolution of HTS was the availability of robust low-volume liquid dispenses. A significant technological advance has occurred in the last decade in the discipline of low-volume dispensing via the development of acoustic droplet ejection (ADE). The term *acoustics* is derived from the Greek *akoustos*, meaning 'hearing,' and it is defined as the science associated with the control, generation, transmission, reception, and the effects of sound. Acoustic waves and their properties have provided innovative approaches to almost all fields and facets of life, and the span of modern acoustics ranges from ultrasonics and infrasonics to the audio range (Figure 11.1). The earliest acoustic references date back to the Chinese dynasty in the twenty-seventh century BC with the invention of the musical pitch pipe [2]. Acoustic applications span multiple fields including medical diagnostics [3], undersea exploration and navigation [4], building design, music, and communications, to name but a few. It was not until recently that the benefits to sample management technologies were realized.

The earliest reference to ADE was in 1927 by Robert W. Wood and Alfred Loomis, who noted that when a high-power acoustic generator was immersed in an oil bath, a mound formed on the surface of the oil and, 'like a miniature volcano,' ejected a

Management of Chemical and Biological Samples for Screening Applications, First Edition.
Edited by Mark Wigglesworth and Terry Wood.

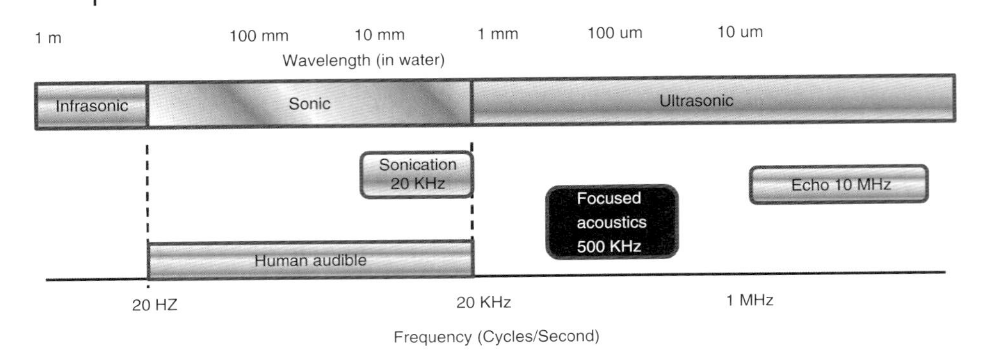

Figure 11.1 Acoustic technology and the sound spectrum.

continuous stream of droplets. This technique was refined in the 1970s and 1980s by Xerox, IBM [5], and other organizations to provide a single droplet on demand for printing ink onto a page. Two California-based companies, EDC Biosystems Inc. and Labcyte Inc., exploit the acoustic energy for two specific utilities: (i) as a liquid transfer device and (ii) as a device for liquid composition auditing.

The technique of moving liquid with sound for liquid transfer is based on a transducer placed below the well of a source plate receiving radio frequency (RF) energy in the MHz range and converting it into an acoustic pulse. If the transducer is focused such that the pulse is just below the surface of the liquid then an acoustic vibration ejects a small drop of liquid. The size of the ejected droplet is tunable, being inversely related to the RF frequency (Figure 11.2). Since 2004, this technology has been adopted and integrated into two distinct liquid-handling devices, the Labcyte Echo™ and the EDC ATS-100™. These devices provide true non-contact, highly precise, low-volume dispensing to facilitate sample transfer in drug discovery applications. The flexible format nature of the acoustic device provides an important attribute. In simple terms, any source sample can be

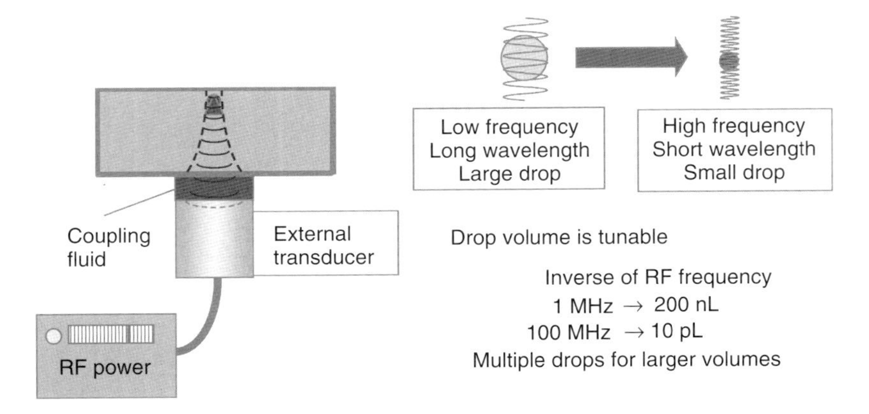

Figure 11.2 Theory of acoustic drop ejection.

transferred to any destination well, in contrast to conventional fixed tip instruments [6]. This provides researchers with the ability to consolidate samples, miniaturize assays and cherry pick compounds to support hit confirmation. Furthermore it has enabled new applications and flexible workflows, including direct dose response profiling and direct dispensing into cell-based assays. The applications of the technology to sample management are described in this chapter together with new and emerging utilities. Another unique aspect of the ADE technology is that successful droplet ejection required certain source well parameters to be derived which in themselves are key quality attributes of the sample. Labcyte have exploited these aspects to successfully provide auditing information, including fluid height and the dimethyl sulfoxide (DMSO)/water ratio of the measured samples.

Although the initial application of acoustic dispensing was focused on compounds in DMSO solutions, the technology has been developed such that many different fluids can be dispensed with the same accuracy and precision. The applications for the technology has consequently broadened significantly to include siRNA screening [7], polymerase chain reaction (PCR), DNA sequencing, and matrix deposition for matrix-assisted laser desorption ionization (MALDI) tissue imaging [8].

Apart from liquid sample transfer, acoustics have also been deployed in applications such as thawing, mixing, and resolubilization of samples in glass containers and microtiter plates. Several technologies are described later in the chapter which use both sonication and ultrasonics. Sonication was first utilized for biochemical applications such as cell lysis and shearing DNA. Recently, the application has extended its use in chemistry and sample management laboratories for solubilization of samples and recently has been adapted for material stored in microtiter plates. Several ultrasonic techniques are described, namely adaptive focused acoustics (AFA) and lateral ultrasonic thrust (LUT) to mix, homogenize, dissolve, and thaw samples. They have some advantages over the sonication methods as they facilitate true non-contact, closed-container, and isothermal mixing, and as a consequence lead to significantly reduced mixing times and limited sample degradation, and are ideal for use with heat-sensitive compounds or materials.

This chapter will review the application of acoustic technology to the scientific discipline of sample management and will describe the current state of the technology and how it is currently being deployed within the biomedical industry.

11.2 Compound-Handling Challenges in Drug Discovery

With the majority of pharmaceutical agents under investigation being dissolved and archived in base solvent, the dissolution of organic molecules in solvents is a fundamental prerequisite of drug discovery and development. Within CM, the bulk of samples supplied for early-phase target-to-hit and hit-to-lead programs are as DMSO stock solutions. As the scale of both HTS and lead optimization increased by perhaps 100-fold, an efficient and rapid method for initial dissolution and subsequent mixing became a necessity. Prior to the introduction of acoustic technologies, compound dissolution was a slow and manual process [9]. Furthermore, methods to ensure efficient mixing in low volumes in microtiter plates were not available. The development of acoustic mixing and dissolution may provide the technical solution to these problems [6].

Until recently, the quality of compound handling in cell-based assays has been largely overlooked and has often been treated as a 'dark art' in the screening workflow. Consequently, any discrepancy in results was attributed to assay variability and poor translation between biochemical end-points and down-stream cell-based assays. Furthermore, DMSO tolerance has often been one of the critical factors in cell-based assays. Generally, final DMSO concentration of 0.1–0.5% v/v is the typical range for most cell-based applications to avoid cell toxicity. This compares to 1–10% v/v in biochemical screening. To achieve such low final DMSO concentration for cell dosing, conventional approaches required the generation of intermediate pre-dilution plates in aqueous-based culture media prior to a further dilution and addition to live cells. No direct DMSO addition was feasible due to the small dispense volume required (e.g., 40 nL DMSO in a 40 μL, 384-well-based assay to achieve 0.1% v/v DMSO) and the need to achieve mixing of the assay reagents within 384-well plates.

In a conventional cell-based screening workflow, all intermediate plates and compound transfers to the destination cell plate is via manual pipetting or liquid handling instruments with 'fixed'-tip configuration. For dose response type studies, the serial dilution of the test agents could be either in aqueous medium or in 100% DMSO before each dilution is subsequently pre-diluted into aqueous culture media. With either of these screening workflows, test agents across a diverse spectrum of chemical classes with a broad range of physical properties were evaluated. Compound properties, including solubility and hydrophobicity, have led to a varying degree of compound loss or carryover in conventional manual and semi-automated compound handling workflows, resulting in deviation from intended test concentration, poor data consistency, and correlation with downstream assays.

Figure 11.3 provides an example of a manual dispense and serial dilution of a test agent with high hydrophobicity. The resulting dilution series were quantified by triple-quad mass spectrometry with standard curves prepared in organic solvent. It has become clear that, due to the high hydrophobicity, significant amounts of the agents were adsorbed onto the plastic or metal surface of the liquid handling devices, resulting in severe carryover of the agent across the dilution series. Conversely, changing tips or acid cleaning of the liquid handling tips reduce the contamination risk but resulting in depletion of the agent in the resulting solution. In extreme cases, the concentration could be less than 50% of the expected concentration at high doses.

In contrast, acoustic technologies provide a non-contact approach to transfer of nanoliter volumes of test agents directly into the destination target plates. This non-contact technology enables the preparation of dose–response series directly in the destination plate. This results in an improvement in the quality

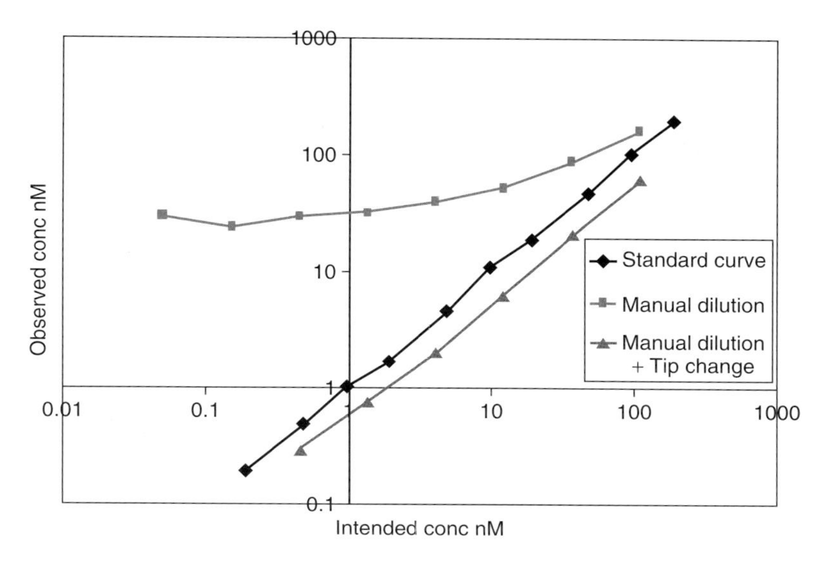

Figure 11.3 Graph comparing the standard curve and conventional manual serial dilution with and without pipette tip changes.

of IC50 analyses when applied to compound screening. In addition, the acoustic approaches reduce compound usage as compared with conventional liquid handling technologies.

11.3 Acoustic Drop Ejection – Performance, Quality Assurance, and Platform Validation

11.3.1 Precision

The two leading acoustic drop ejection instruments on the market are capable of performing high precision liquid transfer at nanoliter volume range. Using a fluorescence carrier, it has been demonstrated that the instruments are capable of delivering repeating droplets of a pre-defined size to the destination well with less than 8% covariance ratio (CV) from 2.5 to 100 nL transfer volume, as illustrated in Figure 11.4. Similar performance on the acoustic instrument has been reported by other investigators [10].

11.3.2 Quality Assurance – Non-Invasive DMSO Hydration Monitor

In order to accurately dispense samples, the acoustic process requires the derivation of parameters about the sample which themselves provide valuable quality information. The amplitude of the reflected energy from the fluid well bottom

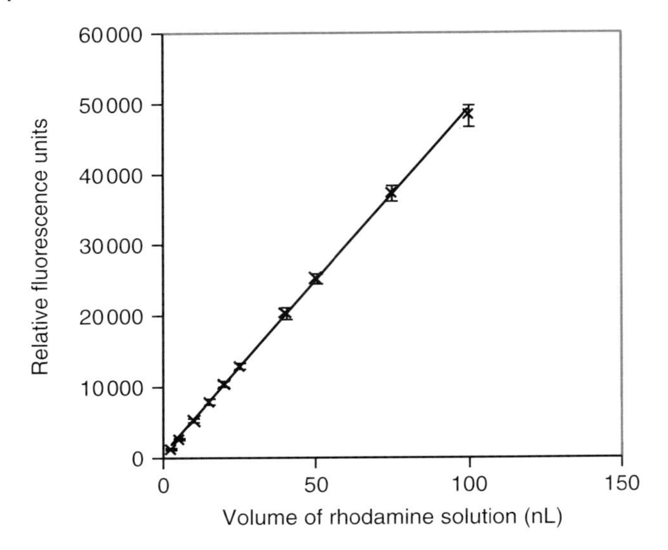

Figure 11.4 Linearity of acoustic drop ejection transfer of rhodamine in DMSO by Echo™ 550 Reformatter. Each data point represents the mean of 72 replicates (±1 SD).

interface provides the hydration level of DMSO and the speed of sound. The time of flight coupled to speed of sound allows the calculation of the fluid height in the source well. These acoustic signals are digitized and processed to generate the volume and DMSO hydration values. Figure 11.5 indicates the measurement of the hydration level of DMSO as compared to the more traditional methods such as Karl Fischer titration and NMR. Unfortunately, the latter techniques are invasive as they require material and they can be time consuming and expensive.

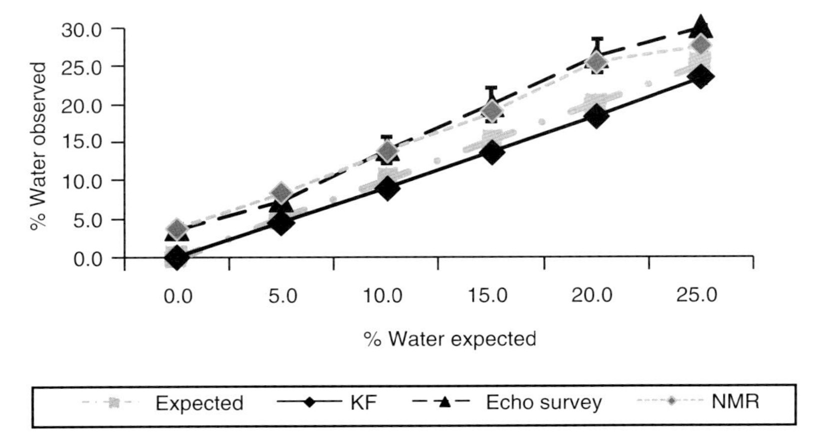

Figure 11.5 Comparison of measurement of water content by different methods. (Karl Fischer, NMR, and Echo™ acoustic dispenser).

Sample management organizations have since exploited this information to improve the quality and integrity of their processes. For example, the hydration levels of source plates are now routinely tracked to indicate appropriate stock replacement time frames. In the past, HTS source plates used for primary plate stamping were not routinely tracked in order to decrement volumes, but by utilizing acoustic dispensing this information can be derived every time a set of samples are dispensed. Furthermore, the acoustic process provides the information that an acoustic pulse has been produced, thus providing for the first time the information that a sample has or has not been dispensed. This signal can be converted in certain CM management software such as Activity Base™ or Mosaic™ software (see Chapter 14 for details) to an electronic event providing dispensing report and flagging missed wells in the process. This therefore provides the opportunity to redispense samples of high value and minimizes false negatives in the screening process.

One of the prevalent quality issues is related to evaporation of small droplets and volumes. Acoustics have provided the capability to cherry pick directly into 1536-well plates, which was not previously feasible. However, the time required to cherry pick into a 1536-well plate can be considerable, particularly with a low percentage hit rate across multiple source plates. In addition, evaporation inside acoustic dispensing instruments can be significant, since in most cases the target plate is being maintained within the instrument for the duration of the cherry pick. Studies have shown that if the humidity and temperature are strictly maintained within the automation system enclosure (45% relative humidity and 21 °C) then plates can be maintained for up to 8 h without significant deterioation. If however the humidity is not maintained, the dispensed compounds will dry out and the compound residues can be difficult to resolubilize.

11.3.3 Platform Validation

Although the Echo™ reformatter has been confirmed to deliver high precision DMSO transfers, it has remained unclear whether or not different compounds affect the fluid dynamics and surface tension of the stock solution, consequently having a detrimental effect on the precision of the acoustic transfer to the destination well. Grant *et al.* [11] have examined the potential effect of a compound's physicochemical properties on the acoustic drop ejection performance by analyzing the dispensing quality of ~100 compounds chosen to cover a wide range of LogD and solubility values. Serial dilution models were used to transfer each of the compounds to their corresponding destination wells and then evaluated by TripleQuad LC/MS to determine the mass of the compound transferred. As shown in Figure 11.6, all replicates are reproducible within a run and measured concentrations for the concentration range 3–300 nM tend to agree quite closely with intended concentrations. For lower concentrations (0.3 and 1 nM intended) the measured concentrations are more variable for many compounds as some of these low concentration samples were near the analytical detection limit of the mass spectrometry instrumentation. This typically leads to a higher variation of the

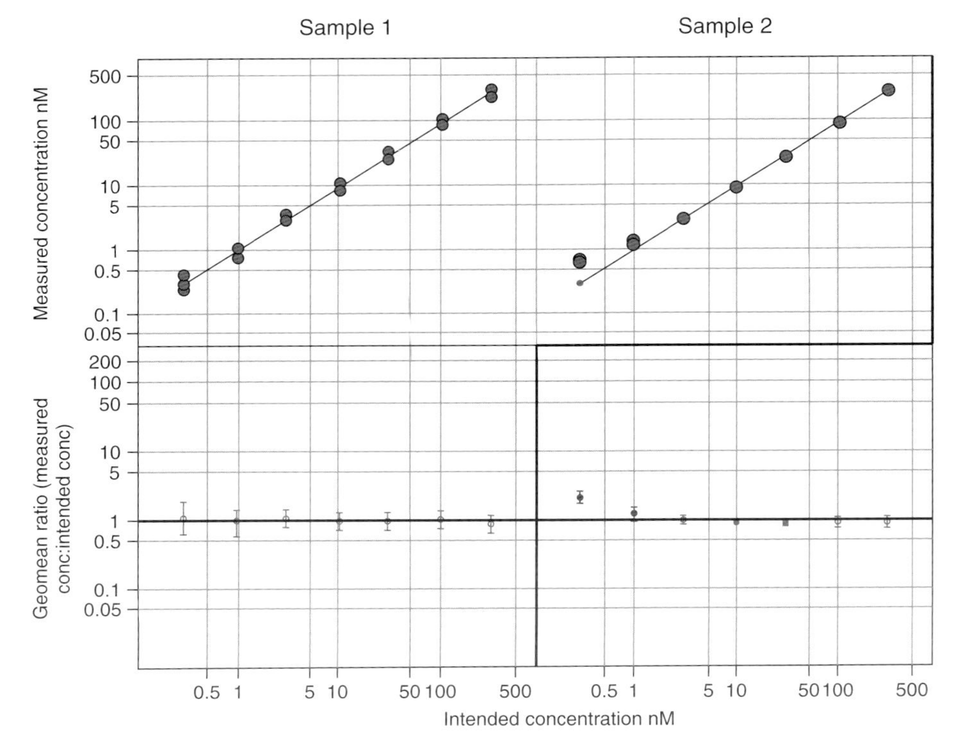

Figure 11.6 Plots for two representative compounds. The top two panels show the intended concentration against measured concentration. A line of agreement is shown – if points fall on this line the intended and measured concentrations are the same. The bottom two panels show the geo-mean ratios (three replicates), which represent the ratio of intended concentration to measured concentration. A ratio of one indicates that intended and measured concentrations are the same.

detection signal. Nevertheless, the results indicate that the acoustic drop ejection platform is suitable for delivery of diverse compounds for drug discovery screening.

11.4 Acoustic-Assisted Compound Solubilization and Mixing

A number of procedures are used routinely in laboratories for solubilization and mixing of compounds in the final destination wells. These include heat treatment and mechanical agitation such as vortexing, orbital shaking, and sonication. Compound solubility is governed by two factors, thermodynamics, which dictates the maximum concentration of the solution, and kinetics, how quickly a compound can attain that concentration. Techniques to enhance solubilization can speed up the process of dissolution, that is, the kinetics, but cannot affect the thermodynamics unless the compound is changed to a more soluble high-energy form. Techniques such as sonication have also been shown to drive precipitated compounds back into solution as a result of acceleration the dissolution process. Compound precipitation and degradation are accelerated in DMSO stock solutions following the introduction of water. Freeze–thawing of these hydrated DMSO solutions can exacerbate precipitation [12]. Prior to the introduction of the septum-sealed tube or single-use tube for compound storage, water uptake and precipitation represent significant issues for many organizations dealing with DMSO sample libraries.

The introduction of miniaturized formats for HTS was a significant breakthrough for drug discovery but provided several technical issues in terms of mixing of compound and assay plates. Conventional mixing techniques were insufficient to overcome the reduced surface area-to-volume ratio and consequent increase in surface tension and capillary forces and thereby ensure adequate mixing. Several technologies have emerged to support solubilization and mixing of compounds in low-volume miniaturized formats. These include the SonicMan™ (Matrical Inc.), the Hendrix LUT™ technology, and AFA. There are also reports describe the development by Advalytix (acquired by Beckman Coulter) of another plate-based mixer using surface acoustic waves (SAWs) technology. SAW devices have been extensively exploited in micro channel devices for mixing and also for activating droplet motion. The technology does not appear to have been adopted for sample management to date.

11.4.1 Sonication

Sonication is often a method of choice to speed up the solubilization of compounds in a solvent such as DMSO. This has traditionally been done with compounds in glass vials within a sonication water bath. There are a number of drawbacks with this technique in that it is a very manual process and consequently has a low throughput. The SonicMan™ (Matrical Inc.) is a plate-based system which utilizes low-energy sonication in the 20 KHz range for high-throughput solubilization and mixing of samples. The technology utilizes matrix of metal pins mounted on disposable microplate lid to transfer these low-energy pulses to the compound solution in each destination well. The technology is applicable to a wide range of compounds and can be used for initial solubilization or to redissolve a compound back into solution following precipitation [13]. Potential drawbacks to this technology include cross contamination due to the contact nature of the treatment and the generation and transfer of heat during treatment. This can be a problem with heat-sensitive compounds, although adjustment of treatment conditions will reduce these side effects.

11.4.2 Ultrasonic Mixing

Ultrasonication generates alternating low-pressure and high-pressure waves in liquids, leading to the formation and violent collapse of bubbles, known as *cavitation*. AFA is a technology developed from medical diagnostic ultrasound imaging and Extracorporeal Shockwave Lithotripsy, a procedure used to dissolve kidney stones since the 1980s. In contrast to sonication, AFA uses higher frequencies, for example, 500 KHz, and low wavelength, typically 1 mm. The shorter wavelength and higher frequency allows the energy to be focused through a container to achieve noncontact isothermal mixing. Specifically, mixing is achieved with pulses of energy which generate cavitation bubbles that collapse producing jet streams that move at speeds of greater than 100 $m\,s^{-1}$, generating strong hydrodynamic shear forces. The AFA technology developed by Covaris Inc. has been deployed in a range of instruments catering for samples in glass vials, tubes, and microplates.

The technology was originally used for DNA extraction from bone and tissue samples [14, 15]. Within CM, AFA was first introduced for the dissolution of compounds in glass vials. In comparison with traditional methods, AFA reduced mixing times and was effective in enhancing the dissolution of poorly soluble compounds [16]. Figure 11.7 shows the concentration of chloroquine achieved across a range of mixing treatments. AFA at 100 s achieved a concentration of 2.1 mM as compared to 0.11 mM with 100 s vortexing and 0.38 mM with 100 s sonication. KBiosciences integrated the acoustic components into an automated platform capable of handling vials in racks of 24. This instrument has been integrated into CM operations in several organizations. Figure 11.8 indicates an analysis of the concentration range achieved for lead optimization samples before and after AFA implementation, demonstrating that the technology was effective in increasing the overall concentration of samples closer to the targeted 10 mM concentration.

In contrast, LUT™ technology utilizes micro-electrical mechanical system (MEMS)-based transducers to convert RF power to ultrasonic waves to mix and solubilize compounds. Unlike AFA, the LUT technology does not cause cavitation but utilizes 384 transducer elements to enable high-throughput parallel processing in 384-well and 1536-well format. The high-energy waves produce a strong LUT to generate a lateral mixing vortex.

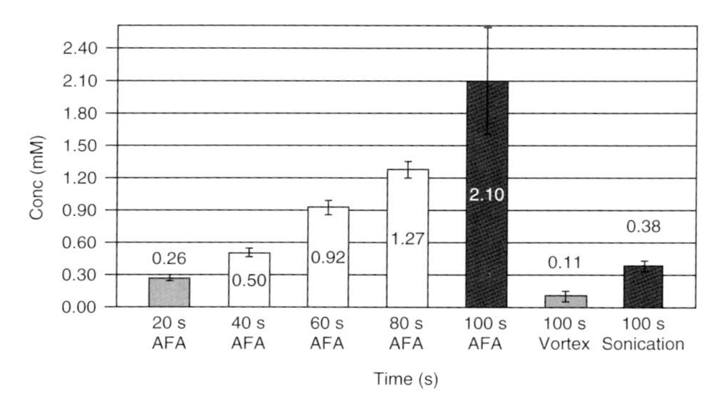

Figure 11.7 Comparison of dissolution of chloroquine by vortex treatment, sonication, and AFA.

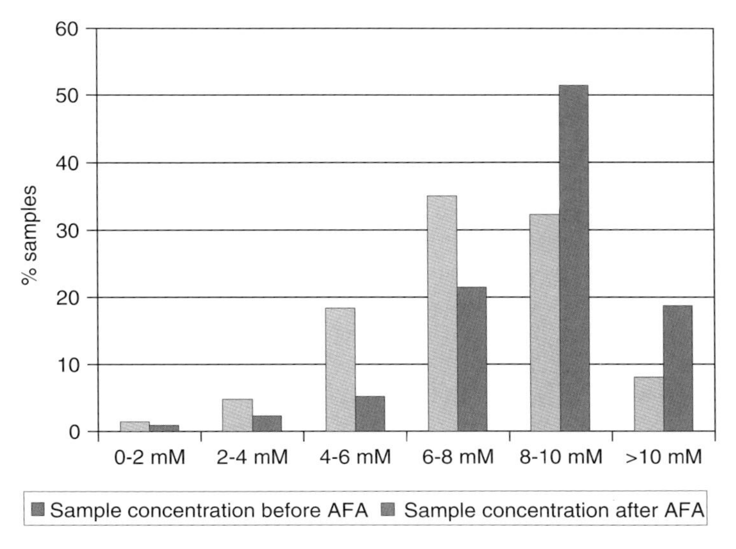

Figure 11.8 The binned percentage of samples within a certain concentration before and after the introduction of AAM to compound management.

The technology incorporated into the Hendrix SM100 ultrasonic mixer has been proven in a number of applications. Several references demonstrate effective resolubilization of precipitated samples. In the first example, data shows enhanced biological activity of a number of poorly soluble compounds after dilution in assay buffer [17]. The second example illustrates the resolubilization of precipitated samples from a fragment library where compounds are normally maintained and screened at much higher concentrations with a concomitant susceptibility to precipitation [18]. It clearly demonstrated that this technology is truly non-contact and non-invasive.

11.5
Acoustic Applications in Drug Discovery

11.5.1
HTS and Assay-Ready Plates – Compound Reformatting and Generic Dose Response Studies

The ability to transfer sub-microliter quantities of material directly to an assay plate has long been an aspiration of scientists in biomedical research. In order to support the increased scale of HTS and the growth of sample collections whilst remaining economically viable, assay miniaturization in low volume 384- and 1536-well format has become an industry-wide goal in drug discovery. Despite significant technological advances, options to enable robust low-volume dispensing directly into assay plates were relatively limited; they included pin tools from V&P scientific and capillary devices, for example, Hummingbird from Genomic Solutions, these have been the instruments of choice until recently. Assay-ready plate (commonly referred as ARP) is defined as a vessel of any format that contains the test material of interest ready to accept biological reagents to complete the assay assembly. ARPs have been used successfully in most biochemical assays and some cell-based applications apart from those involving adherent cell lines.

The delivery of the ARP via ADE can be a manual or an automated process depending on the capacity, location, and budget of the customer group. There are two models for ARP production and distribution. The first is a distributed model in which satellite groups dispense samples from a locally held set of compound source plates. In many cases the acoustic dispenser is integrated directly on an automated screening platform. The second model is a centralized system within mainstream sample management groups in which ARP plates are produced for all customers regardless of the drug discovery phase. Centralization carries many advantages, from process control to budgetary saving [19] including both revenue and capital savings, while providing resource realignment. The choice of implementation strategy is particularly significant with ADE, as the cost of the reformatter often constitutes a significant proportion of the investment budget. Nevertheless, important operational deliverables will determine the necessary cost–benefit projection to determine whether centralization is suitable for individual organizations. Such determinant factors may include a local facility for plate storage, just-in-time production requirements, and sample integrity over storage and physical supply of ARPs between company sites. In recent years, many of these issues have been studied and reported by various organizations. These studies have shown that ARPs can be stored at $-20\,^{\circ}C$ for several months [20] and routinely transported between global sites under dry ice successfully.

Compound screening operations in drug discovery routinely involves two different type of format: single-concentration screening and dose–response studies. HTS starts with single-concentration screening to identify hits across large compound libraries. Dose–response studies will provide concentration-dependent information as part of HTS follow-up and all downstream screening activities in the drug discovery programs. In single-concentration screening, samples are reformatted via a relatively straightforward workflow based on either an ADE or fixed-tip transfer processes. The samples are either transferred in a 1 : 1 well relationship or via a condensing process, that is, 96–384 or 384–1536 reformatting. The subsequent steps of generating concentration response studies for biological testing involve a more complex workflow and are dependent on the assay format utilized downstream.

In order to measure the potency of compounds, most dose–response studies involve diluting the stock compounds at typical concentration at ~10 mM to achieve a concentration range of up to eight log units (e.g., 10^{-4}–10^{-12} M). In contrast, both leading acoustic liquid handlers are only capable of dispensing low nanoliters to sub-microliter volumes in a single transfer operation. This translates to a dilution range of ~2 log units from a single source well, and this limitation is further restricted by smaller assay volumes. As a result, a full dose–response range using acoustic dispensing has to be constructed from multiple source wells containing different pre-dilutions in DMSO. Two different approaches have been developed to generate these mutiple-concentration source wells [21]. The first approach utilizes a separate liquid handler step to prepare quadrant dilutions (10–100-fold in DMSO) of the stock compounds in a 96-well mask over a 384-well-based source plate as shown in Figure 11.9. This method provides a source plate that contains dilutions sufficient to enable the construction of a full dose–response profile from 10^{-4} to 10^{-11} M in a typical assay format. This approach has the benefits that all source plates are pre-assembled prior to dispensing and the acoustic transfer can be operated without the need to swap in different source plates. The second approach is often referred as the *'direct-dilution method'* by the manufacturer. Multiple dilutions of the test agents are created by using several daughter source plates pre-dispensed with diluent (i.e. DMSO), as illustrated in Figure 11.10. Due to multiple source plate handling and the passive diffusion time, this method is only feasible as a fully automated routine within an integrated compound reformatting system. The method also relies on passive mixing within each of the diluted source wells prior to being used as the source material for the subsequent serial dilution step. This approach generates multiple source plates for each test agent, and a full dose–response profile of each agent requires mutiple source plate changes during the execution of the main protocol. The method has the advantage that all source plate preparation and dosing operations can be fully automated within an integrated acoustic system. The down side is the time required for the preparation and passive mixing.

Once the pre-diluted source plate is generated by one of the two methods above, the required volume can be transferred to multiple target plates through standard workflow on the ADE platform. However, depending on the drop size configured on the acoustic instrumentation, the final concentration achieved will need to be recalculated to support downstream data analysis, as the actual volume transferred into each destination well is defined as a function of the drop size offered by the corresponding technology. To complete the dose–response operation, many users will choose to backfill all target wells with DMSO to maintain a final DMSO concentration across all test wells.

11.5.2
Compound Dosing in Cell-Based Screening Applications

In the last decade, there has been a major shift in the screening paradigm, with increasing emphasis on evaluating pharmacologically active agents in cell-based screening. Until recently, cell-based screening was usually confined to small-scale assays in low-well-density formats. With technological advances to enhance cell supply, assay throughput and imaging endpoints, cell-based screening has become a main-stream operation in HTS and secondary cell profiling campaigns. However, employing conventional compound-handling approaches in compound-reformatting and dosing for these assays frequently resulted in throughput bottleneck and data quality issues. The application of acoustic drop ejection to

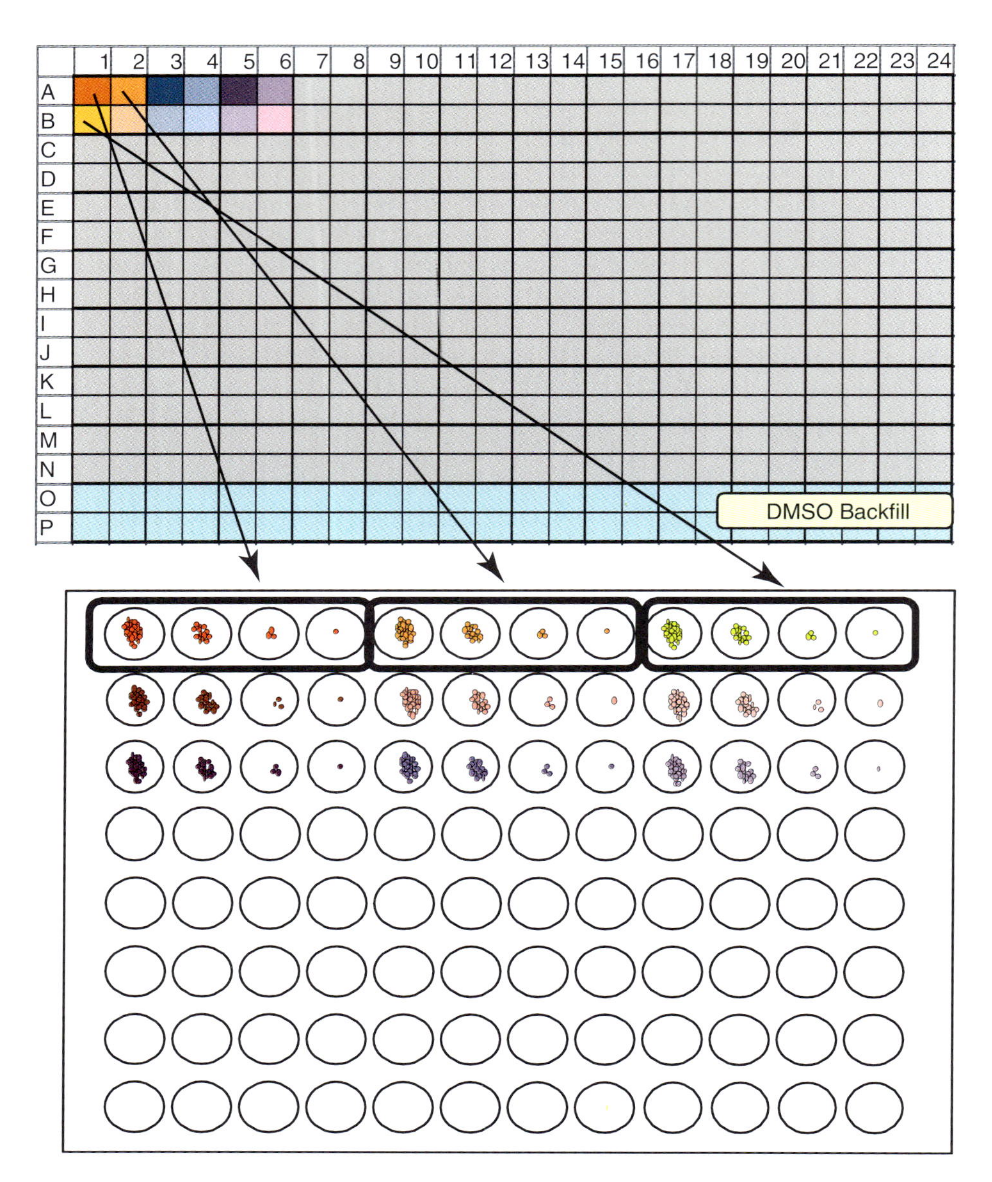

Figure 11.9 Quadrant pre-diluted source plates containing compounds in set of 2 × 2 wells were prepared in typically 100-fold steps by off-line process. In this example, a different number of droplets from each dilution is dispensed into destination wells to form a 12 points dose response in a 96-well plate.

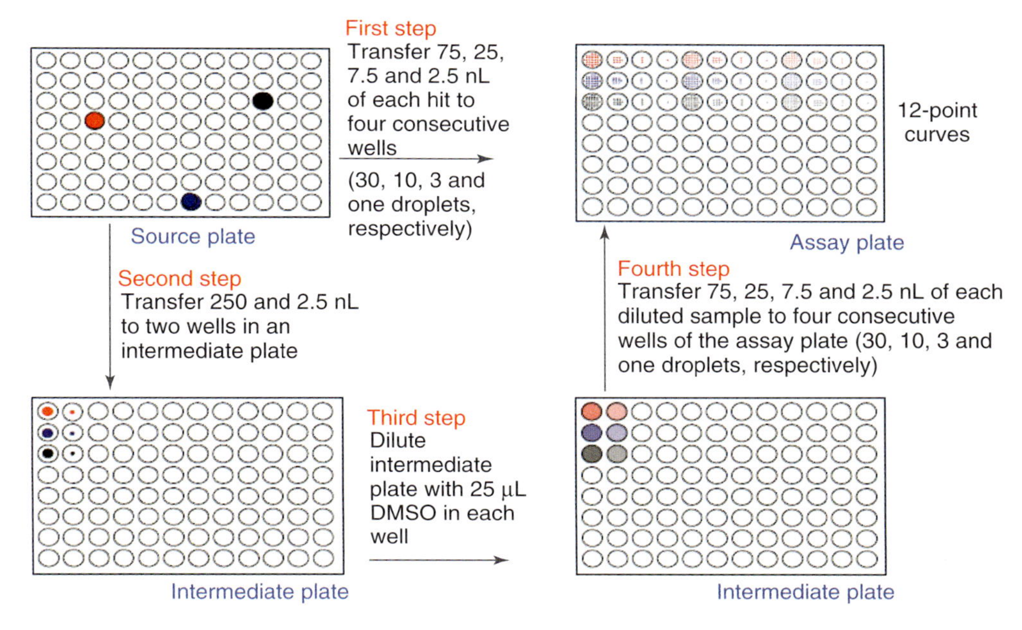

Figure 11.10 'Direct dilution method': the original compounds are transferred from the original source plates to first four locations in the assay plate and to two locations in the intermediate plate. The intermediate plate is then filled with 25 μL of DMSO in each well. This provides a 1/100 and 1/10 000 dilution of the original library material. The intermediate is then used as a source plate to fill in the remaining wells of the assay plate.

direct dosing into cell based assays has provided some opportunity to overcome these issues, albeit not without some other obstacles.

Nevertheless, the benefits of applying the acoustic workflow outweigh the limitations imposed on the screening logistics of the cell-based assay. Most cell-based applications have limited DMSO tolerability (0.1–0.3% v/v), and the ability of direct dosing into the cell-containing destination well negates the need for intermediate dilution with culture media, which less soluble compounds tend to precipitate, thereby affecting the final expected compound concentration. Furthermore, the liquid-handling process is simplified, with reduced run time, as a result of removing the intermediate dilution steps.

Despite the benefits of ADE, one key limitation of this technology is the requirement for the destination plate to be presented to the ADE instrument in an upside-down orientation. With adherence cell-based assays, complete culture media removal prior to acoustic transfer is not feasible due to the toxic effect of high local DMSO concentration formed directly above the culture cells. To protect the potential cell toxicity effect, a work-around has been developed for 96-well-based assays. Acoustic transfer can be successfully performed through partial media removal prior to presenting the cell-containing plate to the reformatter. In our experience, the tight tolerance means that the culture media remaining in the well of a standard 96-well plate after partial removal will be in the range of 30–35 μL. Volumes above this range will result in culture media falling down from the destination wells onto the instrument source stage while the plate is upside-down during acoustic transfer. This obviously compromises the integrity of the compound source plate and causes internal contamination of the acoustic instrument. A lesser volume will be insufficient to protect cells from direct contact with the DMSO/compound droplets and results in cell toxicity. Nevertheless, the acoustic dosing approach has proved to be very successful through careful workflow design and the utilization of robust liquid-handling devices throughout the cell-seeding and media-removal processes.

In contrast, the culture media retention issues within a standard 384-well plate are more complicated. With the small well cross-section area, the ability to retain liquid in 384-well plates while upside down is independent of the retaining volume up to the maximum capacity of the well but dependent on the surface treatment of the plate offered by plate manufacturers. As a result, it is necessary to evaluate the suitability of the microplate product for their application to avoid transfer failures and contamination of the instrument.

Validation of compound transfer by ADE in both 96- and 384-well cell-based assays represents only the first step in defining a successful screening workflow using this new exciting technology platform [11]. The remaining criteria are dependent on the choice of end-point measurement for the given assay design, that is, population-based measurement or a spatial-dependent readout, such as high-content cell imaging. This critical issue is manifested by the fluid dynamics of the upside down meniscus of the retaining volume during the acoustic transfer, as illustrated in Figure 11.11A. At 30 μL of retaining volume, the well surface is adequately covered with fluid in the presence of typical culture media components, such as 10% v/v serum. However, with the mass of the liquid and the surface tension between the liquid and the vertical wall surface, the meniscus is

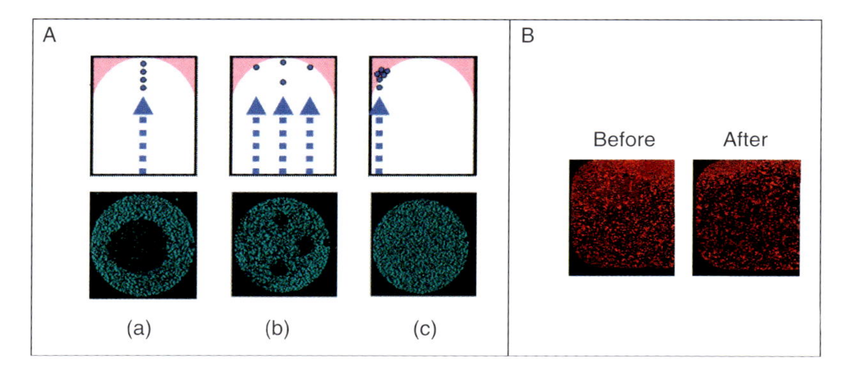

Figure 11.11 Images showing nuclear staining of the cell monolayers in a 96 (A) and 384 (B) well plate after dosing DMSO using acoustic drop ejection. (a) DMSO was dosed with 96-well default well definition setting, that is, in the center of well. (b) The volume of DMSO was distributed between four different target locations within a well. (c) DMSO was transferred into destination well with well offset.

significantly distorted, with the center of the well severely deprived of fluid coverage when the plate is inverted. During the acoustic transfer, droplets containing DMSO-solubilized compounds were deposited directly onto the cell monolayer as illustrated in Figure 11.11A (a, b). Although the droplets are only 2.5 nL in volume, this results in a high localized DMSO concentration at the receiving zone, and this toxicity caused the cells to detach from the vicinity of this area. In our experience, most of these detached cells recovered and re-attached in other parts of the destination well, hence not affecting population-based measurement (e.g., colorimetric assays) in which the total response is independent of the spatial distribution of the cells. However, high-content imaging measurement often focusing on a small area of the well ($\sim$1–2 mm^2) is thus dependent on a consistent distribution of cells across the well surface to represent the biological responses in the assay. This artifact significantly impinges on the quality of the measurement with a detrimental effect on subsequent data analysis if the cell distribution effect was not eliminated as part of the workflow validation.

Nevertheless, this problem was resolved in 96-well plates by adapting the high spatial resolution of the acoustic transfer instrumentation to re-focus the droplet disposition location to the shoulder of the inverted meniscus Figure 11.11A (c). This allows the higher liquid depth of the meniscus shoulder to absorb much of the kinetic energy of the ejected droplets and hold the DMSO-based solution at the lower part of the meniscus until the plate is reverted to the original orientation, and thermodynamic mixing then proceeds rapidly. The mixing can be further enhanced by the re-introduction of culture medium to restore the original cell culture volume.

In contrast, this cell detachment phenomenon does not take place in standard 384-well cell culture plates provided that the volume within the destination well is 30 μL or higher. Additionally, it is important that the culture medium remains adhered to the bottom of the well; the depth of the liquid is at sufficient depth to absorb the energy of the ejected droplets and protect the cell monolayer from being damaged in the process as illustrated in Figure 11.11B.

In summary, acoustic transfer technologies have provided an unprecedented capability in cell-based screening with higher data quality and screening throughput. Through the 'non-contact' property of the technology, it facilitates a 'zero carryover' workflow and allows the transfer of an accurate quantity of compound into the cell-containing wells without intermediate dilution steps. This technology platform has revolutionized cell screening and greatly enhances data consistency and reproducibility (Figure 11.12) in the drug discovery process.

11.5.3
Cell-Based Combination Screening

In large numbers of chronic diseases, clinical efficacy is only achievable with multi-drug therapy. Most of these combination treatments were established through empirical off-label small clinical trials over many decades. However, there is a strong growing demand and expectation from patients and healthcare providers for strong synergistic combination to be identified in the pre-clinical drug discovery phase

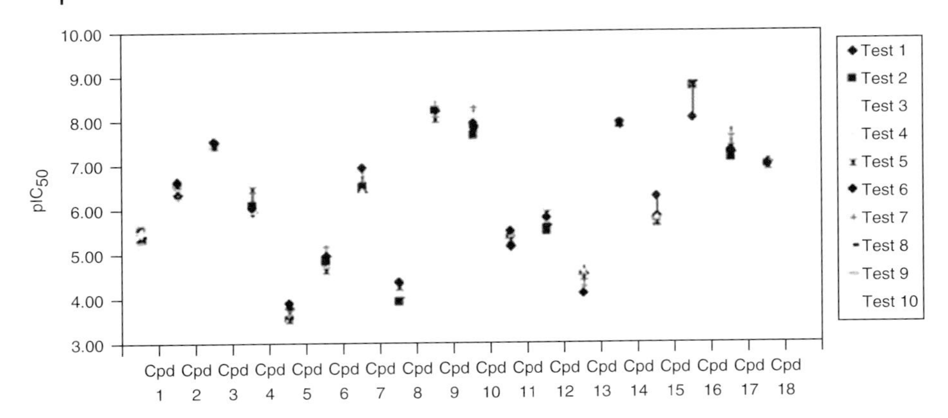

Figure 11.12 pIC_{50} distribution of a compound set in a 384-well kinase cell-based assay.

and translated to higher success rates through appropriate combination clinical trials. This development has become more prevalent in the oncology area due to the complexity of the disease. With the failure of some early target-specific agents, the emergence of a personal healthcare approach demands an early understanding and stratification of patient groups in the development of specific-combination therapeutic regimes.

Combination screening with conventional liquid-handling instruments involves intermediate dilutions and complicated instrument programming and workflows to deliver variable concentrations of multiple agents into each destination well. The workflow is cumbersome and suffers from severe limitations to the resolution of the dose ranges and stability of the compound mixtures in the studies. In contrast, with acoustic transfer, simple software applications can calculate the desirable mixture ratio and the solvent backfill to generate worklist files which directly guide the 'on-demand' direct dosing of each agent into the destination wells. In our experience, the acoustic transfer technology has provided an enabling platform for high-throughput combination screening to evaluate a large number of potential drug combinations in a pre-clinical setting with enhanced data quality, which is impractical with conventional liquid-handling workflow.

11.6 Emerging Applications

11.6.1 Acoustic Transfer of Aqueous Reagents and Biologics

Small-molecule therapeutics has been the focus of drug discovery since the birth of the pharmaceutical industry. With recent advances in molecular biology and antibody technologies, biologics (including antibody therapeutics and silencing RNA) has gradually shared the center stage of blockbuster drug status in certain disease sectors. In order to streamline the evaluation of these agents alongside DMSO-soluble small molecules, both leading acoustic instrument manufacturers have developed capabilities to handle dispensation of biologics, which are often supplied in aqueous preparations. EDC's ATS-100™ platform offered the customer the ability to develop liquid calibrations for any liquids compatible with the acoustic ejection technology. In contrast, Labcyte developed and released the manufacturer-validated 'OMICS calibration set,' which provided support for a defined set of common liquid types. The choice and success of these two approaches is highly dependent on the types of liquids required and the complexity of the liquid class used in the defined operation.

11.6.2 Cell Suspension Transfer

Routine applications of acoustic transfer technology in cell-based screening has concentrated on the transfer of pharmacological agents being evaluated in the process. With the high accuracy and precision of acoustic transfer, it has been demonstrated that, dependent on the cell density of the cell preparation, small volumes or single cells can be transferred successfully to the receiving well plate without cell damage. This approach can greatly facilitate the development of high-density arrays (e.g., 1536-well or 3456-well plates) or single-cell studies in cell culture arrays, as illustrated in Figure 11.13. It is anticipated that further advances of this approach would facilitate the development of tissue scaffolds and micro-tissue structures where precise deposition of specific cell types are required to form functional artificial tissue assemblies.

11.6.3 Matrix Deposition for MALDI Imaging Mass Spectrometry

Imaging mass spectrometry is an emerging technique to measure biomolecules and compound abundance with spatial information across tissue sections from animal studies or human biopsies. Using conventional techniques, a small volume of MALDI matrix reagents were deposited on the specimen manually. This approach limits the spatial resolution and accuracy of the analysis, as the quality of the

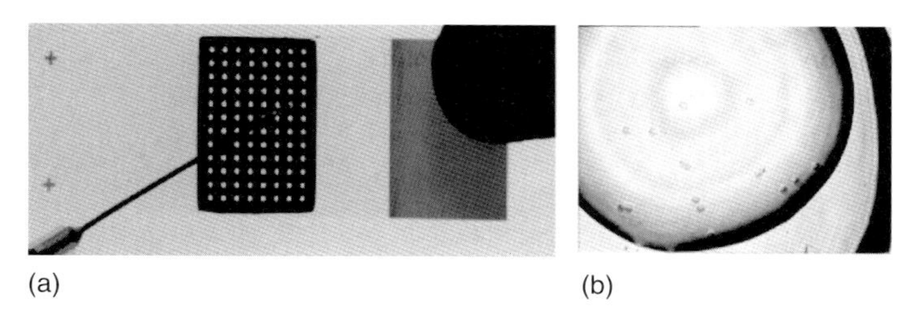

(a) (b)

Figure 11.13 (a) Nanoscale cell-based gel slide. (b) 50 nL droplet containing cells in the microwell formed by gel-based material.

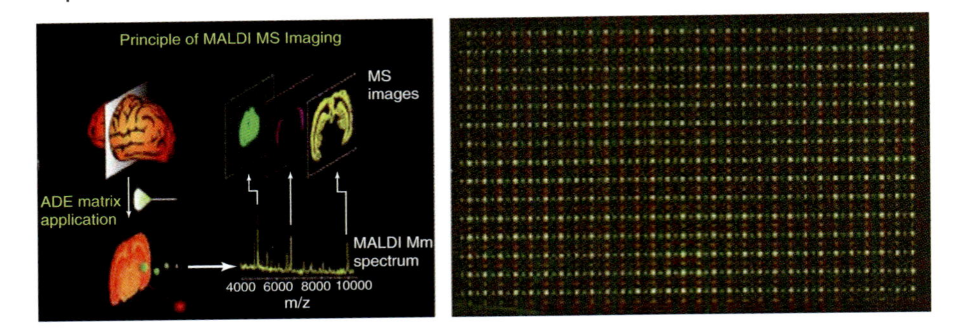

Figure 11.14 Spot-on-spot array created with Portrait® 630 spotter (170 pL sample volume, 800 μm column spacing, and 400 μm row spacing). Average diameter of single spots (red and green) = 168 μm; average diameter of yellow spots (red plus green) = 235 μm.

matrix deposition is compromised by the size of the droplet and reagent diffusion due to the large reagent volume. An adaption of the ADE instrumentation, the Labcyte's Echo Portrait™ 630 spotter delivers precise volumes of different MALDI matrices in droplets as small as 170 pL, with flexibility for a wide range of sample formats and spotting patterns. The high accuracy and precision of the acoustic transfer technology enables the gradual deposition of a MALDI matrix on the imaging sample with significantly less dispersion than that for other conventional techniques [8]. This facilitates a higher sensitivity and spatial resolution for the detection of the analytes by mass spectrometry. The instrument can also 'output' the MALDI reagent pattern coordinates. This data can be used to assist the sampling and analysis of the mass spectrometer to facilitate the reconstruction of spatial distribution of the analyte signal across the specimen (Figure 11.14). This technique has a potential high impact on drug discovery in understanding the drug exposure and cellular response to the treatment by direct measurement of analytes in the tissue sections being examined.

References

1. Fox, S., Farr-Jones, S., Sopchak, L., Boggs, A., Wang Nicely, H., Khoury, R., and Biros, M. (2006) High-throughput screening: update on practices and success. *J. Biomol. Screen.*, **11**, 864.
2. Wood, R.M. and Loomis, A.L. (1927) The physical and biological effects of high frequency sound waves of great intensity. *Philos. Mag.*, **4** (7), 417–436.
3. Whittingham, T.A. (2007) Medical diagnostic applications and sources. *Prog. in Biophys. & Mol. Biol.*, **93**, 84–110.
4. Kious, W.J. and Tiling, R.I. (1996) *This Dynamic Earth*, U.S. Geological Survey, Washington, DC.
5. Krause, K.A. (1973) Focusing ink jet head. *IBM Tech. Disclosure Bull.*, **16** (4), 1168.
6. Ellson, R., Mutz, M., Browning, B., Lee, L., Miller, M., and Papan, R. (2003) Transfer of low nanoliter volumes between microplates using focused acoustics – automation considerations. *J. Assoc. Lab. Autom.*, **8** (5), 29–34.

7. Jarman, C., Bruzzel, V., Davies, A., and Pickett, S. (2009) Miniaturizing RNAi assays: Acoustic Droplet Ejection (ADE) enables efficient and reproducible siRNA library screens at low volumes. 59th Annual Meeting of the American Society of Human Genetics (AFSG), Honolulu, HI.
8. Pickett, S., Chaurand, P., Huang, R., Stearns, R., Browning, B., Hinkson, S., Ellson, R., and Capriolib, R.M. (2006) Acoustic droplet ejection enables precise timing and positioning for deposition of matrix to optimize MALDI tissue imaging. 17th International Mass Spectrometry Conference, August 27–September 1, 2006, Prague, Czech Republic.
9. Spicer, T., Fitzgerald, Y., Burford, N., Matson, S., Chatterjee, M., Gilchrist, M., Myslik, J., and O'Connell, J. (2005) Pharmacological evaluation of different compound dilution and transfer paradigms on an enzyme assay in low volume 384-well format. Poster Presented at Drug Discovery Technology, August, Boston, MA.
10. Harris, D. and Mutz, M. (2006) Debunking the myth: validation of fluorescein for testing the precision of nanoliter dispensing. *J. Assoc. Lab. Autom.*, **11** (4), 233–239.
11. Grant, R., Roberts, K., Pointon, C., Hodgson, C., Womersley, L., Jones, D., and Tang, E. (2009) Achieving accurate compound concentration in cell-based screening: validation of acoustic droplet ejection technology. *J. Biomol. Screen.*, **14**, 452–459.
12. Blaxill, Z., Holland-Crimmin, S., and Lifely, R. (2009) Stability through the ages: the GSK experience. *J. Biomol. Screen.*, **14** (5), 547–556.
13. Oldenburg, K., Pooler, D., Scudder, K., Lipinski, C., and Kelly, M. (2005) High throughput sonication: evaluation for compound solubilization. *Comb. Chem. High Throughput Screen.*, **8**, 499–512.
14. Walsh-Haney, H.M.A. and Coticone, S.R. (2007) Correlation of forensic anthropologic findings with DNA profiles obtained from cold cases. *Proc. Am. Acad. Forensic Sci.*, **XIII**, H37.
15. Takach, E.J., Zhu, Q., Yu, S., Qian, M., and Hsieh, F. (2004) New technology in tissue homogenization: using focused acoustic energy to improve extraction efficiency of drug compounds prior to LC/MS/MS analysis. 52nd ASMS Conference on Mass Spectrometry, Cambridge.
16. Nixon, E., Holland-Crimmin, S., Lupotsky, B., Chan, J., Curtis, J., Dobbs, K., and Blaxill, Z. (2009) Applications of adaptive focused acoustics to compound management. *J. Biomol. Screen.*, **14** (5), 460–467.
17. Allenby, G., Freeman, A., Martin, T., and Travis, M. (2010). Solubility Rules OK! What Ultrasonic Mixing Can Do for You (Poster).
18. Shieh, J., Vivek, V., and Bramwell, J. (2010) Improve the Integrity of Your Fragment Library for Fragment Based Drug Discovery (Poster).
19. Wingfield, J., Jones, D., Clark, R., and Simpson, P. (2008) Assay sciences: a model for improving efficiency through centralization. *Drug Discov. World*, **10** (3), 65–70.
20. Eriksson, H., Brengdahl, J., Sandström, P., Rohman, M., and Becker, B. (2009) Validation of low-volume 1536-well assay-ready compound plates. *J. Biomol. Screen.*, **14** (5), 468–475.
21. Comley, J. (2007) Serial vs direct dilution: time to apply new thinking to IC50 determination and dose-response analysis?. *Drug Discov. World*, Spring, 36–50.

12 Enhancing Biorepository Sample Integrity with Automated Storage and Retrieval

Johann van Niekerk

12.1 The Emerging Growth of Biobanking

12.1.1 The New Face of an Old Practice

The archiving of biospecimens for use in research, or biobanking, is not a new concept. Collections of biospecimens, such as autopsy or histological material, have been in existence for decades, and are often managed in multiple locations around the globe for biomedical research in the public and private sectors. However, this established practice of archiving biological material has recently been touted as one of the ten world-changing ideas in 2009 by TIME magazine [1].

A biobank is an organized collection of biospecimens and associated data. This material can be used for clinical research, criminal investigation, human identification, diversity studies, or other purposes. The collections typically consist of many different types of biological biospecimens (e.g., tissue samples, DNA, and blood) and associated information (e.g., health records, diet and lifestyle information, family history of disease, gender, age, ethnicity, etc.) [2, 3].

12.1.2 Source Material for Post-Genomic Age Research

In addition to molecular information, scientists are also analyzing a vast amount of clinical information from patient records and clinical trials. From this data, it is possible to identify patterns that provide a pathway to understanding disease sub-types, and potential strategies for diagnosing and treating diseases in new and more effective ways.

Human biospecimens can provide a bridge between emerging molecular information and clinical information by enabling researchers to study the molecular characteristics of an actual disease and then correlating those patterns with what is known about the clinical progression of the disease.

Management of Chemical and Biological Samples for Screening Applications, First Edition.
Edited by Mark Wigglesworth and Terry Wood.

The National Cancer Institute (NCI) Office of Biorepositories and Biospecimen Research (OBBR) has identified five critical uses of archived human biospecimens. Specifically, human biospecimens can be used to:

- Identify and validate drug targets
- Identify disease mechanisms
- Develop screening tests for biomarkers associated with certain sub-types of a disease
- Group patients based on their genetic characteristics and likelihood of positive response, for testing of new drugs
- Group patients based on the biomarkers of their disease to determine which treatment is appropriate [4].

Biospecimen science [5], which is dependent on high-quality specimens from biorepositories, and the science of biobanking are developing concurrently. Biospecimen science and biobanking both stimulate and are co-dependent on one another, and it is this mutually beneficial concurrent effect that underlies the emerging growth of biorepositories.

12.1.3
Different Operational Models in Biobanking

Biobanks are developed as repositories of appropriate biospecimens to provide research material for hypothesis testing in a research environment. This implies that each type of biobank will be established according to a specific strategy with specific demands on quality and annotation of the collected biospecimens, resulting in heterogeneous concepts in biobank structures and research goals. Based on collection method and study objectives, human biospecimen-oriented biobanks may be classified as follows [6]:

12.1.3.1 Population Biobanks

The primary goal of population biobanks is to obtain disease biomarkers of particular susceptibility and population identity. Their operational substrate is typically germinal-line DNA from a huge number of healthy donors, representative of a country, region, or ethnic cohort [6].

As such, a population biobank is a collection of biospecimens that has the following characteristics:

- The collection has a population basis
- It is established to supply biospecimens for one or multiple prospective research projects
- It contains biospecimens and associated personal data which may include, or be linked to, genealogical, clinical, or lifestyle data, and which may be regularly updated
- It receives and supplies specimens in an organized manner [7].

12.1.3.2 Disease-Oriented Biobanks for Epidemiology

Disease-oriented biobanks for epidemiology are focused on biomarkers of exposure. These biobanks rely on a large collection of biospecimens, and generally follow a healthy exposed cohort/case-control design. Whereas the cohort study is concerned with frequency of disease in exposed and nonexposed individuals, the case-control study is concerned with the frequency and amount of exposure in subjects with a specific disease (cases) and people without the disease (controls). These studies typically explore germinal-line DNA or serum markers and a large amount of specifically collected data [6].

12.1.3.3 Disease-Oriented General Biobanks

Disease-oriented general biobanks focus on biomarkers of disease as found in prospective and/or retrospective collections of tumor and no-tumor samples and their derivatives (i.e., DNA, RNA, proteins). These biospecimens usually have associated clinical data and may in some instances be associated to clinical trials [6]. Disease-based biobanks (which also may be referred to as *clinical biobanks*) can be used to discover or validate genetic and nongenetic risk factors without having to wait for great lengths of time and spend large efforts on a longitudinal, prospective collection [8].

12.1.4 Why Are Modern Biobanks Needed?

Medicine is moving beyond the 'one size fits all' approach to improving treatment efficacy, minimizing side effects, and reducing the economic cost burden to public and private institutions worldwide. Future healthcare is focused on understanding the complexity of critical diseases and then using this knowledge to develop new diagnostic markers which improve disease prevention and allow earlier, more successful diagnosis and treatment.

Translational research is highly dependent on the availability of large repositories containing high-quality biospecimens and their associated data. Modern biobanks are crucial to supporting and accelerating this medical progress by providing well-characterized biospecimens, accurate annotations, precise validation tools, and advanced technology within a secure, ethically governed arena [9].

12.1.5 The Quest for Biospecimen Quality in Biorepositories

Many biobanks follow their own protocols and generally implement their own quality criteria. Consequently, the integrity, quality, and performance of individual biospecimens and of whole collections can vary widely. Experimental investigations using these resources, therefore, lose their statistical power owing to problems arising from mislabeling, data error, and data loss [10].

The quality and consistency of biospecimens (within a biobank or between different biobanks) will directly impact assay performance and the research performed or treatments developed with them. Proper collection, processing, storage, and tracking of biospecimens are critical components of biomarker-related studies. Rigorous quality control during the life cycle of a biospecimen assists in providing the assurance that the biospecimen is fit for purpose and fully comparable with specimens in the same or in other collections [10].

In many cohort studies, biological specimens are being stored without specific plans for analysis. The collection process typically takes years, and in some cases decades. Therefore, the scope of the end product anticipated as required for research is often kept as wide and as open as possible while endeavoring to maintain the biospecimen quality at as high a level as possible. The advantage gained from this approach is that the biospecimens may prove to be useful for multi-center research efforts that were perhaps not foreseen at the start of the collection [6].

Researchers increasingly require aliquots of biospecimens withdrawn from more than one biobank. They require this aggregation to translate the human genome sequence into health benefits and to increase our understanding of basic human biology. This compilation of resources from multiple biobanks allows investigations:

- to attain the statistical quality required to identify moderate or small risk factors, especially in investigating complex disease,
- to replicate the findings of previous studies,
- to identify gene:gene and gene:environment interactions via the many new large population-based studies that have been planned and started, and
- to investigate less common diseases (or phenotypic/genotypic combinations) [10].

With the biorepository so prominently featured in the genomic age, there is an implicit mandate to assure harmonization of processes and continued research to help ensure that biospecimen quality is maintained at the highest level possible. Prominent institutes and organizations, such as the NCI in the United States of America, the Biobanking and Biomolecular Resources Research Infrastructure (BBMRI) in Europe, and the International Society for Biological and Environmental Repositories (ISBER) on an international level, are promoting quality control and assurance in biobanking practices through the development of best-practice guidelines. The aim of best-practice guidelines is to ensure consistency of all processes for biospecimens and data within each biobank, which will be a critical factor in determining the quality of biospecimens available for research and the utility of the information derived from each specimen [11].

In the next section, the potential use of system and process automation as a means to help maintain and improve the value of biorepository content is explored by examining the beneficial effect that this may have on biospecimen integrity and stored object logistics.

12.2 Automated Storage and Retrieval in a Biorepository

Man has harnessed the benefits of mechanization in many forms to reduce drudgery, repetitive task, and working in inhospitable environments. In fact, it would be hard to imagine modern life without either the familiar domestic appliances or industrial machines that have helped to dramatically reduce personal strain and increase daily productivity. In the following section, the advantages of automated storage and retrieval of biospecimens over manual storage operations are explored.

12.2.1 Inventory Management

The practice of inventory management is a prime example of work consisting of repetitive tasks prone to human-induced error and variation.

In an industrial or supply chain setting, an automated inventory management system comprises an inventory management database, an order control module, and an alarm module. The inventory management database stores inventory information required for inventory management, which includes product descriptions, storage locations, and business rules ranging from the preferred conditions under which products must be kept, to release authorizations. The order/retrieval control module is the main interface between a user and the storage system, and serves to calculate and deliver an order request based on storage location data contained in the inventory management database, and governing business rules. An alarm module generates alerts or notifications when pre-defined storage conditions or other exceptions occur within the storage area.

12.2.2 Automation – Self-Controlling Processes

Automation implies not only a suitable form of mechanization that would reduce human involvement in strenuous tasks, but also refers to the reduction of mental and sensory requirements to accurately perform repetitive tasks. In essence, the operation of the control system of an automated mechanical device is similar to that of both the human cerebellum, which processes input from other areas of the brain, spinal cord, and sensory receptors to provide precise timing for coordinated, smooth movements of the skeletal muscular system, and the cerebral cortex, which stores and processes memory, attention, perceptual awareness, and thought. The obvious advantage of a computerized control system over the human brain is in the way motor actions and reactions, based on sensory input, are executed: computerized control systems are generally hard-coded to continuously perform pre-configured tasks based on established rules, without exception and bias. Automation therefore allows unattended operation of pre-defined processes under both normal and abnormal conditions. In the context of inventory management, once computerized order fulfilment is accomplished, complete walk-away automation is achieved with the aid of robotics, the overall functionality of which may be referred to as an *automated storage and retrieval system*.

12.2.3 Maintaining Biospecimen Value

To understand how the value of biospecimens may be enhanced through automated storage and retrieval in a biorepository, it is important to consider sample integrity and the integrity of the data associated with the sample. The influences impacting on the quality of the biospecimen and the integrity of the accompanying data of the biospecimen during storage operations are considered below.

12.2.3.1 Biospecimen Integrity

Storage-related biospecimen management operations are dynamic actions or events, which may cause changes in the environmental conditions around the stored content. The retrieval of sample containers for reformatting purposes, such as cherry picking, storage space optimization, or sample culling, causes such containers – and invariably the neighboring containers in a container rack – to be moved from their stored location to another area where these operations may be performed more easily.

In the case of manual storage systems, this relocation generally means that biospecimens initially move from a cold storage environment into a relatively warm environment at a temperature to which humans are accustomed. Not only are the retrieved sample containers subjected to a higher ambient temperature, but their relatively cold surfaces cause condensation of water vapor from the surrounding air around them. If sample containers that are moist with condensed water, or dew, are transported back into the cold storage environment, the moisture brought in with them will freeze on their own, or adjoining, surfaces. If such containers are stuck to adjoining surfaces by ice, this may cause difficulty during subsequent attempts to retrieve sample containers. Further complications arise when moist room air flows into the storage area and hoar frost is formed on storage area surfaces that are cooled to below the dew point of the room air while also being below the freezing point of water. Hoar frost on sample containers is notorious for obscuring identification labels. Difficult-to-read labels, in turn, lead to prolonged search time, which sets in motion a cycle of more condensation and frost formation, longer search time, and eventually, increased biospecimen temperature during retrieval from storage. The entire content of a freezer may be subjected to the ensuing temperature fluctuations caused by cumulative inefficiencies in terms of retrieval operations. A resultant loss in trust in the quality and value on the affected biospecimen population, caused by unstable climate conditions, is inevitable.

Stable Storage Conditions The benefits of stable storage conditions for biospecimens are exemplified by dry-state preservation and cryopreservation at $-196\,^{\circ}$C, which share a similar mechanism of protection. Storage at cryogenic conditions can maintain biospecimens in a glassy or vitreous state, which means that the stored material is in an amorphous (noncrystalline) solid form. At such low temperatures viscosity is high and diffusion is insignificant over less than geological time spans [12]. In other words, chemical reactions, and therefore damage due to molecular mobility, are improbable in a time frame of centuries. However, if moisture is added to the biospecimen stored under desiccated conditions, or if a temperature excursion above the glass transition temperature of water (nominally −135 °C) occurs, reactivity is re-established and not necessarily at infinitesimally slow rates [13].

Stability and Homogeneity of Storage Climate While the emphasis is on providing a storage environment with a suitably low temperature for specific biospecimens, maintaining a stable biospecimen temperature is the ultimate goal. Storage conditions that will contribute to maintaining optimal biospecimen integrity are based on a stable, nonfluctuating temperature. This implies that biospecimens required to be stored under cooled conditions must be maintained at an appropriately low but stable temperature, where the temperature of the stored content remains homogeneous around an appropriate set point, whether in different storage locations in the same store or located in different stores.

The storage area temperature is determined by the air temperature in which the biospecimens are stored, which may typically be −20 or − 80 °C in an automated storage system. Sample containers stored in a high storage density robotic-amenable rack can provide a beneficial thermal mass of cold content inside a thermally insulated environment. The thermal mass of the stored content assists in resisting abrupt changes in temperature and will even out fluctuations in biospecimen temperature.

Sample containers with biospecimens traditionally stored at −20 °C, such as extracted DNA in a buffer solution, are robotically handled and stored in the same climate conditions. Sample containers with biospecimens traditionally stored below −70 °C, such as plasma, serum, and extracted protein or RNA in a buffer solution, are typically stored within a dry, frost-free environment at around −80 °C, while the robotic handling of these samples also takes place quickly within a dry, frost-free environment conditioned to −20 °C. This means that sample containers are only briefly subjected to a dry air temperature of −20 °C, which is 60 °C above storage area temperature, instead of normal room temperature, which could be upwards of 100 °C above the storage area temperature.

Given the combined thermal mass of the cold-stored content, the significantly lower difference in ambient temperature between storage area and where sample containers are handled, as well as the dry and frost-free controlled way in which biospecimens are handled in an automated storage and retrieval system, it is clear that a manually accessed storage system operating in a room temperature, non-humidity-controlled environment, cannot provide the same degree of temperature conservation and protection against biospecimen degradation as a dry −80 to − 20 °C humidity-controlled automated storage and retrieval system.

Aliquoting The outcome of a quality-control study to assess the impact of multiple freeze/thaw cycles on biospecimen integrity, such as performed by the Norwegian mother and child cohort study (MoBa) [14], emphasized the importance of storing multiple aliquots of the same biospecimen. The practice of preparing multiple aliquots of the same biospecimen is also central to a conservation strategy commonly employed in biobanks, where such aliquots are disseminated across different freezers and geographically separated locations for disaster recovery. Preparing multiple aliquots of a biorepository's master samples provides the biorepository operator with fast access to these aliquoted samples as well as minimizing freeze/thaw cycling of the master samples as a whole.

Biospecimen Storage under Desiccated Conditions It is important that desiccated biospecimens are stored in a stable and controlled low-humidity environment. DNA, while generally considered a relatively stable macromolecule when purified and stored dry, is very susceptible to water (even moisture) that may cause hydrolysis. Dry, automated DNA storage provides a much higher integrity storage environment than what is typically provided by manual freezer options, which tend to build up layers of frost and ice due to multiple freezer door openings and closings.

Maintenance of Low Humidity Cold Storage Conditions In cold stores, frost formation that may impair mechanical or robotic operation must be actively limited inside the storage area and the robotic handling area. Doing so will maximize the reliability of the robotic sample container transport. Frost build-up also makes it very difficult to identify objects in the store through barcode reading or 2D code scanning. Humidity inside the storage environment can either be passively controlled (through refrigeration) or actively controlled through an automated dehumidification process to assist in maintaining the moisture levels within an acceptable range.

Many of the variables induced by pre-analytical environmental conditions to which biospecimens are often subjected can be mitigated by ensuring that the biospecimens are maintained in highly stable storage conditions. Automated storage and retrieval systems provide the required long-term stable environmental conditions by employing reliable self-regulated and redundant climate control systems. By utilizing redundant environmental control systems, the long-term reliability of automated ultra-cold biorepositories is significantly improved in comparison with the rather poor reliability that is often experienced with conventional single-compressor manual freezers.

12.2.3.2 Data Integrity

In the context of the operational and logistical processes within an automated storage and retrieval system, the focus is on assuring that sample container storage locations and flow of objects within the store are maintained accurately and precisely. This does not include the handling of personal or clinical data, which is linked to the biospecimens themselves. The linking of biospecimen-specific data and the specific storage containers in which they reside is a laboratory information management systems (LIMS) function, where the biospecimen-specific data is intentionally anonymized or pseudonymized.

Biospecimen Tracking The automated reading or scanning of machine-readable codes makes it possible to ensure the completeness and accuracy of biospecimen identification and tracking in an automated storage and retrieval system. Automated storage and retrieval systems reduce human error or omission in terms of content management by consistently tracking specimens from the point where they enter the store until they exit the store. Transcribing errors, occurring when storage containers need to be manually labeled, are eliminated through the use of containers that are pre-labeled with machine-readable unique codes. The facility to read both linear barcodes (e.g., on tube racks) and to scan 2D codes (e.g., on bottom of tubes) inside an automated storage and retrieval system form the basis of sound unequivocal biospecimen management.

This level of pre-defined workflow-based object recognition ensures that the locations of plates, racks, and tubes are verified accurately and repeatedly during storage and before plate reformatting. In the same way, the presence and position of plates and their associated tubes after plate reformatting or cherry picking, before release from the store, is also verified.

Without exception, this consistency is difficult to imagine in operations where humans are responsible for sample tracking, especially in high-throughput systems when large numbers of transactions have to be logged.

Machine-readable codes find themselves at the core of object identity recognition and verification in automated storage and retrieval systems. Usually, sample containers, or racks containing storage tubes, enter an automated storage and retrieval system in a robotic-friendly rack. In this sense, robotic-friendly implies the following:

- the rack is of standard dimensions – that is, conforming to one or more of the ANSI/SBS Microplate Standards (ANSI/SBS 1-2004 through ANSI/SBS 4-2004) [15], thereby making it suitable for use on standard robotic platforms,
- individual tubes in the rack can be robotically accessed and reformatted among several racks, and
- the rack contains a machine-readable code, usually in the form of a 1D barcode, often with human-readable characters included.

It is not essential for individual tubes to each contain a unique machine-readable code, since the content of the tubes will likely be electronically mapped in a grid array position in the rack, which is in turn associated with the unique identifier on the rack. However, once reformatting of tubes take place inside the store, a unique identifier per tube – usually in the form of a 2D code – is highly desirable. 2D codes on multiple tubes can be scanned at the same time as the 1D barcode of the container rack is read. Therefore, 2D codes on tubes assist in both determining and validating the position of a specific tube in a given rack. Uniquely identifiable tubes are essential in verifying the correctness of reformatting actions, such as cherry picking, and act as a redundant means of data security in case of loss of original sample management data files.

As storage containers move through the storage system during sample storage, cherry picking, tube reformatting, or sample retrieval actions, unique identifiers in the form of 1D barcodes or 2D codes can be recognized and interpreted by laser scanners and optical code readers. By recording the unique identifiers on objects and associating it with related data (e.g., user identification, store temperature, date and time, etc.), a complete audit trail based on sample management functions in relation to storage system functions is established.

Audit Trails All functions related to the requirements of user identity and access management, where mandatory (e.g., Title 21 CFR Part 11 of the Code of Federal Regulations, which deals with the Food and Drug Administration (FDA) guidelines on electronic records and electronic signatures in the United States), are generally handled by LIMS or sample administration applications. Given the close connectivity between LIMS (or sample administration applications) and the system controller of an automated storage and retrieval system, it is possible for the storage system controller to inherit the essential user identity and access rules established and fixed in the LIMS. This ensures that unauthorized access to store content or sample location data can be restricted on an individual basis.

Also, user interactions with the store (for example, placing an order for specific specimens or retrieving the specimens from the store) and specimens under the custody of the user may be electronically recorded and archived. Since this happens repeatedly and by design in automated sample management systems, complete audit trails of store-related events are generated dynamically without requiring further thought or input from users.

LIMS Connectivity A key feature of biobank-oriented LIMS is their capacity to integrate with automated systems. Patient-related clinical information is stored in the LIMS and is linked with the biospecimens in a secure way. This ensures that a user may query the LIMS database to perform complex sample requests based on patient demographic and biospecimen type, and then be able to select and retrieve correct biospecimens from the biorepository without violating the privacy of a patient or cohort study participant. The LIMS must manage biospecimens as well as the associated patient consent and the chain of custody. The biobank must respect the patient's consent desires irrespective of where the biospecimen is located in the biobank.

A biorepository user may choose to store biospecimens based on a work list created by the LIMS by providing only a list of sample containers, together with their respective properties and unique identifiers, to the system control database of the automated storage and retrieval system. The automated storage and retrieval system then decides, based on pre-defined business rules, where the sample containers are stored. The LIMS may then be notified of the storage locations of each individual container. Alternatively, sample containers may be stored in clusters or groups according to patient groups or libraries, which in turn implies that more advanced business rules must be observed during storage and retrieval operations, such as conservation of storage location or space. In addition, as part of a conservation strategy or disaster recovery plan, aliquots of the same biospecimen may be distributed across separate storage locations in the same automated storage and retrieval system through reformatting of sample containers once inside the storage system.

Automated space optimization inside the storage system can be performed through a process of defragmentation, or compression, of stored content. Defragmentation can be performed either on rack level or on sample container level. Compression of store content may be required after extensive retrieval of sample containers from the store in cases of cherry picking, culling, and disposing of unwanted or inappropriate content (e.g., biospecimens where consent for their continued use in research had been withdrawn).

The synergistic relationship between a biobank-oriented LIMS and a biorepository in the form of an automated storage and retrieval system provides a complete sample administration solution, where each respective entity is responsible for sample tracking and audit trail creation in its own domain. Figure 12.1 provides a schematic overview of a biorepository user's interaction with an automated storage and retrieval system.

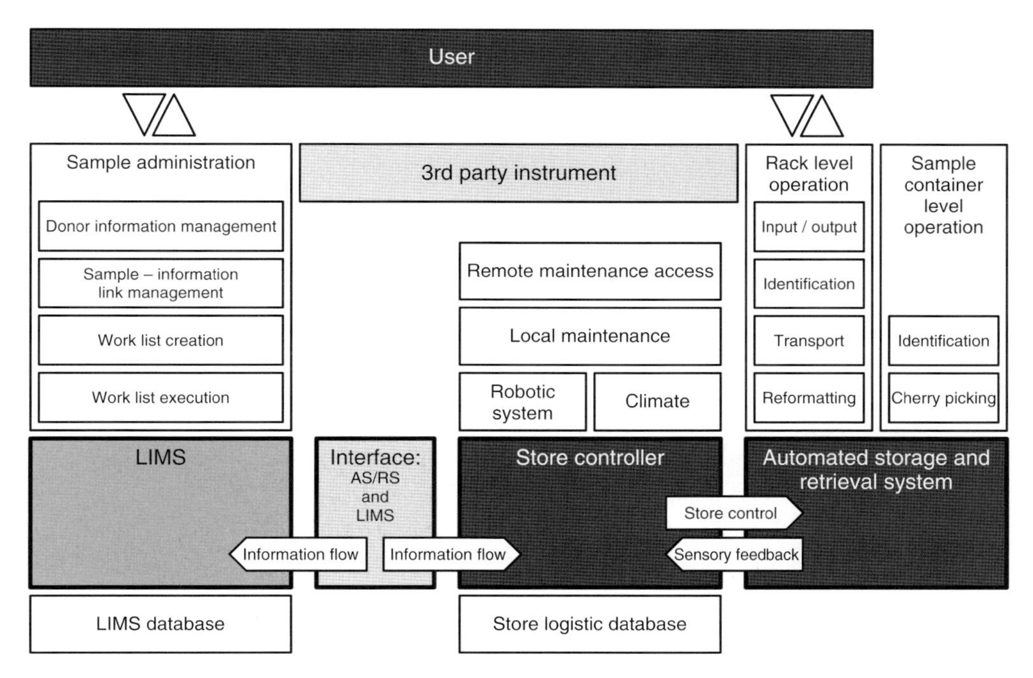

Figure 12.1 User interaction with an automated storage and retrieval system through a laboratory information management system (LIMS).

12.2.4
Advantages of Automated Biospecimen Management

12.2.4.1 Stable Storage Conditions

During normal conditions in an automated storage and retrieval system, storage-related operations are performed according to set procedures and in an environment where storage conditions are maintained close to a reference point and based on a closed-loop sensory feedback to the storage control system. Under abnormal conditions, the storage control system performs error detection and notification by issuing alarms to appropriate recipients.

Automated storage and retrieval systems are capable of providing a controlled climate in the pre-storage area where robotic handling of sample containers takes place. The temperature in the pre-storage/robotic area is typically set to around $-20\,^{\circ}$C. At the same time, the humidity in this area may also be controlled. The combined effect of low temperature and low humidity greatly minimizes the problems encountered with manual biospecimen storage and retrieval. Since stored biospecimens are not subjected to room climate conditions during reformatting procedures, their potential for heating up while briefly held in a frozen state outside the storage area, as well as any unwanted frost build-up, is significantly reduced compared to manual freezers. Generally, sample container reformatting procedures take place at a temperature of at least $40\,^{\circ}$C below room temperature. This ensures that the samples remain frozen during the reformatting steps and reduces the rate at which sample temperature rises while outside of the storage area. Since cold air contains less water vapor than warm air, the low temperature of the pre-storage area further assists in minimizing the amount of moisture entering the storage area. This effect can be enhanced if used in conjunction with active dehumidification of the pre-storage area.

Another advantage of automated storage and retrieval systems over manual storage and retrieval processes, which rely on human operators, lies in the resilience and robustness of robotics. Controlled mechanization is capable of replacing mundane human labor under inhumane arctic conditions and during anti-social hours. Such working conditions could easily lead to lapses in concentration or frustration in a human operator, which may lead to process errors, while robotics consistently perform storage sample handling procedures with precision and accuracy irrespective of adverse working conditions.

12.2.4.2 Operational Reliability

A properly designed and appropriately configured automated storage and retrieval system will ensure that the impact of operator errors, robotic malfunctions, disruption in power supply, and component failures in computers, control systems, and climate control system will not result in jeopardizing the integrity and security of the stored biospecimens.

Availability To reliably maintain storage conditions in a biorepository, robust storage system components are required that will provide an appropriate operational availability (i.e., the ratio of the total time a functional unit is capable of being used during a given interval to the length of the interval, or the proportion of time a system is in a functioning condition).

The notable paper by Groover *et al.* [16] describes a practical approach to evaluating freezer performance by using automated temperature-monitoring systems to identify which freezer units in a mixed population of more than 200 manually operated mechanical freezers have a greater likelihood of imminent failure. The problem of random and frequent component failures faced by many institutions that have large farms consisting of manually operated freezers is also apparent from this paper. Reliability is the cornerstone of a successful biorepository operation. The larger the collection of biospecimens that is stored and the longer the period of time over which they are stored, the more the problem of poor reliability is exacerbated to the detriment of the value of the biospecimen collection.

Redundancy To ensure continuity of storage conditions, biospecimen integrity and productivity maintenance, redundant operation of critical sample management functions is essential. Unattended management of duplicated cooling and system control database components is a common, but hugely beneficial, feature of automated storage, and retrieval systems. The system computers of an automated biorepository are usually supported by an uninterruptible power supply (UPS). As long as the system computers are on UPS, an operator will be able to quickly recover from a power failure, with system software assisting the operator in executing a recovering procedure.

Maintenance Of course, automated storage and retrieval systems are not infallible or impervious to breakdown. However, redundancy of critical functions, such as cooling or data management, as well as planned service and maintenance certainly minimize the risk of catastrophic failure.

If an automated biorepository is designed such that all of the functional modules (e.g., tube selector, input/output (I/O) module) reside within the storage system but are easily accessible from outside the system, maintenance can be performed more readily. Since all of the major system operations are decoupled from each other, each subsystem can be worked on independently. Ease of maintenance may be further increased by a modular design approach of the mechanical components where assemblies can be quickly removed and replaced if required. Equally important, the rest of the system should continue to be usable even when a module is not available. This means that the operator can continue other unaffected operations such as inputting and outputting sample containers even without the available use of a tube selector station, for example. All of these unique features of an automated biorepository work together to maximize system availability.

Remote Diagnostics A significant feature of some automated sample management systems is the ability to provide the user and the supplier's customer support group with remote diagnostic system monitoring and troubleshooting capabilities. With this capability, the supplier can remotely monitor the functioning of their automated systems located around the world and work closely with the system users and the supplier's field service engineers to provide routine maintenance on the system, download new system software packages, and monitor the performance of the user's system remotely with the consent of the on-site system operators.

12.2.4.3 Efficiency and Convenience of Operation

Space Savings Conventional storage systems generally require a lot of floor space in relation to the storage density that is achieved, due to the fact that stored content must be accessible to humans. Automated storage systems provide the benefit of compact design and offer maximum storage capacity per available laboratory floor space, providing variable footprint options and utilizing available room height.

Time Savings As noted earlier, the retrieval of items from conventional storage systems relies on a person-to-goods operation. Automated storage and retrieval involves a goods-to-person process. A goods-to-person process implies an order fulfillment strategy to reduce or eliminate time delays associated with order picking to improve efficiency and productivity. The most pronounced time saving effect experienced with automated retrieval operations comes about when these operations take place without human involvement or supervision. An order for a specified group of objects is placed as a work list from which a control system executes a series of automated and electronically recorded processes.

Storage-related activities, such as cherry picking and order fulfillment, may be scheduled to occur at a planned or convenient time that suits both the requirements of the sample manager as well as those of the researcher. During the output phase, labile biological materials are kept under controlled conditions right up to the point where they are required for analysis. Biospecimen selection or cherry picking is performed under repeatable stable and controlled temperature conditions, which is extremely difficult or impossible to achieve using manual labor. Biospecimens forming part of an order are subsequently maintained within pre-defined climate conditions inside the storage system until they are retrieved by an authorized user.

Ergonomics Manually accessed storage systems have a distinct disadvantage in terms of ergonomics. Users have to bend and stretch and encounter frozen environments when storing or retrieving samples. In this regard, manually accessed chest freezers suffer even more than upright freezers. In addition, users are exposed to cold surfaces of the freezer's storage compartment, as well as to cold content retrieved from the freezer. General purpose laboratory gloves are worn as a protective barrier for laboratory workers for activities involving potentially infectious materials, hazardous chemicals, and biological material, or when hands can come into contact with contaminated surfaces, equipment, or accidental spills. These gloves provide excellent tactile sensation and dexterity, allowing users to work with laboratory equipment and objects in a nonrestrictive way. However, general purpose laboratory gloves are not ideally suited to act as a thermal insulator against extremely cold objects, and are as such unsuitable for handling objects retrieved from ultra-cold storage conditions. Prolonged contact with the cold surface or object may lead to frostnip, which is first-degree congelatio, or frostbite. Second-degree injury occurs when the affected skin is frozen, usually resulting in blisters one to two days after the freezing event. Multi-layer insulated gloves designed to provide protection o the hands and arms from the hazards encountered when working with ultra-cold objects do impede dexterity, especially when working with smaller objects such as cryovials or biospecimen storage tubes.

Automated storage and retrieval systems eliminate the risks associated with the handling of ultra-cold objects experienced by store operators. Objects requested for retrieval from the automated store are cherry picked inside the cold environment and are automatically brought to an ergonomic retrieval height.

Energy Savings A single large storage system has a much lower surface-to-volume ratio, which allows less heat transfer into the low-temperature storage area than occurs with multiple smaller freezers for the same volume of stored content. The relatively small aperture through which samples are loaded into, or retrieved from, automated storage and retrieval systems compared to the doors or lids commonly found on manually access freezers limits the loss of cold air from the storage area. Such automated apertures further contribute to maintaining a stable storage area climate since they are opened and closed in a controlled manner to allow objects to enter or leave the storage area in the shortest possible time. Often, doors or lids of manually accessed freezers are opened for long periods of time when multiple stored containers are retrieved from the freezer. Since the storage area in such an instance is allowed to come into direct contact with the room temperature and humidity, energy consumption is increased during the process of bringing the storage area temperature back to its set point.

Condensers on many larger mechanically cooled storage systems are designed to dissipate their heat either into outside air or into piped chilled water. Smaller freezers, on the other hand, dispose of their heat into the building, putting an additional thermal load on the building air conditioning system.

12.2.4.4 Restricted Physical Access to Stored Content

Physical access to an automated storage and retrieval system under normal operational conditions is restricted. This has a side benefit in terms of improved specimen security. Unauthorized users can neither intentionally nor inadvertently access specimens that they do not have access rights to.

Automated storage and retrieval systems operate on a 'goods-to-person' principle. In this mode of operation, the desired objects are requested from storage in the form of an electronically compiled work list, and the necessary robotic reformatting of objects take place inside the store. Access to the samples is controlled by software and can be protected by security privileges. Once the order is fulfilled according to the work list, the objects are moved to a central collection point as soon as the user is ready to accept them from the store. Since all robotic object movements take place under controlled climate conditions, any adverse effects to biospecimens remaining in storage are minimized. Also, all object movements are electronically tracked. This means that both biospecimen and data integrity are fully maintained during an automated sample container retrieval process.

This is in sharp contrast to the usual 'person-to-goods' way of collecting objects from one or more manual stores. Not only does it take longer for the user to find and collect the individual objects in a work list, but the biospecimens – those to be collected as well as those that remain in storage – are subjected to both detrimental climate condition changes as well as intentional or unintentional removal of inappropriate samples. Again, the potential for error rises with the number of objects that are manually handled.

12.2.4.5 Redeployment or Reduction of Full-Time Employees

With the reduction in the number of full-time employees (FTEs) required to manage an automated biorepository, biobanks may realize a significant annual cost savings or cost avoidance. Alternatively, FTEs may be redeployed to another function in an organization as the organization transitions from manual storage of biospecimens to an automated biorepository.

12.2.4.6 Configuration Flexibility – Fit for Purpose

Automated storage and retrieval systems based on intelligent modular functional components allow expansion or changing of storage capacity at any time without any problem. Functional automated storage and retrieval units can be expanded step by step at a later date to meet growing storage or throughput needs. The flexibility offered by automated storage and retrieval systems based on modular, scalable components allows for an automated biorepository to be custom tailored for a specific storage application and then easily expanded at a later date to store a larger number of samples. The following sections give an overview of the general configuration of an automated biorepository.

12.3 Configuration of an Automated Biorepository

Ideally, an automated storage and retrieval system should be configured according to storage temperature, sample container type, and required storage capacity. Such a system should be capable of expansion, both in terms of increased storage capacity or enhanced performance.

12.3.1 Modularity

The term *modularity* is widely used in studies of technological and organizational systems. Product systems are deemed modular, for example, when they can be separated into a number of components that may be mixed and matched in a variety of configurations. The components are able to connect, interact, or exchange resources (such as energy or data) in some way, by adhering to standardized interfaces.

In general, decoupling of components or processes in a modular system is highly desirable. Unlike a tightly integrated product whereby each component is designed to work specifically (and often exclusively) with other particular components in a tightly coupled system, modular products are systems of components that are loosely coupled. In the inventory management processes that typically take place in automated storage and retrieval systems, decoupling allows economy of scale within a single system, and permits each process to operate at maximum efficiency rather than having the speed of the entire process constrained by the slowest component. In the context of automated storage and retrieval, economy of scale relates to the situation where a module required to perform a specific function is available at a time and in a place where a user needs it.

A well-designed automated biorepository will have many of its internal robotic systems and operations decoupled, making it possible to perform most operations in parallel. This leads to significantly enhanced sample access times and cherry picking rates. Simultaneously, the system can perform operations such as transferring storage trays while at the same time cherry picking tubes, plates, and vials.

In contrast to the beneficial decoupling of modular systems, close coupling of inter-dependent systems ensures that seamless communication between the related systems take place. An example of this beneficial coupling between inter-dependent systems is the LIMS – store controller connectivity. Such close integration between the biobank LIMS and the control system of the automated biorepository ensures complete and accurate sample tracking and creation of an audit trail without depending on user intervention.

12.3.1.1 Storage Chamber

In an automated biorepository, the storage chamber is usually a discrete unit comprising a freezer with a specific storage capacity and temperature and an adjoining area where a robot responsible for the transport of stored items can move. The basic requirements to be fulfilled by the storage chamber include providing a sufficiently insulated enclosure that will withstand temperature fluctuations during normal operation. The storage module should further provide a high storage density both per unit footprint, to make as efficient as possible the use of floor space, and per available height, to further maximize system sample capacity. The storage chamber should ultimately also provide a high storage density per unit storage volume, which serves as an indication of the energy efficiency of its operation.

12.3.1.2 Robotic Object Handling

Material storage at $-80\,^{\circ}$C presents some challenges that are not present in $-20\,^{\circ}$C applications. One challenge is the inability of robotics and electronics to perform well in a $-80\,^{\circ}$C environment. To overcome this, an automated biorepository typically separates the storage area into two distinct and physically separated areas. The actual storage chamber where the biospecimens are stored at $-80\,^{\circ}$C is typically a separate sub-compartment. No robotics or electronics are present in this compartment. The biospecimens are stored in trays or stacks that remain in fixed positions.

Transport robotics that are used to retrieve the sample containers from the −80 °C storage chamber are completely insulated from the −80 °C environment and are typically kept, instead, in a −20 °C environment. Robotic performance and longevity are well characterized at this higher −20 °C temperature.

A robot capable of transporting stored items between their storage locations and either an interface module outside of the automated biorepository or a sample container picking module performs its activities under the command of the store control system in the pre-storage area of a storage chamber. Usually operating at −20 °C, this robot ensures that the storage area temperature is maintained close to its set point by eliminating the need to expose the storage area to room temperature conditions. This robot is thus an essential component of an automated biorepository, ensuring that biospecimen integrity is maintained during storage and retrieval operations.

Sample containers are generally transferred into and out of an automated biorepository by a two-axis gantry system. This transport robotic system moves in both the horizontal and vertical plane allowing it to access any of the fixed storage locations in the storage area and move stored items between the storage locations and the various functional modules (I/O module, tube picker, etc.). The entire storage robotics operates independently from the functional modules. By being able to perform parallel operations, maximum throughput is achieved. Just as importantly, each robotic system can still operate even if for some reason the others cannot. This means that in most cases, a single robotic failure cannot disrupt the operation of the entire storage system.

A video system normally forms an integral part of an automated biorepository. A camera mounted on the storage robotics system enables monitoring of all crucial tray transfers into and out of storage. The video signal may be relayed to a video display or to a web-based application, which makes it possible to view store operations via a secure internet connection. This allows the user, and even the automation provider through remote diagnostics, to easily observe the storage system robotics in operation, and clearly detect any potential problems.

12.3.1.3 **Functional Modules – Execution of Specific Tasks**

Input/Output of Storage Containers The I/O module serves as the main sample I/O location. The module usually has a door to allow the user access to a buffer compartment which can hold multiple items to be stored or retrieved from the store. Usually, all sample container identity verification is performed in this module.

A conventional barcode scanner identifies tray and tube rack barcodes, while a vision system scans sample container 2D codes. All sample container 2D codes and their positions within their host racks are scanned when they first enter the system as well as when they exit via the I/O module to ensure the highest level of database integrity and to verify that the appropriate biospecimen was indeed retrieved. The results of this redundant scanning can be compared to the data that is stored in the biobank LIMS database to ensure the accuracy of biospecimens selected by the user. In standard configuration, the storage container identification scanner can distinguish between a tube present with an unreadable barcode, a tube present with a readable barcode, and a missing tube. This prevents unnecessary errors in the case when tube racks are introduced that are not completely full.

Sample Container Selection Cherry picking is performed by dedicated functional modules that are specifically designed for a particular container type. This design philosophy ensures a high level of reliability and performance. Specialized robotic hardware components and software algorithms allow for multiple, simultaneous container picking, look-ahead tray positioning, and tray-to-select-station delivery, to ensure secure high-speed retrieval of sample containers. This, in conjunction with the thermal mass of the stored content, limits unnecessary exposure of biospecimens to temperature fluctuations during cherry picking operations. The modules are connected to the same atmospheric conditions as the storage system, so biospecimens are not unnecessarily defrosted or exposed.

An automated biorepository is typically a completely pre-configured, computerized system that can operate unattended for an extended period of time. This is often due to the unique way that cherry pick jobs are handled. The user can typically create an unlimited number of cherry pick work lists with an unlimited number of requested sample containers. The way in which work lists are automatically processed, the use of parallel operations, and exceptionally fast cherry pick rates make this unattended, unlimited operation of a biorepository possible.

Once sample container selection and reformatting according to the work list is complete, the destination tray/rack can either be delivered immediately to the I/O module for automatic defrosting or placed back into storage for retrieval at a later time. The operator can then manually request via the operator interface that the tray/rack to be delivered to the I/O module. Once a storage tray is delivered to the I/O module, the sample container identifiers are scanned and verified against the database to ensure that the correct biospecimens were selected.

12.3.2
Scalability

Scalability refers to the characteristic of a system, module, or function that describes its capability to cope and perform under an increased or expanding workload. A system that scales well will be able to maintain or even increase its level of performance or efficiency when tested by higher operational demands. An automated biorepository whose design is based on standardized units or dimensions is typically able to expand due to its modular design, allowing flexibility and variety in use (Figure 12.2). An important benefit of scalability in an automated storage and retrieval system is the fact that a biobanking operation can invest in an automated biorepository with confidence that future storage or performance requirements will not outgrow the system's capacity for productive and efficient expansion.

Figure 12.2 An automated biorepository based on a modular, scalable design.

12.3.2.1 **Expansion of Storage Capacity**

A well-designed modular automated biorepository is typically able to expand fairly seamlessly to allow for increased storage capacity. Additional storage units can be added, when expansion is required, with very little disturbance to the ongoing store operation. Additional expansion storage units do not often have to be of the same size or capacity as the existing system. Multiple storage units can be bridged to an existing system while sharing any number of cherry picking and I/O functional modules, as well as a common software and inventory database.

12.3.2.2 **Increased Storage System Performance**

An automated biorepository offers functional flexibility that can adapt to changing user requirements. All major process operations such as barcode or 2D code scanning, cherry picking, and liquid handling are all performed in separate functional modules that are independent though integrated within an automated biorepository. If all the functional modules are completely self-contained and operationally independent from each other, the user can selectively incorporate the desired level of process functionality at any time. Operations such as barcode or 2D code scanning, cherry picking, and plate replication are all performed in separate functional modules that can be added at any time. This modular approach provides a high degree of flexibility while allowing the system to evolve as needed. As new technologies and process requirements develop, new functional modules can be developed and added, ensuring that the automated biorepository remains 'future proof.'

12.4 Conclusions

In the scope of a biobanking enterprise, industrial-scale automated inventory management solutions provide tremendous benefits compared to manual freezer storage in terms of preserving the long-term value of stored biospecimens and providing high-integrity access to the entire biospecimen collection at any time.

The larger the collection, the more important the beneficial contributions of automation become for biospecimens that are subjected to the differing events and conditions during storage.

The longer a well-maintained collection is in existence, the more valuable automation becomes since the biospecimens become more and more irreplaceable. Failure of the storage operation in such a collection is not an option, and therefore, highly robust and reliable storage systems should be considered and implemented to safeguard the long-term content of the biorepository. Reliability and built-in redundancy of the climate control system of an automated biorepository ensures higher sample integrity, sample availability, and usefulness over the expected long-term period during which biospecimens are stored. Therefore, a robust and reliable cooling mechanism that minimizes the risk of irrevocable damage to unique biospecimen collections is another strong advantage that automated sample management systems will provide.

References

1. Park, A. (2009) 8. Biobanks: 10 Ideas Changing the World Right Now. *Time Magazine*. 12 March 2009. *http://www.time.com/time/specials/packages/article/0,28804,1884779_1884782_1884766,00.html* (accessed 13 December 2010) [Online].
2. Lexicon (2005) P3G Observatory, *http://www.p3gobservatory.org/lexicon/list.htm* (accessed 13 December 2010) [Online].
3. FAQs Biobanking in British Columbia – a Deliberative Public Consultation (2009) *http://biobanktalk.ca/faqs/* (accessed 13 December 2010) [Online].
4. Biospecimen Research Network: Lifecycle of Biospecimens (2007) NCI Office of Biorepositories and Biospecimen Science (OBBR), *http://biospecimens.cancer.gov/researchnetwork/lifecycle.asp* (accessed 10 January 2011) [Online].
5. Biospecimen Research Network: Biospecimen Science. NCI Office of Biorepositories and Biospecimen Science (OBBR) (2007) *http://biospecimens.cancer.gov/researchnetwork/bs.asp* (accessed 3 November 2011) [Online].
6. Riegman, P.H.J. *et al.* (2008) Biobanking for better healthcare. *Mol. Oncol.*, **2** (3), 213–222.
7. Council of Europe – Committee of Ministers (2006) Recommendation Rec(2006)4 of the Committee of Ministers to Member States on Research on Biological Materials of Human Origin. CHAPTER V: Population Biobanks – Article 17, 15 March 2006, *https://wcd.coe.int/wcd/ViewDoc.jsp?id=977859* (accessed 15 December 2010) [Online].
8. BBMRI Work Package 3 – Disease-oriented Biobanks. BBMRI. (2008) *http://www.bbmri.eu/index.php/workpackages/wp-3* (accessed 9 December 2010) [Online].
9. Integrated BioBank of Luxembourg (IBBL) About the Integrated BioBank of Luxembourg (IBBL) (2010) *http://www.ibbl.*

lu/fileadmin/media/backgrounders/IBBL_-_Fact_sheet.pdf (accessed 13 December 2010) [Online].

10. Yuille, M. *et al.* (2010) Laboratory management of samples in biobanks: European Consensus Expert Group report. *Biopreserv. Biobanking*, **8** (1), 65–69.
11. Baust, J.G. (2007) Biospecimen research: enabling translational medicine. *Cell Preserv. Technol.*, **5** (4), 177.
12. Manzur, P. (1984) Freezing of living cells: mechanisms and implications. *Cell Physiol.*, **16**; *Am. J. Physiol.*, **247**, C125–C145.
13. Baust, J.G. (2008) Strategies for the storage of DNA. *Biopreserv. Biobanking*, **6** (4), 251.
14. Paltiel, L. *et al.* (2008) Evaluation of freeze-thaw cycles on stored plasma in the biobank of the Norwegian mother and child cohort study. *Cell Preserv. Technol.*, **6** (3), 223–230.
15. Learning Center: Microplate Standards. (2004) ANSI/SBS Microplate Standards, developed by the American National Standards Institute and the Society for Biomolecular Sciences (now Society for Laboratory Automation and Screening). *http://www.slas.org/education/microplate.cfm* (accessed 3 November 2011) [Online].
16. Groover, K., Drew, K., and Franke, J. (2007) The use of compressor cycle patterns: the ability to predict freezer failure. *Cell Preserv. Technol.*, **5** (4), 225–228.

13
Information Technology Systems for Sample Management

Brian Brooks

13.1
Sample Registration

13.1.1
Why the Need for Registration?

In the journey to find a new drug, pharmaceutical companies will synthesize and test thousands of chemical compounds. Potential drug candidates are tested in many 'screens' – experiments to test the compounds – in order to evaluate them for efficacy and undesirable effects. In order to relate a particular compound's results across the screens used, the compound is assigned an identity – a 'compound number.' This process of assigning a compound number is known as '*compound registration.*'

A compound number has two main roles:

1) A human-friendly label that is used to link a compound with its results
2) A grouping identifier which gives a common identifier for multiple preparations of the same compound.

Figure 13.1 illustrates the terms commonly used to describe the compound as it is synthesized, subdivided, tested, and re-synthesized. The exact meaning of these terms will vary from company to company, so it is always wise to check on local interpretations. However, the general concepts are always the same, so these terms will be used in this chapter.

A compound that shows promise as a drug candidate will end up being synthesized many times over; these are called new '*preparations*' or '*batches.*' Often a different synthetic route is used for different batches, and different salts are prepared in an effort to improve properties such as solubility. There will inevitably be some variation in composition between these different batches, but when they are tested, the research scientist is assuming that the major ingredient of the batch is responsible for the screening activity being measured. When the results are recorded, scientists across a company want to be sure that they are referring to the same

Management of Chemical and Biological Samples for Screening Applications, First Edition.
Edited by Mark Wigglesworth and Terry Wood.

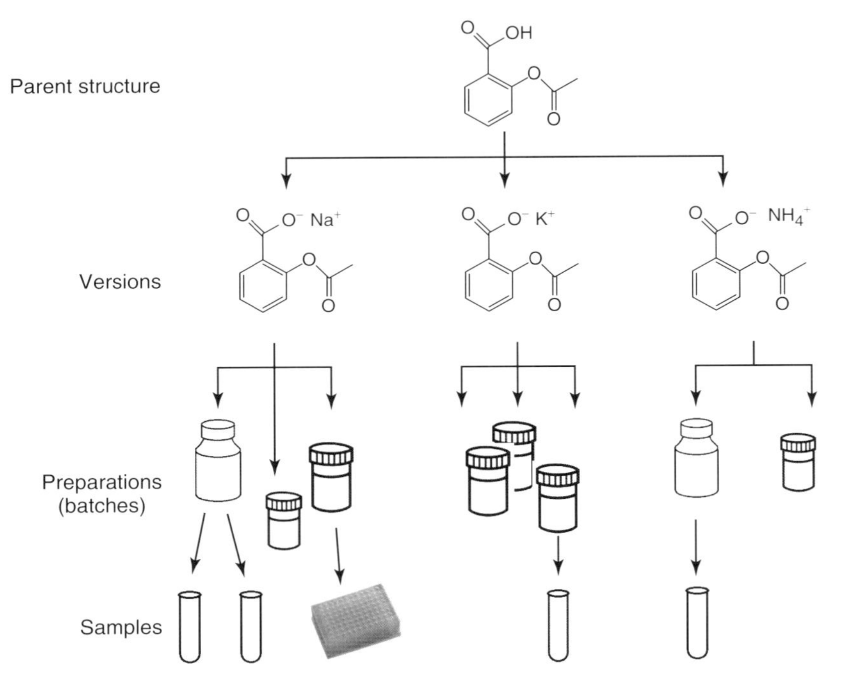

Figure 13.1 Different levels of the compound hierarchy: parent structures, different versions, preparations (batches) of those versions, and samples of those preparations.

structure, and this is why a compound identifier is used as a common usable name that groups together the results from all the different versions and preparations.

To register a compound the scientist must provide the information required to describe the compound to the registration system. One of the key tasks of a registration system is to check a structure for novelty; this needs to be quick, even if the database has many millions of structures. The details specified are checked against the data already in the registration database to see if this compound has been registered before (see Figure 13.2). If it has, the new batch of compound is given the appropriate existing compound identifier and a new preparation identifier. If the compound is new, then a new compound identifier is issued. In this way, registration builds up a database of all the structures, prepared or acquired, known to the organization. The registration database is the authoritative reference system for structural information used by all other systems in the company.

A state-of-the-art system for handling screening data results would allow tracking at any level of the compound hierarchy – that is, the ability to track individual samples, batches, or compounds. In practice, this is difficult; newer systems such as highly automated high-throughput screening are typically capable of recording results at sample level. For older systems and for screens further down the screening cascade, it is common to record results at the parent or version level, with some level of capability at the preparation level.

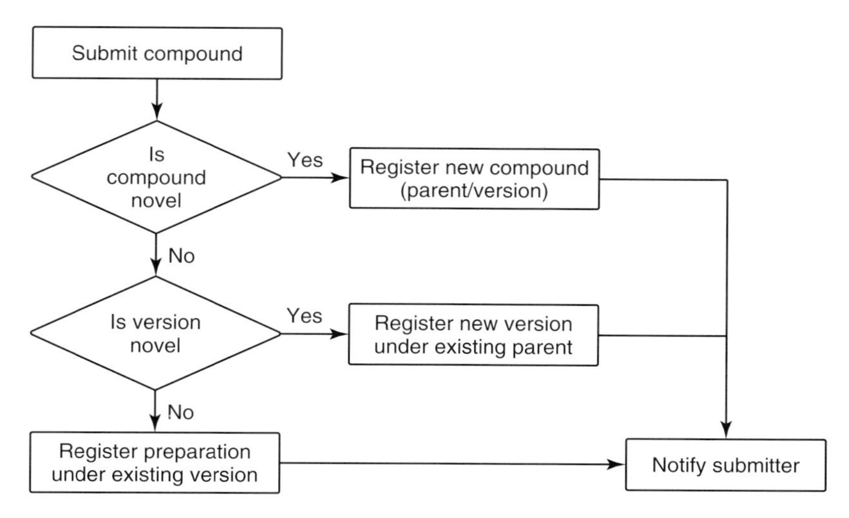

Figure 13.2 A flowchart for registering a preparation of a compound.

The ability to track screening results at the preparation level is a significant help in interpreting apparently anomalous results for assays performed on different batches of compound. A batch of compound prepared using one synthetic route could have different impurities from a batch prepared using a different synthetic route. It may be that the impurities themselves have an effect on the screening assay being run, so the ability to tell which preparation (batch) was used in testing can help interpret screening results.

13.1.2
Assigning and Using Identifiers at Different Levels

As can be seen in Figure 13.1, there are three levels at which a compound can be registered: the parent structure; the versions; and the different preparations of each version. Each level can have its own identifier, though not all of them will be seen and used by the scientists themselves (i.e., some levels may be for purely internal computing use). The designers of a registration system must decide whether to have three different identifiers, one for each level, or whether they want to merge two of the levels, using a single identifier for both levels. They must also decide which identifier is to be used by scientists to identify structures and store data against. A common solution is to merge the top two levels, so that one Compound ID conveys information about the parent structure and the salt version. The parent structure is indicated by a numeric portion on the Compound ID, and a suffix is added to indicate the version. A completely separate identifier, typically the laboratory notebook reference, is then used for the different preparations. Sample IDs are typically assigned locally within a screening system.

It can be an advantage to be able to assign a Preparation ID to a new compound without assigning a Compound ID at the same time. Some companies wish to be able to screen a compound without knowing the exact details of the compound's structure and composition. For example, they may be screening combinatorial mixtures, or extracts from natural samples (such as plants). The screening is done just using a Preparation ID; data is collected and processed using just this identifier. If the compound then produces interesting results, the structure may be determined and a Compound ID then assigned. The advantage of this screening approach is that the effort involved in determining a compound's structure is only made when you know you're interested in it. However, the smaller set of structural data generated by this approach limits the opportunity to use molecular modeling to further characterize the target.

13.1.3
Preparation Numbering

As discussed above, it is advantageous to name preparations using a different number from the compound identifier. Because the details of how a preparation was made are recorded in a laboratory notebook, the preparation ID is usually based on the laboratory notebook numbering system. Commonly, the Preparation ID will use the Laboratory Notebook ID plus some indication of the page number, with perhaps some extra digits to allow multiple structures on one page to be distinguished.

13.1.4
Sample Numbering

It is uncommon to find a sample numbering system which is standardized across the whole company. The main reason for this is that screening systems within companies tend to have grown piecemeal over time so that there is no one optimum opportunity to define and introduce a standardized sample numbering system. Another reason is the sheer difficulty of keeping track of all the individual samples which are taken from a batch, duplicated, sub-sampled, transformed into different physical forms, particle sizes, solvates, and so on, and assayed in different places. Consequently, sample numbers are almost always local to an individual screening system, with no common numbering approach being defined.

13.1.5
Methods for Naming Compounds

At first sight it appears straightforward to number a compound. You know the chemical structure that you made, so give it an identity number and call it job done! However, in practice there are many complexities involved in describing a compound, making it necessary to define rules to follow when naming a compound. These are commonly referred to as the '*business rules.*'

Business rules vary from company to company, and, once defined, they are built into the registration system. This variation between companies means that, if you were to register a test set of compounds in two different registration systems, the pattern of compound numbers assigned would be quite different. The implication of this is that if two companies merge, it is a significant task to stitch the two registration systems together. When two pharmaceutical companies merge, one of the first things to be done is to analyze the overlap between the two companies' compound portfolios, and look at the corresponding biological targets and results. Merging the registration database is an early requirement, and it is not an easy task. During the integration phase after a merger, the definition of the new company's business rules can be debated with almost religious fervor.

13.1.6
Some History to Compound Registration

The need for compound registration starts early on in the life of a pharmaceutical company. Some companies have been registering compounds for more than 70 years, long before the advent of computer databases capable of storing chemical structures. (The first commercial database system capable of storing chemical structures was the MACCS database, developed by MDL Information Systems in the early 1980s. Introduction of a chemical registration system was one of the earliest uses of a computer database in a pharmaceutical research laboratory.)

Consequently, many compound numbering systems were defined before the advent of computer databases. These older registration systems were operated by hand, using filing systems based on paper cards. Because of the difficulty of searching and even accessing the data in this format, it was common to make the compound number format convey as much meaning as possible to help the biologists and chemists understand the specific batch of the compound being tested. Compound number formats therefore tended to be quite complex, encoding various pieces of information to classify the substance being tested.

With the advent of computerized registration systems, giving the ability to quickly look up a compound's information, compound number formats have been simplified. Typically, a modern compound numbering format is much simpler, often comprising a prefix to indicate the company, a sequential integer to identify the parent structure, and a sequential alphabetic suffix to indicate different salt versions. Nowadays it is generally considered best-practice to keep identifiers a simple as possible, and not to add meaningful values into the identifier. Table 13.1 gives some examples of different compound numbering approaches.

The introduction of computer-based registration systems, plus the maturing of the functionality of the chemical registration software, has not only made the retrieval of information easier, but has allowed additional information to be stored with the structure, such as atom-centered stereochemistry.

13.1.7
Business Rules for Compound Registration

There are many things to consider when defining the business rules for a registration system. The decisions are not in themselves technically difficult, but they do

Table 13.1 Some examples of different compound numbering business rules. Best practice is now considered to be to keep them as simple as possible, and not to load the compound number with various coded meanings.

Compound identifier	Comments
RDM100916W	RDM = company prefix; 1 00 916 = sequential number for structure; W = salt or version code which could have meaning, that is, represent a given salt form. Lack of a suffix could indicate a free acid/base
RDM-L-3916 RDM-S-3917	L and S could indicate stereochemistry of the compound. Other isomers, for example, cis/trans, could also be encoded
RDM-26801AO-A	AO represents the project/target for which the compound was prepared, for example, AO = anti-obesity

require a position to be taken to adopt one or other approach. Often the pros and cons of competing alternatives are quite small and subtle, which can lead to long debates on the approach to adopt. Strong leadership (and perhaps a coin to toss) is a real benefit here. The following sections give examples of areas to consider.

13.1.7.1 Number Format

The business rules must define the overall format of the compound number. Compounds are things that a company cares about, so the numbering format is usually easily recognizable as a compound number (i.e., they are obviously different from, for example, a batch number, or a laboratory notebook number). The number format needs to define such things as whether the compound number is zero-filled, how many digits it has, whether it contain dashes or spaces, what the suffix should be, whether it uses curved or square brackets, and so on.

13.1.7.2 Compound Number Prefix

Many companies prefix their compound numbers with a code that identifies the company, for example, GSK109984, but this is not always the case. Some companies may use just a letter unrelated to the company name, for example, l-698532. Prefixes may also change over a compound's lifetime. A company may use one prefix for research compounds and a different prefix for development candidates.

13.1.7.3 Purity

Decisions need to be made as to how the compound numbering should treat compounds of different purity. If the compound is not 100% pure, should the compound number reflect this in some way? If the purity of your compound is below a certain boundary, should it be given a new compound number? If you know the structure of the impurities, should these be drawn in the registered structure?

An advantage of including all known structures in the registration is that they will then show up in structure searches of the database.

13.1.7.4 **Salts**

Promising compounds will be prepared in a range of salt forms. These different versions all have the same parent structure, so most registration systems have the same compound number, with a suffix to indicate the salt version. Some companies make this salt suffix 'meaningful' whereby, for example, an 'A' suffix always means the hydrochloride salt. The alternative is to use sequential salt codes – they have no meaning other than that they are incremented whenever a new salt form is prepared (e.g., 'A' is used for the first salt form made, 'B' for the second, etc.).

13.1.7.5 **Stereochemistry**

A compound's three-dimensional structure is vital to its biological activity. The registration system must therefore represent as much as possible about the exact configuration of the molecule, distinguishing between different stereoisomers. When molecular modelers are mining biological results from large numbers of compounds, they rely on stereochemistry from the registration database. The more accurate the representation of the compound's 3-D structure, the better will be the results from mining the data.

Many registration systems were created before chemical databases were able to represent stereo information, with the result that many compounds are recorded without stereochemical markings. Frequently, the exact stereochemistry is unknown or is only partially known; it may be absolute or it may be relative between two stereo centers. The business rules must define the approach taken in addressing these situations, and the solutions chosen need to be guided by the capabilities of the chemical database software to represent and search stereochemistry.

Stereochemical registration requires a lot of skill. For example, consider the following typical structure that might be submitted by a chemist:

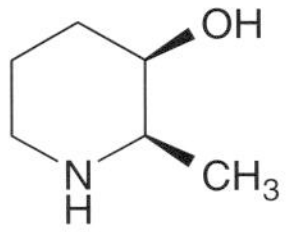

It looks clear enough; but the chemist might in fact be meaning any of the following structures:

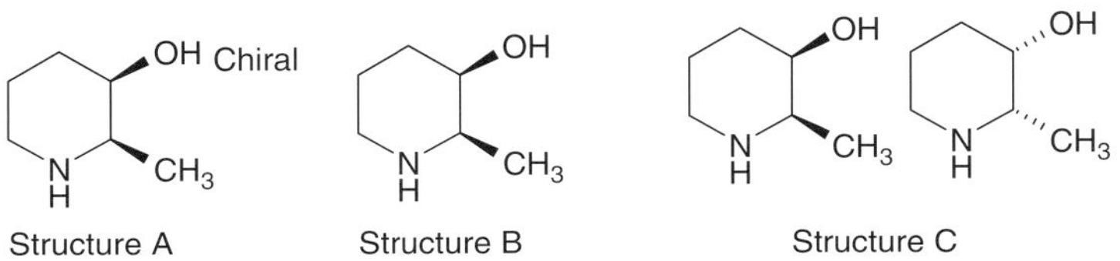

Structure A

Structure B

Structure C

The chemist might be saying that he (or she) knows the orientation of all the stereocenters (Structure A). Alternatively, he may mean that he knows he has a chiral molecule but doesn't know which exact stereochemical composition, R-S or S-R (Structure B). A third interpretation is that the chemist is saying that he has both stereoisomers. It requires training and awareness by the chemical registrar to spot these interpretations and to work with the submitting chemist to resolve them.

13.1.7.6 **Enantiomers and Racemic Mixtures**

Enantiomers are mirror image stereoisomers of each other. If they are equally present in a compound, it is a racemic mixture. The business rules need to consider how the stereochemistry of this mixture should be represented. For example, should both stereoisomers be drawn in the registered structure? The capabilities of the chemical database software often dictate the solutions that are chosen.

13.1.7.7 **Standardization of Charge Form**

The registration system also needs to have rules defined for how functional groups that may be represented in different ways are actually stored within the database. The most common example is the nitro group, which can be represented in either the pentavalent or ionic form:

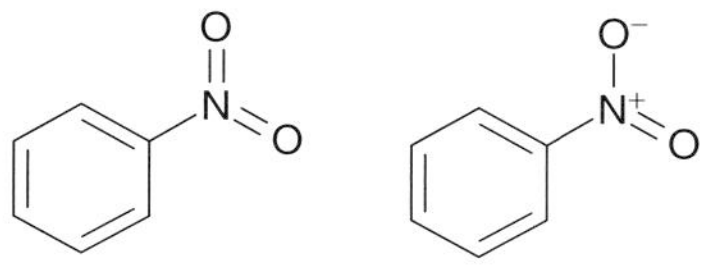

Ideally, the registration system should automatically convert one form to the other, depending on the storage rules.

13.1.7.8 **Tautomerism**

Tautomers are isomers of compounds where the different forms readily interconvert (see Figure 13.3, for example). It is common for the registration business rules to specify that one form of the tautomer be used for registration. This allows chemists to search more quickly – so long as they know only one form will be represented in the registration database.

Some registration systems can automatically convert one tautomeric form into the other, thus removing a possible source of error on registration (and searching) by ensuring that the structure is always stored in the desired form. Richly functional chemical database software can automatically match both tautomeric forms while searching, thus simplifying the business rule on tautomers.

13.1.7.9 **Radioactivity**

Compounds can be synthesized with radioisotopes at defined parts of the molecule. Some compound-naming systems indicate this by including the isotope in brackets at the end of the name. Other systems rely on different Preparation IDs to distinguish between different radioactive forms. The business rules need to be able to distinguish between two preparations of the same compound, where the same isotope is in different locations of the molecule, or the site of the labeled atom is random or unknown, or there is more than one isotope present.

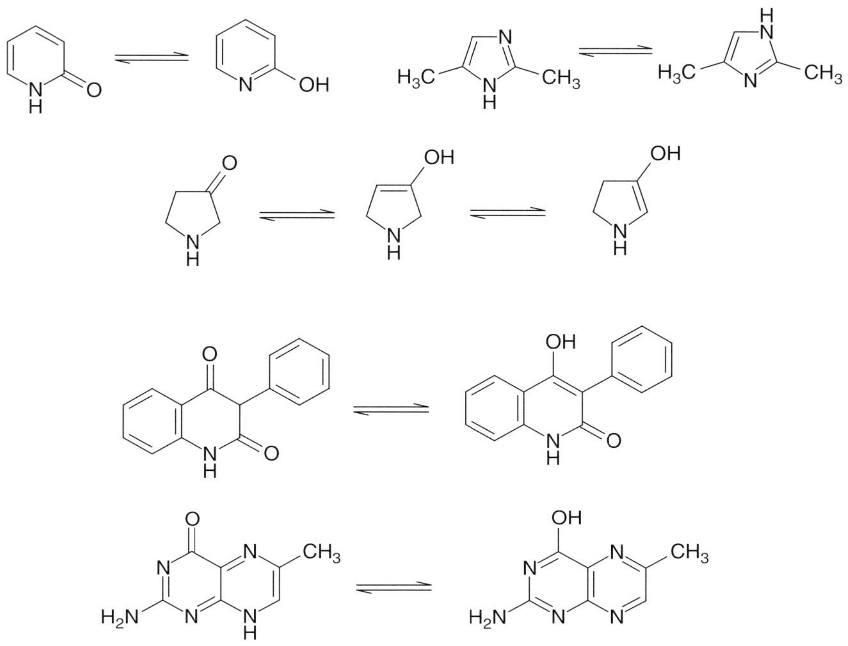

Figure 13.3 Examples of different types of tautomers.

13.1.7.10 Project Codes, Site Codes, Country Codes

Pharmaceutical companies commonly create a project team to work on particular diseases. Some compound-numbering systems have a way of identifying the compounds produced by each drug project by including a project code as part of the compound number. Similarly, some companies encode into the Compound ID an indicator of the site or country where the compound was originally synthesized. However, modern compound-numbering systems rarely have such a complex compound-numbering form.

13.1.7.11 Larger Molecules – of Known Structure

The smaller molecules that are the majority of compounds used by pharmaceutical companies are easy to draw in their entirety. However, as the molecules get bigger, into the realms of polymers, poly-peptides, and short DNA or RNA sequences, the question arises as to whether the exact atom-by-atom structure needs to be drawn. Drawing out the exact structure with hundreds or thousands of atoms is painful, unnecessary, and error-prone. In such cases, the prime purpose of registration is to create an identifier for tagging biological results – chemical searching to match the molecule is a lesser consideration. Hence it is sufficient to represent the compounds using abbreviations of each unit such as A, C, G, T for the nucleic acid bases, or the amino acid abbreviations Gly, Ala, Val, Leu (or their single-lettered equivalents, G, A, V, L), and so on. However, note that the system still needs a way to perform novelty checking so as to ensure that the same molecule does not get multiple identities.

The approach taken to registering these compounds will vary depending on the capabilities of the software on which the registration system is built. If it is able to understand abbreviations for the subunits, then the compound can be coded as such. If not, then the system designers have to be more creative, or even use a separate registration system, with suitable precautions to ensure that numbers created in the two systems do not conflict.

13.1.7.12 Larger Molecules – of Unknown Structure

At some point the size of the molecule being registered becomes too big to be drawn in any exact representation. Registration of very large molecules, such as biological macromolecules, natural products, and antibodies, should not require an exact knowledge of the compound structure. If such molecules need to be registered, then the registration system must be able to cope with these sorts of molecules. One approach is simply to register an empty parent structure and assign preparations to it as usual. Note, however, that the business rules must allow this, as must the capabilities of the registration software. Also, there must be some way of novelty checking to prevent multiple identities for the same molecule. Such novelty checking need not necessarily be done inside the registration system itself; novelty could be ensured using procedural methods in the group preparing the large molecule.

13.1.7.13 Combinatorial Mixtures

With combinatorial mixtures, batches for screening are deliberately made as mixtures of compounds. One possible way to register this sort of mixture is to use a Markush-type representation of the structures present in the sample. Another method is not to register any chemical structure at all and to use a Preparation ID during screening. When a compound of interest is found it can then be registered in the usual way.

13.1.7.14 Inorganic Compounds

Software developed for storing chemical structures was typically aimed at the pharmaceutical market, catering for the large diversity of organic compounds produced when searching for new drugs. For materials science companies, often working with inorganic compounds, the capabilities of the software to represent the chemical complexes developed can cause difficulties in registering the exact compound. Particular care is therefore needed in the choice of chemical database software for registering inorganic compounds.

13.1.7.15 Development Compounds, Outside Publications, Generic, and Trade Names

It is common for development compounds (e.g., batches prepared in development stages such as toxicology or formulation) to be used in experiments in the earlier research screens. Ideally one registration system would be used for both research and development compounds. Often, however, two different systems are used, because the needs of research and development differ; research has a few batches of many compounds, whereas development has many batches of a few compounds.

As a compound progresses through the R&D stages, at some point the company will start releasing documents (such as scientific research papers) to the outside world. Some companies have a separate compound-numbering system for external use. This publicity compound number is eventually superseded by the generic and trade names. Another common practice for referring to compound numbers in external documents is to use the parent compound number, dropping the more esoteric extra codes used.

13.1.8 The Role of the Chemical Registrar

Inevitably, mistakes get made. A compound may not be what it was originally thought to be, so the structure will have been incorrectly registered, or other registration details may have been incorrectly specified. The research process is fairly fault-tolerant, and extensive effort does not need to be expended to update old data and documents. However, because the registration database is a fundamental reference point for many systems, it is necessary to correct any mistakes so that subsequent interpretation of screening results is correct. It is also important to keep a log of corrections for future reference in case issues arise.

Chemical software is increasingly user-friendly, and chemists are now familiar with using computers and chemical drawing packages. It is now common for chemists to directly register their compounds into the registration system. However, correction of mistakes can be very complex, and can involve changing the structure or compound number, or retiring it, and moving batches of compound between different parents. The complexity involved is such that specialist skills are needed, and in larger organizations these skills may reside in a specialist chemical registrar.

Registrars are the experts who understand the chemistry, the business rules, the representation of the chemical structures, the history of the different chemical numbering systems used, the data in the registration database, and the tools for data entry and querying. They oversee the registration process, correcting mistakes as necessary. Many companies still have a registration process in which the registrars are the people who do the actual registration of new compounds. Other activities that registrars' skills are vital for include: testing of new releases and versions of software, batch loading of chemical structures from large compound purchases or from mergers and acquisitions, specialist searching for legal purposes (for example, patent challenges), and providing expertise to and training of scientists.

13.2 Intellectual Property and Laboratory Notebooks

For a pharmaceutical company it is vital to document their research in a way that protects their intellectual property (IP). Until recently, the paper laboratory notebook ruled supreme as the method to establish IP ownership. Increasingly, electronic laboratory notebooks are taking over the role of collecting experimental data and results, and of proving a company's IP. The chemist draws the desired compound structure and the reaction used to make it in the electronic laboratory notebook. This information is then sent to the chemical registration system to achieve an end-user registration system. The registration database is typically not the system used to prove a company's IP; this is still the job of a paper laboratory notebook or an electronic laboratory notebook with a digital signature facility for witnessing by another scientist.

13.3
Some Observations on Information Technology

The chemical database software market is now fairly mature. Many software companies now offer integrated suites of products to meet the needs of a research company. These include registration systems which can be tailored to different business rules designs and which integrate with other software applications such as electronic laboratory notebooks, inventories, screening systems, data processing tools, biological results databases, analytical software, and so on.

The most common database technology used for a chemical registration system is the relational database. Some relational databases can be extended in functionality by adding a 'database cartridge.' Database cartridges enable relational databases to handle specialist data types such as chemical structures. The cartridge makes the relational database 'chemically intelligent,' allowing storage of chemical structures, and enabling searches such as exact chemical match or sub-structure searches.

Despite the maturity of the chemical software tools, with chemical data cartridges now (2010) becoming (almost) a commodity, it is still not easy to move a registration system from one chemical vendor to another. Chemical databases from different vendors have slight differences in behavior for chemical searches and matches. When registering a compound, a registration system must give the same search results each time, so that when a re-synthesized compound is registered, the system will detect that the compound has previously been registered and thus assign the new preparation the correct compound number. This makes it difficult to move a registration system from one software supplier to another, because the chemical searching that is necessary during registration will differ.

Research data systems can fall into a trap similar to the Y2K (year 2000) issue when the length of the Compound ID increases by one character. The advent of combinatorial chemistry led to an explosion in the number of compounds being registered, and it is easy to rollover to millions or tens of millions of compounds registered. Systems built a long time ago might assume a certain number of characters when collecting Compound IDs, and might not accept an ID with an extra character.

When designing a new application which requires validating a compound number during data entry, it is best to validate by looking it up in a list of valid values (i.e., registered compounds), rather than using a format-based approach. In early computerized registration systems it was common to validate a compound number being entered into the system by validating the input against the format for the compound number as defined in the business rules. This works well as long as there are no changes to the compound number format. However, if there is a change such as a merger, then this validation technique does not work with the compound numbers from the other company. To accommodate the new compound numbers it is necessary at the very least to update the program. By validating new input by looking it up in a list of valid compound numbers, maintenance overhead over the lifetime of the system is reduced. Current best-practice software design is to perform such validations using a central server which provides a validation service to any software that needs it. In this way, updating the validation is easier because there is only one location that needs altering.

13.4 Biological Data Management

13.4.1 The Corporate Biological Screening Database (CBSD)

After synthesis, a compound is tested in biological screens to evaluate it for activity. A drug research project will have a variety of screens to assess desirable and undesirable properties of the compound. In a large pharmaceutical company there are many scientists working on a wide variety of diseases. They can be located on many sites all around the world, and as years pass, drug projects and the personnel in them will come and go. This creates a challenge to know what has been done and what the results were.

There is therefore a need for a Corporate Biological Screening Database (CBSD) to make screening results available in a format that allows scientists across the company to find and use the information. The CBSD serves the following needs:

- Helps a scientist build a profile of a compound by showing where in the organization it has been tested and whether it was active
- Provides the scientist with tools for storing, processing, analysis, reporting, and mining their own project's data
- Captures organizational memory of previous research activities, facilitating future analyses.

The CBSD is not the same as a high-throughput screening database. All high-throughput screens (HTS) systems are computerized – they have to be in order to handle the very large amounts of data – but their specialty is collecting and processing the raw experimental data. HTS systems deal with plate layouts, duplicate and triplicate samples, standard curves, sample dilutions, edge effects, calculations, and statistics. They understand experiment design, and process the raw results to produce the final result. Most of the data in an HTS system is of no interest to those scientists who just want to know the final result that describes the effect of a compound on a target. This final result, together with supporting metadata, is what is passed to the CBSD for storage. The CBSD contains the final results from the high-throughput screens, plus the results from the lower-throughput screens (LTS) which come later in a project's screening cascade.

CBSD also differ from HTS databases in the length of time that data is held in them. Data are only stored in HTS systems for a limited length of time; after a year or so the results from an experiment are deleted. The main driver for this is performance; the accumulation of raw data in the HTS database leads to a progressive decline in performance. Because the final results have been passed on to the corporate database, there is no need to keep the raw data in the operational HTS database.

Another reason why HTS data is not stored for long periods is the effort that would be needed to keep old data compatible with new versions of the screening systems. As the science, techniques, operating systems, database, and software versions change over time, new database formats evolve which are not backwards compatible. It is therefore a large effort to keep older HTS data in a readable format. The CBSD, in contrast, will guarantee that all data will at all times be readable.

It is worth noting that when raw data are removed from the HTS database, due consideration must be given to whether the raw data should be archived for protection of IP. The interaction of a compound with a target in an HTS screen is usually the first indication of a company's interest in that compound for the disease, so recording this fact could be vital for establishing a date of invention.

The CBSD must have good quality contextual data (often called '*metadata*') to support the screening results. A CBSD can become very large, with data from thousands of screens, and many millions of screening results, resulting from decades of research. It is a challenge to organize all of this data consistently and to a standard that is useful to the scientists across the company. Imagine that you are a scientist who discovers that a compound you're interested in has been tested elsewhere in the company. How do you understand what the results in this unfamiliar screen mean? For the data to be useful there needs to be sufficient documentation so that the results can be understood by other scientists in the organization in addition to those intimately involved with the screen.

After high-throughput screening, a compound is tested in LTS. These screens are used to characterize a compound's activity profile and to evaluate activity in more intact, physiologically representative screens. They are therefore tailored to the particular disease being researched, and they create smaller amounts of much more diverse data. Even though there is only a small amount of data produced, they are much more 'information-dense' than HTS data, the vast majority of which is simply a negative result. The LTS data contains the results that are used to progress along the path of lead optimization, and they tell the story of how a final candidate is reached. LTS results therefore define a drug project's research path, and it is vital to capture them in the CBSD.

Because of the variety of types of low-throughput data, they are more of a challenge to store than HTS data. In HTS, due to the large numbers of samples involved, much of the data collection, processing, and analysis has to be done using computers. The formats of these experiments are fairly well-defined, and are re-used across many different targets. Data collected can be rigorously validated and checked for quality. However, in low-throughput screening, systems are typically manual, which presents a challenge for data collection and validation.

The CBSD needs expert staff to manage it. However, it is remarkably difficult to manage scientific data consistently across many people. Scientists are of course skilled in managing their own data for their own purposes. However, managing data so that it is understandable by other scientists takes time, effort, and dedication. The analysis of the screen and its data storage needs, configuring the metadata to a high standard, and understanding the needs of the scientists who need to query the data requires skill and experience. It also requires suitable processes to ensure consistency across the organization, especially if this is geographically dispersed. For a good quality database you need dedicated staff to manage it.

Firm governance is needed in order to ensure that data is stored in the CBSD. Human nature is a wonderful thing, as is having a diversity of opinions. However, diversity of opinions will inevitably mean that some scientists will consider the need to record results in a central database as a lower priority than other activities, with the result that the data doesn't get stored. Similarly, other frequently encountered pleas are that 'other scientists will misunderstand the results,' or 'my data is not of interest to other scientists.' Some highly computer literate scientists may create their own databases. However, these are not accessible by scientists across the organization, and they rarely get managed over longer term. The company therefore needs to issue clear rules on the requirement to store results into the CBSD; ideally they should be mandated by the head of research.

13.4.2 Data Entry Tools

Because it is a corporate database it is likely that there will be more than one way of entering data into the database. Loading data from HTS databases is likely to be automatic, with an interface that allows the scientist to control when an assay's results are loaded. For the more manual LTS, a different method of data entry is needed. This may be via a dedicated application, or it may be via applications such as Excel. During data entry, it is important that the data being entered is validated against known values of identifiers such as the compound number and the batch number. Additionally, as much validation as is practicable should be done on the data; for example, checking the range of data entered, or ensuring that the value exists in a list of valid values. The easiest and cheapest time at which to fix errors is during data entry, so it is worth building good quality data validation procedures that run at this time.

In addition to validating data, the CBSD can save scientists significant time and errors by doing simple transformations. It is common for a compound to be repeat-screened in an assay. Normally the biologist has to find the previous results obtained and then average them out. The database can do this automatically; the

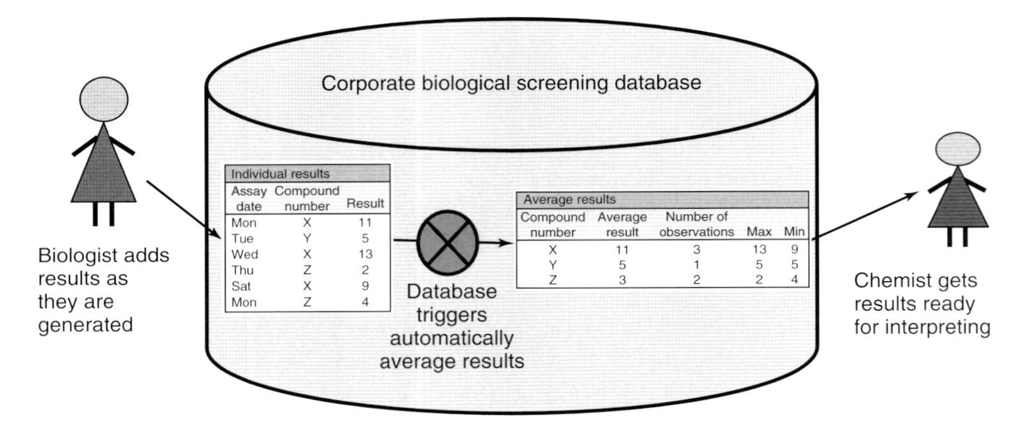

Figure 13.4 Added data-entry functionality saves time and reduces errors. Here the biologist enters individual observations, and the database calculates an average for the chemist to use.

biologist enters the individual screening results, and the average is automatically calculated for the chemist (see Figure 13.4).

13.4.3 Database Querying

The application used for querying the CBSD is key to enabling scientists to make the best use of the data stored. Ideally, the query application should provide two main types of functionality: analysis of datasets and navigation between datasets.

Each research project will have a set of assays with which it assesses potential compounds as candidates for treating the disease. The scientist will scrutinize the results from these assays in great detail, preparing structure-activity relationship analyses, and balancing desired and undesired activities. The software to perform this rich data analysis will be used by all the research projects in an organization, so usually it is linked to the CBSD so that everyone can use it. This stage of data querying is characterized by data processing and manipulation, chemical structure analysis, and comparison of different datasets. Table 13.2 gives examples of the typical tasks that the query tool should aim to do.

The CBSD will contain data that is related to other data in a variety of ways. For example, there may be graphs or instrument traces that show the detail of each experiment collected. The ideal client would make it easy for the scientist to navigate up and down these different levels of data, thus making it easy to browse the database contents (see Figure 13.5). Navigation in this way requires that the client has the appropriate functionality built into it, but also that the database itself has the necessary rich metadata that describes the relationships between the different types of data collected for a screen.

Table 13.2 Typical database querying activities.

Activity	Example
Querying based on identifiers	Show me results for this Compound ID. Where has this Batch ID been used?
Querying based on a biological property	What results are there for pKa > 7? Show me the compounds with the lowest IC50.
Querying based on chemical criteria	Run this sub-structure search and show me results for the compounds found using my local visualization tool. Do an exact-match search; see if we have any results.
Structure–activity relationship analysis	Do a sub-structure search for this structure, showing results for this biological parameter. On the results returned, do an SAR analysis to show how biological activity varies with structural components.
Working with sets of compounds, batches, screens, and so on	Show me the results for this set of compounds in these screens. Save this set of compounds so I can use it in future searches. Show me the compounds that are at least 10-fold more active at Target A than they are at Target B.
Querying a single assay	What were the results for compound X? Which were the most active compounds in this assay? Display them in my chemically-aware spreadsheet.
Querying a group of assays	How active is this compound in this assay as compared to these other two assays?
Querying all assays	Where has this compound being tested and was it active?
Time-based querying	Give me results collected in the last week.
Reporting	Prepare a monthly report for results collected in the screen
Understanding assay	Show me the documentation and methodology for this screen. What was the rationale for running the assay in that way? Show me which assays have been run recently using a specific technology.

When a project scientist identifies a compound of interest, they will then wish to find out more about it. This stage of investigation is characterized by fact-finding, involving navigation to datasets outside a research project's screening cascade. The query tool should therefore provide hyperlinking functionality, enabling the scientist to navigate from a data point of interest to further sources of information on the compound. This sort of navigation requires a well-constructed information architecture and software infrastructure.

In hyperlinking-based activities, the result that the scientist wishes to browse could be within the same organization, or it could be external, out on the internet

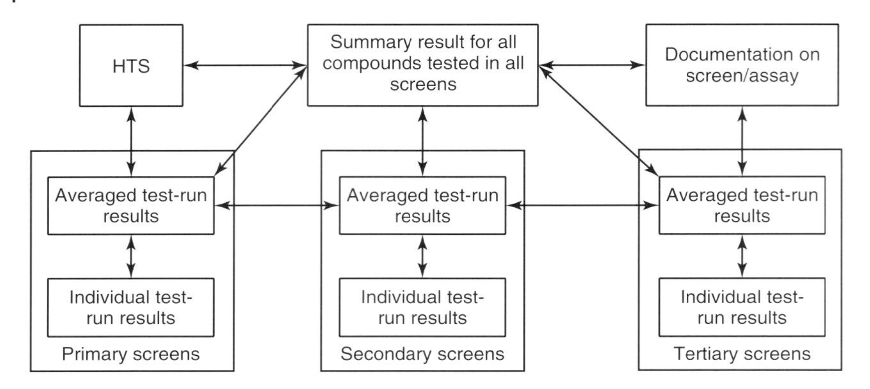

Figure 13.5 Examples of typical navigations between types of data within a corporate biological screening results database.

Table 13.3 Examples of data sources external to the corporate biological screening database that the scientist may wish to navigate to from the CBSD.

Browsing destination	Example
Corporate compound inventory	How much of this compound is available?
Corporate ELN	Show me the details of this experiment.
Corporate HTS database	How was this value derived? Show me a graph of the results.
Corporate search engine	What documents do we have on this compound?
Genomics database	What targets are similar to this one?
Patents database	What patents are there on this target? Has this compound been patented?
External compound inventory	Can we buy this compound?
Metabolic pathway database	Where is this target expressed within the body?

(see Table 13.3). For optimally integrated navigation it is necessary to use standardized identifiers and terms across the different data sources. Even within an organization this is surprisingly difficult, because often different groups are responsible for developing different databases, and lack of understanding of other areas can mean that corporate identifier schemes are not followed sufficiently to achieve the identifier quality necessary for facile navigation. Ideally, an organization needs a data architect who understands the scientific needs and who has the responsibility for ensuring data consistency and quality across the organization, plus the authority to implement any recommendations.

An issue that makes standardization of metadata difficult is that, by definition, we are venturing into the unknown. This is Research; new things are being discovered, and the names that get associated with the new knowledge can change over time, or be found to be the same as previously discovered items. Different identifiers may be used by scientists external to the organization. It is very difficult to find the time to update older metadata keep things consistent over time; the pragmatic solution is to build IT systems that allow use of synonyms, allowing the different names to be associated. However, even when such functionality is available, it requires time, understanding, and expertise to maintain.

13.4.4 Special Data Types

One of the interesting problems encountered when storing biological results is values of the form, for example, '<1' or '>50.' These 'soft numbers' occur, for example, when a result lies outside the limits of sensitivity of an assay, or is outside the range of calibration standards used. Soft numbers cannot be stored in a database field of type 'number,' which is the normal way of storing numeric values in a database system. The database designer therefore has a dilemma about how to record a soft number. One solution is to store it as a string, which has the advantage that you can store the value exactly as observed but the disadvantage that it is not numerically searchable (and for example, '1' and '11' will sort before 2). Another solution is to store the numeric part as a number, and store any modifier separately as a string. When querying and reporting, one can then refer to the modifier field where appropriate for the query being run. However, this does make the data entry and the querying applications more complex.

Graphs and images are common outputs from biological screens. Images can be such things as microscope slide sections, scans of gels, photographs, and so on. Graphical output is generated in many ways, from standard curves, to dose-response curves, to time-based chart outputs. Storing these sorts of datatypes in a relational database usually involves storing the result as a 'blob,' which is a large bit-bucket that is handled as one chunk of data. The interpretation of the structure of the bits within the blob is done by the client applications – the relational database simply delivers a lump of data-bits to the calling application.

13.4.5 Database Designs

There is a spectrum of detail that can be used to report biological screening results. In some assays, the result could be given simply as 'Active.' Other assays could consist of a set of pairs of dose-response results, and could be reported as a set of values, or as a derived value such as an EC50. For some needs, 'Active' is sufficient; for others, detail is needed.

When designing a corporate biological database, therefore, it is important to decide what levels of detail of screening data the system is required to store. Will the CBSD simply be a basic record of the screening activity across the whole company, or should it be a system that project scientists can use for storing and analyzing their project's data? The simplest possible design would be to store a single table of results, with fields in the table being something such as Compound ID, Screen Name, and Result. A simple design such as this would meet the minimum needs of scientists wanting to know where a compound has been screened and what the result is. However, this would not be of much use to a scientist on the program itself, who would probably want to make use of several parameters being measured and perhaps to relate them to results from other assays. A more ambitious database design would add data such as all the parameters measured in the experiment, plus graphs of experimental results, images of slides or traces, statistical results, and so on.

Another challenge for the designers of a corporate biological results database is to provide functionality that links the biological results to the results of chemical searches run against the chemical structures database. As discussed in Section 13.3, chemical structures are a specialist data type that are typically handled using a data cartridge. Joining together the results from a chemical search to results from multiple assays requires quite complex software techniques if the application is to speedily return the results in the form that scientists require.

Performance (i.e., speed) is a factor that must always be in the forefront of the database designer's mind. Performance is necessary not only for data searching and recall, but also for data entry, editing, and deleting. It is not possible to design a database that is optimal for both data input and querying. A common solution is to design a database in two parts; a transactional part which is used for data input, and a warehouse part which is optimized for querying. The degree of logical and physical separation between these two varies between the possible extremes, with hybrid implementations being a pragmatic solution. Data is moved between the various parts, sometimes being transformed en route, using various data pumping tools such as ETL (extraction, transformation, loading) or bespoke applications (Figure 13.6).

One solution to the design of the data warehouse is to use a star schema design. A star schema has a Facts table at its heart and various Dimension tables in support. The Facts table holds the values that are the main measured results of experiments, and the Dimension tables hold contextual information such as the assay in use, experimental conditions, time values, and so on. Star schema designs are good at querying widely across results from a set of screens. However, for such wide querying, it is vital to keep metadata consistent across the database; star schema is ideal for querying high-throughput results. As the number of types of assays increases, the star schema approach becomes less useful.

Another database design approach is to use 'flat' tables – where the design of the table looks much as you would lay out the data in a spreadsheet. The advantage of this is that the interactions with that table are straightforward, and can make use of the powerful capabilities of the SQL query language. Many commercial querying tools are built with this sort of database design in mind. Searching within a single screen is extremely fast. However, this approach is at a disadvantage when you need to search across a set of screens, or across all screens. A way to solve this is to build dedicated tables that are designed to optimize the performance of these multi-screen queries. A common desire is to compare the results for a compound in the set of assays used by a drug project, often looking for increased activity in

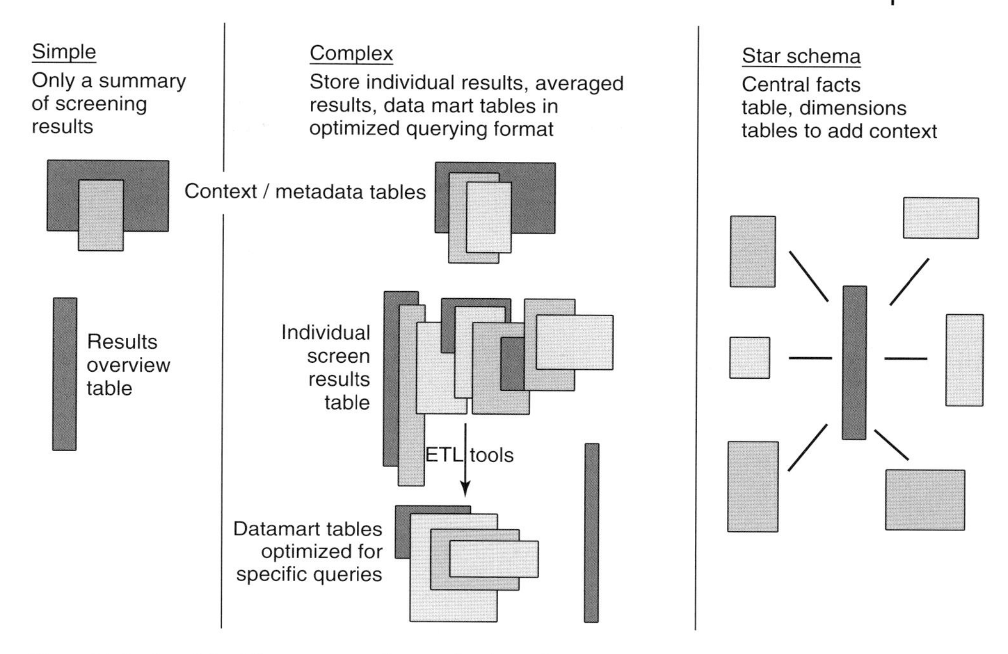

Figure 13.6 Possible different approaches to database design.

one screen with decreased or no activity in other screens. A way to tackle this is to use tools such database triggers or ETL tools to build a wide, flat table specifically for this query. Because SQL is a row-centric query language, with this sort of wide flat table it is easy to create fast queries that compare values in different columns.

A common desire for a scientist is to find where a compound has been tested in the whole database, and to get an indication of the activity of the compound in each screen. This query can be answered quickly by maintaining a single table which has a summary description of every result in every screen.

The essence of a pharmaceutical company's business is knowledge of the interaction between a compound and a biological target. To capture the knowledge of these interactions is to secure the crown jewels of a company's research efforts. Research data has its own entropy; without at least some energy expended to organize it, data chaos will ensue. Organizing data costs time and effort from dedicated data shepherds. As the adage goes – you get out what you put in.

Dedication and Acknowledgments

My thanks are due to Jenny Caldwell, Dr Gill Cooper, Ian Mawer, and Colin Wood for their help in the preparation of this chapter. The chapter is dedicated to Dr. Merrie Wise, who was a passionate advocate of the benefits of managing research data – and the most wonderful boss.

14
Key Features of a Compound Management System

Clive Battle

14.1
Why Do We Need Compound Management Information Technology Systems?

The growth of high-throughput screening (HTS) over the last 15 years has placed greater demands on compound management resources to increase both productivity and quality of samples supplied to scientists. This has led to the widespread use of automated hardware systems, capable of producing many thousands of samples per day.

The day-to-day management of all the physical and logical processes that contribute to this high-throughput supply of samples would scarcely be possible without an effective software system to oversee the process. Companies have addressed this issue either by developing compound management software systems in-house 1–4 or by acquiring commercially available systems configured to their specific requirements 5–7.

A compound management IT system has many requirements. Even the most basic system must enable the sample manager to hold an inventory of samples. It should track receipt, creation, movement, and despatch of these samples. There must also be a means of viewing the inventory and placing requests against it. The more advanced systems will also provide processes that efficiently guide the sample manager through a workflow when fulfilling requests. These systems will interact with automation and manual equipment to control and track the processing of samples, providing data integrity to the complete sample management process. They may also exchange information about samples with other software, ensuring continuity throughout a system and providing accurate stock management. Within the following sections of this chapter we describe a more advanced system to cover all of these elements. Although it should be noted that as these systems are now available, they are as applicable and accessible to small inventory stores as they are to the large inventory capacities required within the pharmaceutical sector.

Management of Chemical and Biological Samples for Screening Applications, First Edition.
Edited by Mark Wigglesworth and Terry Wood.

14.2
Compound Management Software

Software tools and applications that undertake specific functions in compound management may be developed by pharmaceutical companies or purchased from specialist suppliers. In either case, a compound management system represents the integration of these systems, communicating via application-programmable interfaces (APIs), database interfaces, and web services (see Figure 14.1).

14.2.1
Inventory Management

All of the labware items (bottles, tubes, and plates) that contain compounds for distribution make up the physical inventory. A compound management system must therefore have an inventory management subsystem comprising the system inventory, inventory viewing, and order management components that are integrated with the inventory system, as shown in Figure 14.1. These components then exchange data with the corporate inventory, the corporate ordering system to facilitate storage of sample data, and inventory tracking, providing the facility to import, view, create, and edit inventory, and enable the logical organization of the inventory, all of which functions are described in detail below.

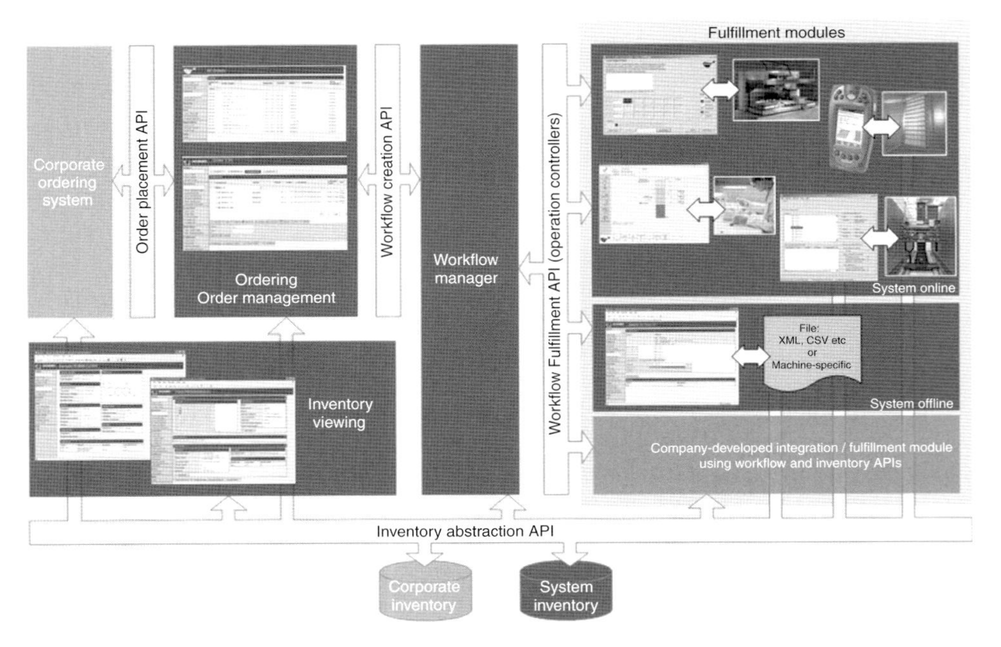

Figure 14.1 A schematic representation of the key functional components of a commercial system and its interfaces with corporate systems.

14.2.1.1 Data Storage

Compound collections vary in size from a few hundred thousand to several million items. The inventory database stores information about the labware items that contain compounds. The database recognizes labware items by their barcode or two-dimensional barcode and associates this with the type of labware, the identity of the compound it contains, the stored amount of compound, the location of the item, and so on. There may be any number of properties associated with labware items.

14.2.1.2 Inventory Tracking

The inventory is a dynamic entity, as samples are depleted and created as orders are placed on the compound management system. The inventory database keeps track of the inventory items, maintaining and updating information as items are processed. Tracked information includes all of the processes that items undergo, their locations, changes in the amount of compound they contain as a result of processing, and so on. The inventory database tracks the life of a labware item in the compound management system and maintains the relationship between the parent item (the labware item which is the source item) and any child items that are created from it. It is therefore possible for the user to identify the child of a parent item or the parent of a child item. Tracking labware items in this way enables the compound management system to generate a full audit trail for each item in the inventory.

14.2.1.3 Inventory Browsing

The user can browse the inventory to view information about labware items or the compounds they contain. A wide range of information about inventory items can be viewed including the volume or mass of sample present in a labware item, the physicochemical properties of the compound, its structure, safety data, and so on.

14.2.1.4 Importing Inventory Items

The compound management system needs to provide a means of importing labware items into the inventory database. There are several ways in which labware can be imported, depending on the process that the labware will undergo when it is received. For example, a vial containing a sample that will be weighed might be conveniently imported at the weighing operation, whereas a set of plates for storage would be more efficiently imported using a bulk import file-based process.

14.2.1.5 Editing Inventory Items

Under some circumstances, it may be necessary to edit the details of an inventory item. If, for example, a labware item has undergone an offline process, such as an *ad hoc* manual transfer of solution from a tube to a plate, the user would want to update the volume of solution in the tube. Such activities are rare, but provision for them is required.

14.2.1.6 Organizing the Inventory

The inventory is logically organized in a way that enables items which belong together to be easily selected for use in an order; for example, plates that belong to a particular library may be ordered for a particular screen. In some systems this is done by organizing labware items in a series of 'sets,' where the entire contents of a set can be ordered.

14.2.2 Ordering

Labware items for screening, quality control (QC) or other testing purposes are created as a result of orders. A compound management system must therefore have an order management subsystem to enable scientists across the company to search for items and place orders, defining precisely which outputs they want for their screen or test and receiving notifications about the status of their order. This interface could be directly via the order management component of the compound management system or via this system and an API to a corporate ordering system, as shown in Figure 14.1 and detailed below in external ordering systems.

14.2.2.1 Web-Based Ordering

Requesters who place orders for labware items are often physically remote from the compound management facility, and so web-based ordering systems are commonly utilized to allow orders to be placed by users in multiple sites in remote locations.

14.2.2.2 Sample Naming

Every organization has its own nomenclature for sample naming (see Chapter 13 for detailed discussion), so the ordering subsystem needs to be configured to use the names that the users of the system understand. Items may be known by a multi-part name, a typical example being the compound ID (indicating the molecular structure), the lot number (indicating the batch of substance made at a particular point in time), and the salt form. Items may be ordered by a combination of all three name parts or by compound ID and one other name part.

14.2.2.3 Definition of an Order

An order defines the sources (which may be specified by a batch/compound number or by a labware item barcode) and the outputs that the requester wants to receive. The requester will specify the attributes of the outputs, such as the type of labware, the volume of solution, concentration, and so on. In some systems, the creation of orders is simplified by using templates that have pre-populated fields. This is particularly useful when repeat orders are made for the same or similar outputs as it avoids having to re-enter order details every time the same type of order is placed.

14.2.2.4 Order Validation

The compound management system checks the order, validating that it can be processed to completion; for example, it checks that there is sufficient volume in the source vessels to achieve the output and it checks that adding diluent to a vessel will not cause it to overflow.

14.2.2.5 Order Approval

In order to manage the fulfillment resources effectively in the compound management facility, orders that are submitted by requesters are received for approval by sample management supervisors. The ordering subsystem therefore requires a submission function and an approval or release for fulfillment function before processing can commence.

14.2.2.6 Restrictions

If a compound has limited availability or is particularly valuable, a requester may choose to place a restriction on it. There might be several effects of placing restrictions on items. It might be that the compound is not available for use by other requesters until the restriction is removed, or there may be a limit on the amount of substance that can be requested. The ordering system notifies the originator of the restriction if the item has been requested. The originator can choose to allow the item to be used or reject its use (thus upholding the restriction).

14.2.2.7 Queries

The ordering system may provide a list of pre-defined parameterized queries that can be used to search for items. For example, a low-stock query can be used to identify all the stock labware items in the inventory that have less than the minimum pipetting volume in them. Any number of queries can be used to identify items for use in orders.

14.2.2.8 Order Status Notifications

As an order progresses to completion, the scientist may want to know what stage the order has reached in the workflow (see Section 14.2.3 below) and may want to receive email notifications informing of the status of the order. As a minimum requirement, the scientist will want to receive email notification that the order is complete and awaiting collection.

14.2.3 Workflow Management

The compound management system needs to translate an order into a series of process steps that can be fulfilled by an operator. Having knowledge of the precise processing steps enables the system to validate an order before physical processing begins. The workflow manager subsystem understands the process steps that are required to fulfill an order and packages-up the work in such a way that it can be processed by each workflow step in the most efficient manner.

14.2.3.1 Workflow Steps

The workflow manager knows the precise details of what happens to each labware item in a workflow step. Typical workflow steps include dispense, solubilize, dilute, replicate, and so on. The workflow manager is therefore able to calculate what should be done at each workstation to achieve the desired outputs. The workflow manager creates instructions for the operator (or a workstation) to undertake the process as specified. When interacting with workstations the success or failure of each operation is tracked and logged within the system (see Sections 14.2.4.1 and 14.2.4.2 below). Similarly, when instructions are completed by an operator they log the completion of each task as they move through the workflow. Hence the workflow manager provides a means by which an operator or a requester can easily identify the progress of an order or the status of a particular operation, with obvious benefits when operators hand work to a colleague or when requesters wish to plan downstream events. In some systems, the workflow steps are represented graphically, with visual indicators for items in progress.

14.2.4 Fulfillment

We have seen how orders are created and how these orders are translated into workflow steps by the workflow manager. The physical processing of work, or fulfillment, uses the instructions provided by the workflow manager to create the items requested in an order.

The way in which the workflow manager interacts with the workstations to fulfill an order will vary depending on the instruments in use, the availability of an application to 'drive' an instrument, and the operator's preference. There are two general strategies for fulfillment of physical processes; 'offline' and 'online' processing.

14.2.4.1 Offline Instrument Integration

When the fulfillment process is offline, the system typically produces a file describing the operations that need to be performed for a specific step in the workflow. In this mode of operation, the operator will instruct the instrument to perform the process as it is described in the file. When the process is complete, the operator reports what was actually done (in most cases this will reflect what the file asked to be done), by importing another file into the compound management system. It is possible that the operator did not follow the instructions of the file precisely for some good reason, and so the inventory database is updated with data that reflects the physical reality.

14.2.4.2 Online Instrument Integration

When the fulfillment process is online, there is a direct communication between the compound management system and the instrument that is fulfilling the workflow step. So, for example, the instrument will receive exact information on how it will pipette a solution or what barcode it will apply to labware items. This communication may be via a fulfillment application which acts as a user interface to the instrument, instructing the operator to load both source and destination labware, for example, and prepare the instrument in accordance with the instructions passed down from the workflow manager.

14.2.4.3 Offline vs Online

Whether a fulfillment process can be run offline or online depends mainly on the availability of an appropriate compound management system application or 'driver' for the specific instrument in use. In general, an online fulfillment approach offers a lower risk of labware tracking errors because there is no file manipulation by an operator. Most online applications have robust user interfaces to ensure that movement of labware by the operator is properly tracked and the risk of human error is minimized. However, the number of different sample processing instruments in use by the industry today is such that suppliers of compound management systems tend to provide the more commonly used instrument drivers, so that in some cases offline sample processing is still the only fulfillment option.

14.2.4.4 Despatch

The final step in the workflow is to despatch items. This workflow operation finishes off the processing of an item, by despatching it from the fulfillment system. The despatch operation can involve both local delivery (where the item is being delivered to a location in the same area or building as the compound management facility) and remote shipping (where the item is being shipped to a different site from that of the compound management facility). When an item is despatched, its new location is updated in the inventory database. Despatching an item would typically involve sending the recipient an email with an accompanying delivery note giving full details of the item(s) in the order.

14.2.4.5 Reports and Metrics

The compound management system will produce a wide range of reports to assist the facility supervisor in planning and routine management activities. The kinds of reports which might be generated include process throughput, stock and re-stock metrics, turnaround time for specific orders, specific machine utilization, and so on.

14.2.5 Interfaces with External Systems

The compound management system forms part of the informatics infrastructure supporting the drug discovery process. It must interface with both upstream (chemical registration/corporate inventory) and downstream (results analysis) systems, and may be required to interface with other existing corporate systems.

14.2.5.1 Chemical Registration System

The chemical registration system may store a wide range of data on a chemical compound. These data are not duplicated in the compound management system. On receipt of an item into the compound management system, an association is made between the barcode of the labware item and the batch it contains; it is this association that enables the compound management system to look up any data associated with the batch that is stored in the chemical registration system.

14.2.5.2 External Ordering System

The ordering system is the front end of the compound management system, known and understood by potentially many end users. For this reason, a company that is implementing a new compound management system may prefer to retain their ordering system and use this to place orders on the compound management system. Under these circumstances, the compound management system provides an API which enables orders to be placed using an ordering system that is external to the compound management system itself.

14.2.5.3 Results Analysis

If the compound management system provides labware items for use in screening, it is likely that information about the compounds provided will be used by downstream results analysis software. In this case, the compound management system may be required to pass plate maps, in a form that the analysis software can use, into an external database.

14.3 Benefits of Commercially Available Compound Management Systems

Commercially available compound management IT systems can offer significant benefits over such systems developed in-house. For example, apart from being economically accessible, they

- Provides a means of organizing, viewing, and editing the sample inventory.
- Optimize the allocation of work to maximize efficiency.
- Enable the efficient placement of orders, utilizing commonalities in ordering requirements, order tracking, and control.
- Provide simplified order fulfillment through the use of online applications that guide the operator through each processing step.
- Provide flexible integration with corporate systems through the use of APIs, database interfaces, and web services.

Some additional benefits of a commercial product are given below:

- Commercial products benefit from a development focus on industry-wide issues, offering standard functionalities at lower cost compared to in-house developments.
- Commercial products offer continued product development, enabling companies to benefit from enhancements and new functionalities.
- Product vendors offer a support model that is more commercially viable than continued maintenance by in-house resources.

References

1. Sander, T., Freyss, J., von Korff, M., Reich, J.R., and Rufener, C. (2009) OSIRIS, an entirely in-house developed drug discovery informatics system. *J. Chem. Inf. Model.* **49** (2), 232–246.
2. Warne, M. and Pemberton, L. (2009) Informatics in compound library management. *Methods Mol. Biol.*, **565**, 55–68.
3. Gobbi, A., Funeriu, S., Ioannou, J., Wang, J., Lee, M.L., Palmer, C., Bamford, B., and Hewitt, R. (2004) Process-driven information management system at a biotech company: concept and implementation. *J. Chem. Inf. Comput. Sci.*, **44**, 964–975.
4. Feuston, B.P., Chakravorty, S.J., Conway, J.F., Culberson, J.C., Forbes, J., Kraker, B., Lennon, P.A., Lindsley, C., McGaughey, G.B., Mosley, R., Sheridan, R.P., Valenciano, M., and Kearsley, S.K. (2005) Web enabling technology for the design, enumeration, optimization and tracking of compound libraries. *Curr. Top. Med. Chem.*, **5**, 773–783.
5. (2001 The Automation Partnership: Merck Orders the Automation Partnership Compound Repository. Press release, January 29.
6. (2009) RTS-life Science: RTS Life Science Automation Chosen to Increase UK Biobank Sample Processing Capability. Press release, August 18.
7. Richard, F., (2009) Titian Software: 10 Years on and the Future Continues Bright for Titian. Press release, August 4.

15
What Does an HTS File of the Future Look Like?

François Bertelli

15.1
Introduction

The screening of chemical compounds for pharmacological activity has been ongoing in various forms for at least 40 years. High-throughput screening (HTS) is the process in which small-molecule compounds are tested for binding activity or biological activity against target molecules. Test compounds act as inhibitors of target enzymes, as competitors for binding of a natural ligand to its receptor, as agonists or antagonists for receptor-mediated intracellular processes, and so forth.

The screening paradigm says that when a compound interacts with a target in a productive way, that compound then passes the first milestone on the way to becoming a drug. Compounds that fail this initial screen go back into the library, perhaps to be screened later against other targets. HTS is a key link in the chain comprising the industrialized drug discovery paradigm (Figure 15.1). Positive HTS results are usually called *hits*. Compounds resulting in hits are collected for further testing, in which, for example, the potency of an enzyme inhibitor or the binding affinity of a ligand for a receptor may be determined.

Today, many pharmaceutical companies are screening several million compounds per screen to produce only few hundreds hits. After this second level of triage, hits become lead compounds. Further synthesis may then be required to provide a variety of compounds structurally related to the lead. These sublibraries must then be screened against the target again in order to choose optimal structures. At this stage, some basic indicators of toxicity or bioavailability may be considered in an attempt to eliminate potential failures as early in the discovery process as possible. This is why a new focus has been developed on improving lead generation and therefore original hit. The better the starting point, the more quickly can medicinal chemists improve the lead into a drug-like molecule. This lead-likeness concept is crucial when thinking of the HTS collection of the future. Compounds which have no chance of being followed up, though they bring diversity, have to be removed from the file. Also, generating large libraries containing very similar compounds is counterproductive to the drug discovery process at HTS stage. The aim of the

Management of Chemical and Biological Samples for Screening Applications, First Edition.
Edited by Mark Wigglesworth and Terry Wood.

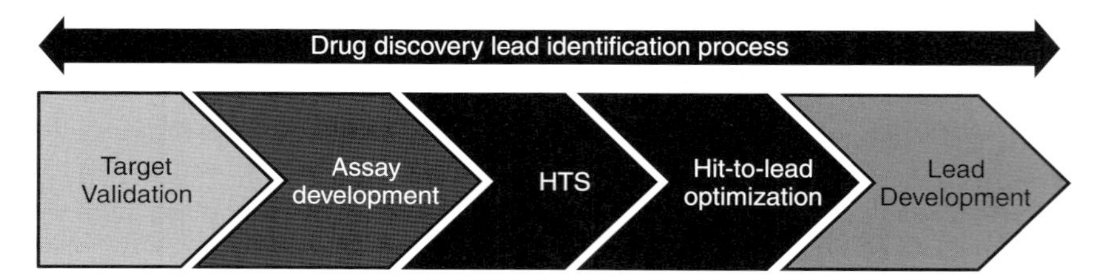

Figure 15.1 Chart showing the different stages of the early exploratory drug discovery process from target identification and validation through preclinical with the lead optimization.

HTS is to discover specific series; screening similar compounds displaying similar activity is part of the lead optimization process (Figure 15.1). To design the file of the future it is therefore important to include only compounds from the same library which are dissimilar enough for the series not to be missed during the initial HTS. Methodologies explaining how these compounds are compared and how to define the maximal size of each library integrating the corporate collection are described. With such drastic modifications of the future file, the impact of future hit identification strategy is evident. From a default full file screening we are moving toward a tailor-made target screening.

15.2
History of Compounds Collection for HTS

The first mass screening reported took place at the Terre Haute laboratories of the Pfizer Company, when 56 scientists tested almost 100 000 soil samples over a period of approximately one year. This program led to the discovery of the novel antibiotic product terramycin, which eventually captured 25% of the American broad-spectrum antibiotic market. It was first found near the Pfizer laboratories in a soil sample yielding the soil actinomycete, *Streptomyces rimosus*. In 1950, Robert B. Woodward, worked out the chemical structure of oxytetracycline, enabling other scientists at Pfizer to mass produce the drug under the trade name, terramycin [1]. This discovery by Woodward was a major advancement in tetracycline research and paved the way for the discovery of an oxytetracycline derivative, doxycycline, which is one of the most commonly used antibiotics today.

Once established as a technique, this practice was expanded to other sources of chemicals. Initially this included naturally derived extracts from plants, fungi, and microbial sources for the fixed therapeutic indication of infectious disease treatment, but it later expanded to include more complex diseases such as hypertension, metabolic diseases, and psychiatric disorders. Much of this expansion was driven by advances in medicine and the associated biological sciences that led to a better understanding of the molecular causes of disease. This methodology, which prefigured the screening in pool, proceeded with only incremental improvements until the late twentieth century, when several forces conspired to bring about significant changes.

The discovery of DNA in 1953 [2] soon gave rise to the new fields of biotechnology and genetic engineering, which, in turn, made possible the large-scale production of proteins and the creation of engineered cell lines such as chinese hamster ovary (CHO) through cloning technologies [3], which enabled the production of large quantities of protein without following cumbersome activity-based purification processes. This was followed closely by breakthroughs in the chemical synthesis technique of parallel synthesis (also known as *combinatorial chemistry*) for the large-scale production of synthetic medicinal chemistry compounds [4]. All the drug discovery tools were there; automation was then implemented, and this led to the formation of specialized departments to perform screens and, subsequently, the formal creation of HTS groups.

15.3
Impact of High-Throughput Chemistry on Corporate Files

Traditional corporate compound collections prior to HTS consisted of thousands or tens of thousands of discrete entities, usually in the form of analytically pure powders. It became clear in the late 1980s that the capacity to screen compounds for biological activity would greatly outstrip the compound collections of most pharmaceutical companies. Industry leaders thus started to build their compounds collection through purchasing of compound sets and by investing heavily in combinatorial chemistry technologies. These new advances built upon the Nobel Prize-winning work of Merrifield in solid-phase synthesis [5] by showing that not just one but many molecular entities could be prepared through a logical sequence of chemical reactions. Today those principles have changed little since the 1980s; what has changed is the technology to enable the science by increasing the efficiency and quality and by decreasing the timeframes and costs. Screening of those mixtures was thus similar to screening natural products in the old days except that no organism will ever produce so many similar entities for itself. Natural diversity will never be matched by synthesized mixtures. Creating molecular diversity in a test tube is a dogma with the drug space estimated to be between 10^8 and 10^{19} unique compounds [6]. Furthermore, combinatorial chemistry, not being able to create genuine diversity, was in contrast creating something even worse, which is now a plague in all large corporate collections: Redundancy. Thus, combinatorial chemistry produced hundreds or thousands of compounds in libraries where they ended up looking like each other and, even better, showing very similar biological activity profile. However, combinatorial chemistry is still one of the pillars of drug discovery as it is the quickest way to generate structure–activity relationship (SAR) in order to select the best lead compound for project progression.

Corporate collections contain molecules from a variety of sources, with a core set of compounds representing the synthetic efforts of all their chemists over years. These compounds usually represent close analogs of series that were created

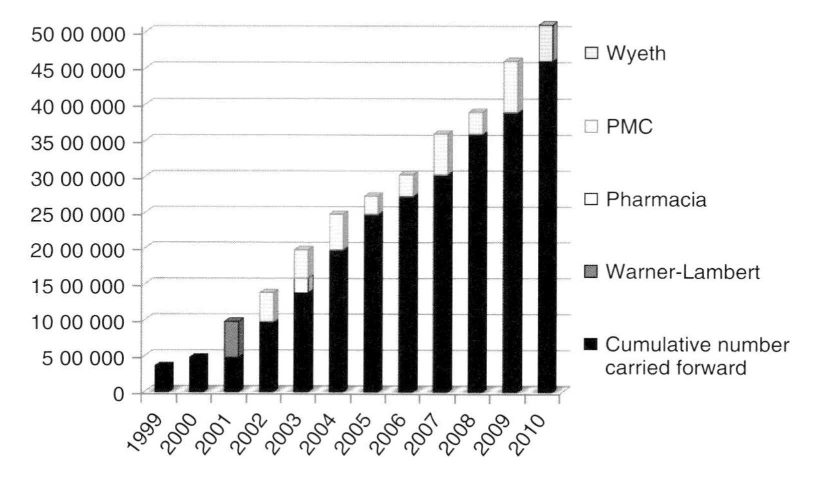

Figure 15.2 Chart showing the evolution of the number of compounds in the Pfizer corporate file for the past 10 years.

for specific programs and are not very diverse. Typically, the Pfizer collection has grown from historical central nervous system/cardiovascular/anti-infectives programs to the incorporation of a more biased kinase-focused, ion channels and metabolic collection with the acquisition of Warner-Lambert/Agouron, then moving more into oncology and pain with the pharmacia/sugen acquisition. Only then was a major initiative to expand library diversity from custom synthesis triggered. Aggressive efforts in library expansion from five major suppliers (Arqule, Chembridge, ChemRx, Tripos, and Wuxi) in parallel have led to the creation of a corporate collection of more than 5 million compounds (Figure 15.2). But breaking the records as the biggest corporate file in the industry has forced Pfizer to re-think drastically its primary screening collection, otherwise HTS operations would no longer be a viable option.

15.4 Chemical Library Management

The assembly and management of chemical compound libraries has evolved from a secondary consideration to one of central importance in HTS. In most companies, the compound library is treated as a one of the most valuable corporate assets and is managed centrally. This allows global corporations to ensure consistency across many testing sites. In an effort to increase efficiency, most of the major pharmaceutical companies have now centralized and co-located compounds management and HTS operations.

It is virtually impossible to handle hundreds of thousands of samples per day in a powder format. Even with automation, the task of accessing and retrieving that many individual samples, let alone weighing and solubilizing them, could not be managed cost effectively. But that many samples must be fed into every HTS operation every day. The solution to this problem is the creation of presolubilized compounds stored in microplates. The solvent dimethylsulfoxide (DMSO) is the industry standard for compound solubilization. Although DMSO is widely regarded as a universal solvent and can solubilize most compounds, it does not, by any means, dissolve all chemicals, and has properties that can make it difficult to handle [7]. Nevertheless, no viable substitute has been found, and the operational concerns of HTS prevent the use of multiple solvents, so DMSO remains ubiquitous in HTS. Once solubilized, the compound library is normally stored in separately addressable tubes or directly in microplates.

Because HTS laboratories have robotic systems already in place for handling and pipetting into and from microplates, having the compounds in the same format facilitates the transfer of compounds into the screening plates. Library management must, therefore, work in three distinct phases; dry sample management, solubilization, and plate-based or liquid-phase library management. Although any of these steps can be accomplished in a basic laboratory setting, each of them requires specialized equipment for high-throughput implementation. Automation systems supporting library management are the largest and costliest systems in HTS. A typical system that can store several million samples with random-access storage and retrieval capacity can be roughly the size of a gymnasium, with robotically accessed sample carousels or stacks 3–4 m high in a controlled environment, a robot around which a building has to be put together (Pfizer's Remp AG system, Sandwich, UK). It is therefore crucial that the solubilized samples (liquid bank) is co-located with the HTS operations, whereas the dry samples bank could be remotely located or even outsourced to a third party like Sigma, Milwaukee. Because weighing is the most labor-intensive step in the process, enough sample will be aliquoted to support an operation for a period of several years.

The most basic concept in building a chemical library is chemical diversity. This is a mathematical representation of how different chemical compounds are from one another. These calculations are quite complex and usually involve sets of descriptors, which are variables describing different attributes of a chemical [8, 9]. They will range from very simple factors such as molecular weight to complex calculations describing the distance between specific types of molecular bonds. The goal is to create a collection of molecules which represents as broad coverage as possible of potential drugs with the minimum number of samples. Once an active molecule or series of molecules is found, they can be expanded by further chemical synthesis.

15.5 The Concept of Drug-Likeness and the Lipinski Rules

The fast identification of quality small-molecule lead compounds in the pharmaceutical industry through a combination of high-throughput synthesis and screening has become more challenging in recent years. Although the number of available compounds for HTS has dramatically increased, large-scale random combinatorial libraries have contributed proportionally less to the identification of novel leads for drug discovery projects. Therefore, the concept of 'drug-likeness' of compound selections has become a focus in recent years [10, 11]. In particular, the concepts of lead-likeness that followed drug-likeness propose that it is advantageous to use small, hydrophilic molecules to start drug discovery projects [12, 13]. This allows the processes of chemical optimization, which frequently increase physical properties, to work within drug-like space. These strategic developments, which clearly emphasize 'small is beautiful,' have yet to be fully accepted or implemented as standard practices in some medicinal chemistry departments.

In parallel, the low success rate of converting lead compounds into drugs, often due to unfavorable pharmacokinetic parameters, has sparked a renewed interest in understanding more clearly what makes a compound drug-like. Various approaches have tried to address the drug-likeness of molecules, employing retrospective analyses of known drug collections as well as attempting to capture 'chemical wisdom' in algorithms.

15.5.1 Drug-Like

Still, the meaning of 'drug-like' is dependent on mode of administration. The original Rule of five (Ro5) deals with orally active compounds and defines four simple physicochemical parameter ranges (highlighted in gray in Table 15.1) associated with 90% of orally active drugs that have achieved phase II clinical status [11]. These physicochemical parameters are associated with acceptable aqueous solubility and intestinal permeability and comprise the first steps in oral bioavailability. The Ro5 were deliberately created to be a conservative predictor in an era where medicinal and combinatorial chemistry produced too many compounds with very poor physicochemical properties. The goal was to change chemistry behavior in the desired direction. If a compound fails the Ro5 there

Table 15.1 Typical range for parameters related to drug-likeness, with the Lipinki's rule of 5 highlighted in gray.

Parameter	Minimum	Maximum
LogP	−2	5
Molecular weight (MW)	200	500
Hydrogen bond acceptors	0	10
Hydrogen bond donors	0	5
Molar refractivity	40	130
Rotatable bonds	0	8
Heavy atoms	20	70
Polar surface area (PSA) (AA^2)	0	120
Net charge	−2	+2

is a high probability that oral activity problems will be encountered. However, passing the Ro5 is no guarantee that a compound will become a drug. Moreover, the Ro5 says nothing about specific chemistry structural features found in drugs or non-drugs, hence the extension of those physicochemical characteristics described in Table 15.1. Rotatable bond count is now a widely used filter following the finding that greater than eight rotatable bonds correlates with decreased rat oral bioavailability [14]. An analysis of small drug-like molecules suggests that a filter of LogD > 0 and <3 enhances the probability of good intestinal permeability [15]. If a drug has a low molecular weight (MW), does the method of delivery have to be oral? Can the Ro5 be by-passed by delivering the drug by a non-oral route (e.g., pulmonary, intra nasal, or dermal)? Oral drugs are lower in MW and have fewer H-bond donors, acceptors, and rotatable bonds. Pulmonary drugs tend to have a higher polar surface area (PSA) because pulmonary permeability is less sensitive to polar hydrogen-bonding functionality [16]. Figure 15.3 shows that some approved drugs can violate the Ro5, especially with respect to the MW and cLogP, and still be a blockbuster, as shown with atorvastatin, whereas kinase inhibitors like sunitinib and imatinib are well within the Ro5. However, some compounds in development are still stretching those boundaries significantly; this is illustrated by BMS-790052, which needs to compensate for its MW (739) by having exceptional affinity to its target, and produced miraculous results in its early clinical trials. BMS-790052 is an NS5A inhibitor obtained from Hepatitis C Virus (HCV) replicon screening. Indeed, EC_{50} values obtained ranged from 9, 50, and 146 pM for 1b, 1a, and 3a genotypes respectively [17].

Atorvastatin Lipitor atorvastatin	Sunitinib SUTENT capsules sunitinib malate	Imatinib gleevec (imatinib mesylate) tablets	BMS-790052 phase II/III
MW 558	MW 398	MW 494	MW 739
cLogP 5.9	cLogP 3.0	cLogP 4.5	cLogP 4.7
HB A/D 5/4	HB A/D 3/3	HB A/D 6/2	HB A/D 6/4
Rot bonds 12	Rot bonds 7	Rot bonds 7	Rot bonds 15
PSA 112	PSA 77	PSA 86	PSA 175
2 Ro5 violations	No Ro5 violation	No Ro5 violation	1 Ro5 violation + 2 drug-like

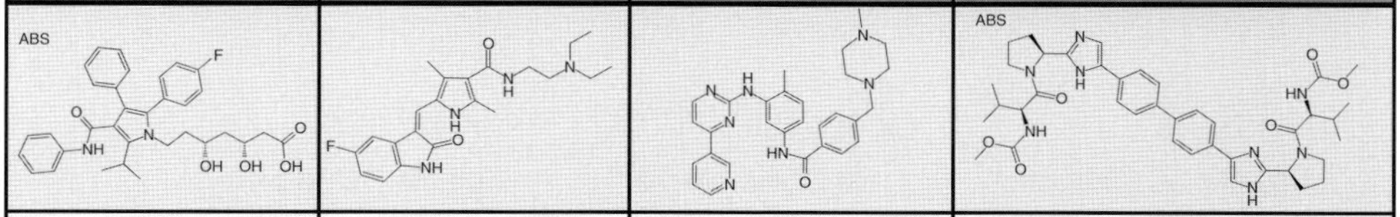

Figure 15.3 Drug-likeness and rule of 5 violation for three approved drugs and one drug in development with several rule violations.

15.5.2
Lead-Like

However there is a clear difference between drug-like and lead-like, which has been extensively described [13]. Although the HTS hit criteria are getting much more like lead criteria and because it is understood that a quality lead as opposed to a flawed lead is more likely to result in a real drug, there is now a strong structural resemblance between starting lead and drug. There is a pressing need to go back to the original drug discovery process, which is to find a specific small chemical template as a starting point. In one lead-like definition, compounds have reduced property range dimensions compared to the drug. In another definition, lead-like discovery refers to the screening of small MW libraries with detection of weak affinities in the high micromolar range, which is closer to the hit discovery, therefore HTS. Leads are less complex in most parameters than drugs, which is understandable in that medicinal chemistry optimization almost invariably increases MW and LogP. The discovery of Maraviroc followed the ideal drug discovery pathway, moving from weak hit to lead-like molecule then to drug within 10 years, with a clear improvement of most of the key drug-like properties recommended for an oral drug (Figure 15.4).

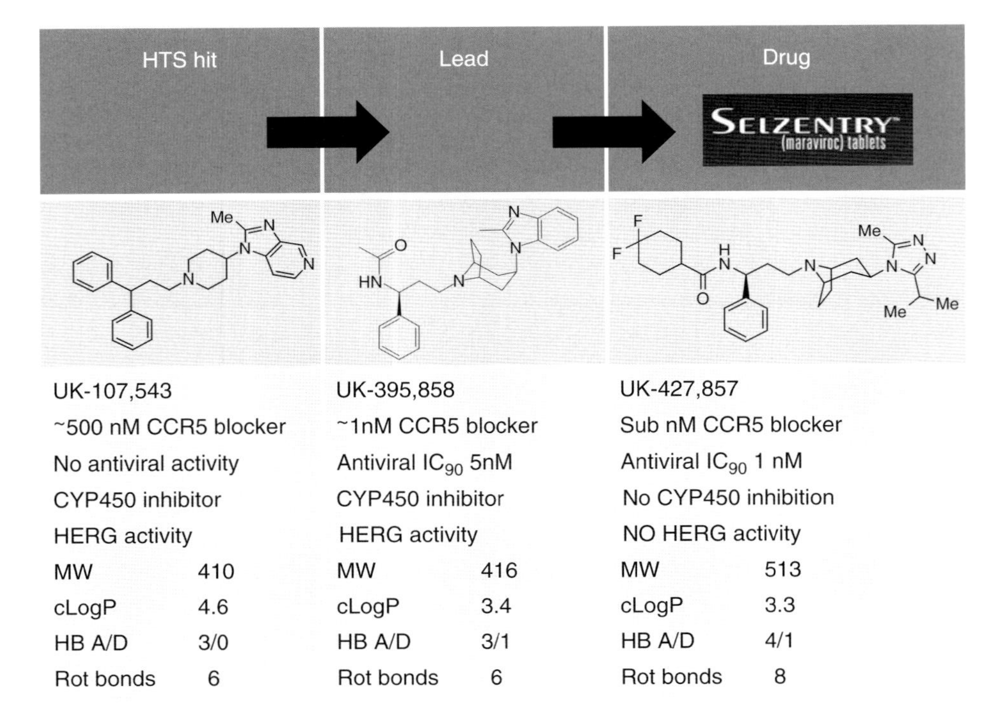

Figure 15.4 Hit-to-drug improvement of chemical properties for CCR5 antagonist Maraviroc.

However, looking at the physico-chemical properties of the patented compounds of the four companies with the most prolific contributions to small-molecules discovery over the last few years, there are clear differences [18]. The rank order of cLogP and molecular mass values agree for both the Prous Science Integrity and GVK Bio patent data [19]. For median cLogP, GlaxoSmithKline > Merck > AstraZeneca > Pfizer; for median molecular mass, GlaxoSmithKline = Merck = AstraZeneca > Pfizer. Taking the average property values for GVK Bio compounds aggregated from all companies confirms that the movement of chemical space in current chemistry to higher molecular mass and cLogP has progressed even further from historical oral drugs, recent oral drugs and development compounds (3.1–4.1 and 432–450 Da for cLogP and MW respectively).

In order to complete the picture around corporate medicinal chemistry practices, all four companies have been pursuing the chemokine (C–C motif) receptor 5 (CCR5) as the target for HIV or rheumatoid arthritis, and they all arrived to the same class of CCR5 antagonist containing the same common phenylpropylpiperidine pharmacophore [18]. The interesting feature is that the spread in cLogP (2.3 units) and especially the molecular mass (110 Da) reflects exactly the trend described above for the four companies. Not surprisingly, the company with the lowest compound molecular mass and cLogP is Pfizer, the originator of the Ro5, making a conscious choice to stay within those boundaries. The differences observed with the other companies are important and the risks taken could be reflected in their development pipelines. It is down to the medicinal chemist to stay in or out of approved physico-chemical properties and therefore to have a direct impact on the development attrition rates.

15.6 Quality versus Quantity

Besides the debate about how large a corporate compound collection should be, the questions of how to judge the quality of the inventory, and how to ultimately improve it, are important issues. The collections of large pharmaceutical companies are over several millions of entities, which represents historical collections, integration of acquired companies collection, and efforts to enrich the file by outsourced custom-synthesized file enrichment (FE) libraries.

This is around an order of magnitude higher than 15 years ago, when HTS and combinatorial chemistry first emerged. Although this number is somewhat arbitrary, logistical hurdles and cost issues make this inventory size an upper limit for most companies. Many research organizations subsequently scaled back their large compound production units after the realization that the quality component needed to get reliable and information-rich biological readouts cannot be obtained using such ultra-high-throughput synthesis technologies favored in the early 1990s. Today, instead of huge internal combinatorial chemistry programs, purchasing efforts in every pharmaceutical company are directed toward constantly improving and diversifying the compound collections, and making them globally available for random HTS campaigns [20].

Although the chemical integrity of compounds can be checked by various analytical techniques, determining whether the chemical entities are useful in general as starting points for a hit-to-lead program is far more complex. Besides the variable perceptions of medicinal chemists of what makes a valuable hit (lead-like versus drug-like), the issue concerning structural similarity, and in particular the overlap of chemical space between libraries, is frequently debated.

Various computational algorithms can be applied for the validation of compound collections to be acquired in terms of their lead-likeness, chemical diversity, and similarity to the existing corporate compound inventory [21]. Although prediction tools for physicochemical properties and aggregators (frequent-hitters) are successfully applied in a routine fashion, the issue concerning diversity is largely unresolved [22]. When combining two large compound collections, the question is always raised 'how much extra chemical space coverage is gained?' but what constitutes a valuable gap in the lead-like chemical space is still very empirical. The value of structure-oriented diversity descriptors is not in question, although determining pharmacologically relevant similarity is far more complex and cannot be described accurately by any single metric, such as biological activity of affinity parameters. Conversely, it is difficult to describe the similarity (or dissimilarity) of two compounds that display the same activity, but possess, for example, different functionality, selectivity, toxicological liabilities, and so on. Similarity is a context-dependent parameter and therefore the context must define the appropriate metric, otherwise it is meaningless. In any case, to increase the quality of a compound inventory, certain filtering techniques have to be applied for weeding out compounds that contain unfavorable chemical motifs. Database-searching tools have been developed that allow the differentiation between desired and undesired compounds. These computational algorithms are often based on sub-structural analysis methods, similarity searching techniques, or artificial neural networks. Besides the application of these for filtering physically available compound collections or vendor databases, such algorithms can of course also be used to screen and validate virtual combinatorial libraries. It is in this setting that computational screening can have the greatest impact, owing to the overwhelmingly large number of compounds that are synthetically amenable using combinatorial chemistry technologies.

The motifs for undesirable compounds have been classified, and establishing filters for these is relatively straightforward [23]. In an attempt to discard the corporate file from those templates, more than 600 'chemical alerts' filters have been implemented in order to 'clean' the file but also to prevent any undesirable compounds from entering the collection.

As a consequence, there is now a clear trend to move away from huge and diverse 'random' combinatorial libraries toward smaller and focused drug-like subsets. Although the discussion of how focused or biased a library should be is still an ongoing debate, the low hit rate of large, random combinatorial libraries, as well as the steady increase in demand for screening capacity, has set the stage for efforts toward small and focused compound collections instead.

15.7
The Emergence of the Subsets: Fragment, G-Protein-Coupled Receptor (GPCR), Ion Channel, Kinase, Protein–Protein Interaction, Chemogenomics, Library Of Pharmacologically Active Compounds (LOPAC), Central Nervous System (CNS), and Diversity

Typically, screening collections of pharmaceutical companies contain more than a million compounds today. However, for certain HTS campaigns, constraints posed by the assay throughput and/or the reagent costs make it impractical to screen the entire file. Therefore, it is desirable to effectively screen subsets of the collection based on a hypothesis (targeted-subset screening) or a diversity selection when very little or nothing is known about the target. How to select compound subsets is a subject of ongoing debate. The 'combinatorial explosion,' meaning the virtually infinite number of compounds that are synthetically tractable, has fascinated and challenged chemists ever since the inception of the concept. Independently of the library designs, the question of which compounds should be made from the huge pool of possibilities always emerges immediately, once the chemistry is established and the relevant building blocks are identified. The original concept of 'synthesize and test,' without considering the targets being screened, was frequently questioned by the medicinal chemistry community and is nowadays considered to be of much lower interest because of the unsatisfactory hit rates obtained so far. The days in which compounds were generated just for filling up the companies' inventories, without taking any design or filtering criteria into account, have passed. In fact, most of the early combinatorial chemistry libraries have now been largely eliminated from the standard screening sets because of the disappointing results obtained after biological testing. Primary screening has changed from mass screening to smart screening, implementing knowledge based upfront and screening less but with high quality in order to identify series instead of singleton hits.

The marriage of HTS with computational chemistry methods has allowed a move away from purely random-based testing toward more meaningful and directed iterative rapid-feedback searches of subsets and focused libraries and performing staggered screening instead one-go full file. The prerequisite for success of both approaches is the availability of the highest quality compounds possible for screening, either real or virtual.

Another widely used approach in the generation of targeted compound collections is ligand motif-based library design. This is particularly relevant for targets for which very limited or no biostructural information is available. It is here that elements of known biologically active molecules are used as the core for generating libraries encompassing these privileged structures. Especially in the area of GPCRs, such design tactics have been applied successfully to create a subset based on all corporate historical data for all three GPCR families in agonist, antagonist, and allosteric modulation. An inherent issue linked to this approach is the fact that these motifs can show promiscuous activity for whole target families, so selectivity considerations have to be addressed very early on. The restricted availability of privileged structures and resulting issues concerning intellectual property clearly limit the scope of this ligand-based approach to some extent. As a result, there is a continued need to identify novel proprietary chemotypes, and computational tools have already shown their potential in this area.

Subsets could be classified into two main categories: the knowledge-based subsets, which consist of a significant gathering of information from corporate database around specific type of targets in order to select compounds with a high likelihood of hitting the same type of targets, and the diversity subsets, which are purely based on chemical properties and similarity. The first category consists of the targets with a long history such as the kinases and the GPCR, which have constituted the majority of portfolio targets in the last 15 years. Also in this category is included the ion channel subset, although there is still a lot to learn about the difference between the modes of action of the ion-based (Ca, Na, K, H, ⋯) and ligand-based channels, and some companies like Wyeth have started to gather all this information into a subset [24]. Some subsets could be designed based on the backbone structure of the compounds such as peptides, peptoids, or macrocycle for protein–protein interaction targets or based on the physicochemical characteristics of compounds such as the CNS-penetrant subset, for which it is not the target which matters but its location; there are some properties such compounds need to have in order to reach the target within the CNS. A PSA value of less than 60–70 tends to identify CNS-active compounds [25]. A very simple set of two rules predicts CNS activity: If $N + O$ (the number of nitrogen and oxygen atoms) in a molecule is less than or equal to five, it has a high chance of entering the brain. The second rule predicts that if $\log P - (N + O)$ (the CNS (MPO) multiparameter optimization score) is positive, then the compound is CNS-active [26]. A computational prediction cannot be better than the underlying experimental data set. The log–brain to blood–drug concentration ratio is almost universally the experimental measurement that is used for predicting CNS-active compounds.

15.7.1
'Cherry Picking' from Virtual Space

A highly sophisticated way to avoid the synthesis of trivial analogs is the application of virtual screening tools in order to search through chemical space for topologically similar entities using known actives (seed structures) as references. In addition, biostructural information can also be applied if available [27]. Principally, one can subdivide such a virtual screening exercise into three main categories, namely virtual filtering, virtual profiling, and virtual screening. The first focuses on criteria that are based on very fundamental issues concerning pharmacological targets in general. In this filtering step, all candidates are eliminated that do not fulfill certain generally defined requirements. These elimination criteria can either be based on statistically validated exclusion rules, substructural features, or on training sets of known compounds. Virtual screening technologies help to filter out the unfavorable combinations and predict actives out of such a library proposal if particular target and/or ligand information is available.

15.7.2
Diverse Subsets

The second main category of compound subsets is the diverse subsets. They are not constructed from any target knowledge but are only based on lead-likeness and similarity scores. Nature itself constructed one of them. Indeed, natural products are the most diverse compound class, with significantly higher hit rates compared to the compounds from the traditional synthetic and combinatorial libraries. But unfortunately nature can go as wild as possible and the templates observed could go well beyond any lead criteria and are therefore most of the times disregarded by the medicinal chemists. Nevertheless, as described below, natural product templates have taken a significant share of the drug landscape market especially in anti-infectives and oncology.

Natural products are undeniably the best source for diversity in chemotype for the discovery of novel therapeutics. One-third of the top selling drugs in the world are natural products or their derivatives. Roughly 60% of the antitumor and anti-infective agents that are available commercially or were in late stage clinical development from 1989 to 1995 are of natural products origin. Even recently discovered natural products may have still an impact in the oncology, CNS, and cardiovascular therapeutic areas [28, 29].

However, natural products have been de-emphasized as HTS resources in the recent past, certainly at Pfizer, in part because of difficulties in obtaining high-quality natural products screening libraries, or in applying modern screening assays to these libraries. In addition, natural products programs based on screening of extract libraries, bioassay-guided isolation, structure elucidation, and subsequent production scale-up are challenged to meet the rapid cycle times that are characteristic of the modern HTS approach. Fortunately, new technologies in mass spectrometry, NMR, and other spectroscopic techniques can greatly facilitate the first components of the process – namely the efficient creation of high-quality natural products libraries, biomolecular target or cell-based screening, and early hit characterization. The success of any HTS campaign is dependent on the quality of the chemical library. The construction and maintenance of a high-quality natural products library, whether based on microbial, plant, marine, venom, or other sources is a costly implication. The library itself may be composed of samples that are themselves mixtures – such as crude extracts, fractionated or semi-pure mixtures, or single purified natural products. Some cone snail venoms have been described to contain more than 100 different toxins. Each of these library designs carries with it distinctive advantages and disadvantages. Crude extract libraries have lower resource requirements for sample preparation, but high requirements for identification of the bioactive constituents alike the deconvolution applied to the compressed screening. Pre-fractionated libraries can be an effective strategy to alleviate interferences or toxicity encountered with crude libraries, and may shorten the time needed to identify the active principle. Purified natural product libraries require substantial resources for preparation, but offer the advantage that the hit detection process is reduced to that of synthetic single-component libraries; this is the current preferred option within HTS. The integration of the Wyeth collection into the Pfizer corporate file included its purified natural products library; moving forward this set will be part of Tier 2 screening, described later. As we move to more intractable targets, whether by cell-based or biomolecular target-based assays, screening of natural product extract libraries continues to furnish novel lead molecules for further drug development. Despite challenges in the analysis and prioritization of natural products hits, they can always be a new chemical source in the generation of a specific template more within the 'lead-like' environment.

Another type of diverse subset consists of the monomers or fragments. These low-MW libraries are often referred to as *fragment libraries*. Small-fragment screening can be performed by NMR, X-ray or in theory any method capable of detecting weak interactions, which includes conventional screening at high concentration. A rule of three has been coined for these small-molecule fragment screening libraries; $MW < 300$; $\log P < 3$; H-bond donors and acceptors <3 and rotatable bonds <3 [30]. The experimental compared with the theoretically achievable diversity is much higher in a low-MW fragment-based library. Combining some active low-MW chemical features together into a single molecule could provide interesting starting points to some intractable targets, especially in the protein–protein interaction arena [31].

The last type of diverse subsets is made from gathering physico-chemical properties and by calculating a similarity parameter in order to cover maximal chemical space.

15.7.3
Creation of the Global Diversity Representative Subset (GDRS)

The growth of the Pfizer compound file has led to the need to identify subsets of compounds for use where it is not feasible or desirable to screen all compounds, whether prepared using singleton or combinatorial chemistry methods. The non-feasibility or desirability may be due to a time constraint, for instance, the urgent need for a tool or lead compounds for a target that does not have an assay ready for large-scale HTS, or a practical one such as the supply or toxicity of reagents. Using clustering methods for our data, the cluster-based subsets have higher mean hit rates than those of the random ones. In addition, subsets comprising random plates are compared with subsets of random compounds. While the mean hit rate of both is the same, the former demonstrates more variation in hit rate. The choice of compound file, rational subset method, and ratio of subset size to compound file size are key factors in the relative performance of random and rational testing. To design the global diversity representative subset (GDRS), 'lead-like criteria' from the Pfizer file have been used [32].

The Tanimoto coefficient (T_c) which is most widely used for binary fingerprints between two molecules has been used to select the compounds for the most diverse set construction [33].

$$T_{(A,B)} = \frac{c}{a + b - c}$$

where a = bits set to 1 in A, b = bits set to 1 in B, c = number of 1 bits common to both. The dissimilarity is then defined as $1 - T_{(A,B)}$. A higher T_c value means that the two molecules have higher similarity. As far as the fingerprint descriptor is concerned, a T_c of 1 means that both structures have the same fingerprint, while a value of 0 indicates that there is no overlap between these two fingerprints.

Series in the full file have been clustered and selected using a 0.5 similarity Tanimoto radius. The legacy Pfizer had many compounds that contribute to its diversity but are highly undesirable in terms of lead matter and were therefore deselected. The designed subset has been designed to be a good balance between legacy medicinal chemistry efforts and FE chemistry space. Using the Tanimoto radius calculation to define structural similarity, the ratio of structurally diverse clusters for the legacy file versus the FE library is 1 FE for three legacies, emphasizing the fact that there is a significant amount of redundancy within the FE component of the file. >95% of FE series have been captured between 0.5 and 0.6 Tanimoto radius and will therefore be represented in the GDRS subset. This translates to between 20 and 40 K compounds selected and can represent >95% of the FE chemical space. A complementary FE set which contains alternative compounds that cover from the same chemical can be used to confirm the activity. Screening both complementary sets provides a little more SAR and maximizes chances of finding series. The GDRS contain an equal proportion of FE and legacy compounds (~75 k each), with most but not all unattractive chemistry filtered out. The compounds are nested in tranches according to physical properties for ease of screening (see Figure 15.5). The usage of this subset is clearly defined; it provides:

- A chance to screen a non-HTS compatible target or assay format (e.g., high-content screening) with a diverse representation of the file
- A cost- and resource-effective alternative to a full-file HTS under specific constraints (e.g., reagents too costly)
- Access to selected compounds within specified property ranges relevant to your project (e.g., fragments or acidic compounds).

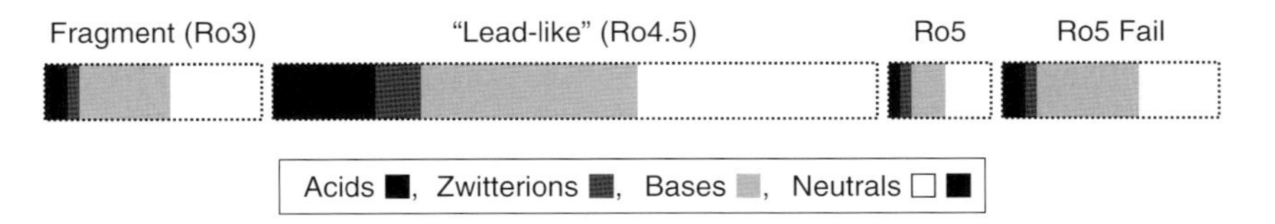

Figure 15.5 Composition of the global diverse representative subset (GDRS). Fragments are defined with cLogP < 3 and MW < 300. Lead-like Ro4.5 have MW < 450 with cLogP < 4. Ro5 are defined in the Table 15.1, and Ro5 fails where the compounds failed Ro5 for one parameter only.

- The opportunity to facilitate confidence in mechanism (CIM), confidence in rationale (CIR), and 'druggability' through the quick identification of 'tool' compounds as part of the target validation package.

On the other hand, the GDRS does not provide a guarantee that failure to find lead matter implies that lead matter is not present in the file, as it does not represent a full coverage of the file chemical space. Also, finding a tool from the GDRS does not mean that this particular compound is the most active of the series; there should be an iterative screening process in place in order to enrich the series obtained from screening the GDRS.

The success of the GDRS prompted a search for an alternative. Because this subset was largely distributed to all local screening groups and widely used, it was apparent that it would become depleted very quickly, and replenishment could be a limiting step in the process. Needless to say, the disappearance of these compounds, especially some singletons in the legacy set, could create significant chemical space gaps. It was therefore urgent to think of another diversity strategy which was less demanding regarding logistics and maybe more restricted to HTS groups instead or widely spread globally. In 2005, the plate-based diversity strategy was implemented to leverage GDRS usage and to provide extra chemical space and SAR during the primary screening.

15.7.4 Plate-Based Diversity Set (PBDS)

The compounds in the Pfizer liquid store center of emphasis are currently stored in 384-well plates. For the purpose of selecting a subset, automated 'cherry picking' of individual compounds from specific plates is possible. However, it becomes a nontrivial, resource-consuming endeavor when the total number of compounds to be cherry picked runs into tens or hundreds of thousands. So, when it comes to replenishing the GDRS subset this becomes unsustainable. In addition, companies' screening collections constantly change, some existing compounds being depleted and new ones being regularly added. Performing a chemical diversity selection on a per plate basis instead of a per compound basis offers a simple yet effective *in silico* alternative to select a subset of the collection for screening. If we were to implement a plate-based approach for subset selection, the next question would be how we measure the plate diversity, or how we determine the chemical diversity of compounds on a per plate (instead of per compound) basis. Measuring diversity of compounds is often tricky in practice, as the definition of a scaffold or chemotype is a matter of ongoing debate [34]. Current methods to measure diversity can be broadly divided into two categories, namely, rule-based and chemical descriptor-based methods. The rule-based methods usually reduce the compounds down to a core molecular framework, obtained using a set of prioritization rules. This reduced representation of compounds is then used as a diversity metric to classify them.

Murcko scaffolds, as defined by Bemis and Murcko [35], represent a well-established example of a rule-based method. Chemical descriptors, on the other hand, use numerical vectors, usually represented as discrete binary values, to abstract the structural features of compounds. The pairwise Tanimoto coefficient between the chemical descriptors is most commonly used to quantitatively measure the similarity or dissimilarity between compounds [33, 36]. The variability in the results obtained from these methods, perhaps because of the differences in their 'definition, calculation, and uses' of diversity, underscores the challenges of expressing a medicinal chemist's intuition in a computer algorithm [36].

In previous work from the Novartis group [37], a plate-based diversity selection method has been described, based on Murcko scaffolds, to select a medium-sized subset of the screening deck. The objective was to select the combination of in-house plates that would maximize the diversity, defined by the number of unique Murcko scaffolds, in the selected subset. The same group two years later [38] then used 2D chemical descriptors – namely, ECFP-6 (extended connectivity fingerprints with a diameter of 6) as another definition of diversity, to perform a chemical diversity selection of compounds measured by a suitable objective function. The dissimilarity between the compounds in a plate was measured by calculating the average pairwise Tanimoto coefficient between their ECFP6 descriptors. The two methods, Murcko and ECFP6, were then compared on retrospective analysis of historical HTS. They recommended using ECFP6 fingerprints for selecting subsets of compounds to be screened in screening campaigns [39, 40].

Based on these methods the most diverse plates were selected and cut-offs for plate properties were derived to 90% of the 'best' plates, based on three criteria listed below and referred to as the '*Rule-of-40*' (as they are mostly all divisible by 40): more than 200 unique compounds on a plate, more than 160 Ro5 compliant compounds (MW cut-off >200 as well as <500), and more than 120 compounds free of chemical alerts. The size of the subset analysis identified 1200 plates which permit 95.4% double coverage of a defined BCUT chemical space within the HTS file.

However, screening a subset containing the most diverse plates is unlikely to yield all the hits that could have been identified by screening the full library, but it offers a useful alternative to screening the GDRS. Figure 15.6 recapitulates the features of both main Pfizer diverse subsets and their recommended usage.

15.8 Re-designing the Corporate File for the Future

15.8.1 Pooling Compounds Moving Forward

Advances in genomic biology and combinatorial chemistry have led to an increasing number of targets and compounds to screen. The past trend of reducing assay size is aiding HTS to keep pace with the increasing target lists and corporate file size, but miniaturization alone is insufficient. One of several ways to improve the efficiency of HTS is compound pooling (or compression), where mixes of

Property	GDRS	PBDS
Sub-set basis	Compound level	Plate level
File size	150,480	429,067
384-well plate number	418	1,200
Representative fraction of chemical space defined by 3MM HTS compounds covered *(as defined by BCUT analysis)*	56%	95.4%
Coverage overlap by compound identity	17.5% GDRS in PBDS	4.1% PBDS in GDRS
Logistics		
Amount compound available per plate set	0.1 μmoles	0.15 μmoles
Replenishment process time	1-2 months	Few days
Availability	Global	HTS groups only

Figure 15.6 Metrics of the two main Pfizer diverse sub-sets: the GDRS (compound-level diversity) and the PBDS (plate-based diversity).

compounds are tested in each well of an HTS assay rather than individual testing of each compound.

The central rationale of compression is that most compounds can be quickly identified as inactive via a negative result in a pooled test. The concept of testing pools of samples in biological experiments was conceived during World War II (WWII), when blood samples from cadets were pooled to optimize syphilis testing [41]. Dorfman sought an efficient method to screen WWII recruits for syphilis, since testing everybody individually was expensive. He proposed pooling large numbers of samples. If a pool tested negative for syphilis, all the samples could be declared negative with no further work. If the pool tested positive, then individual samples within the pool would be tested.

The review by Kainkaryam and Woolf [42] discusses the complexities and solutions to pooled drug discovery, and focuses on the following five topics: (i) design of pooling schemes, (ii) experimental success of pooling, (iii) mixing constraints and cutoff selection of pooling, (iv) additive, synergistic, and antagonistic effects of pooling compounds, and (v) recent advances in pooling.

Currently most HTS groups still employ the 'one compound, one well' approach. This approach is simple in terms of both the physical implementation of the assay and the analysis of results. However, given the usually small number of active compounds in a library that is typically comprised of a much large number of compounds, this approach can be wasteful. Also, HTS screens are prone to several forms of experimental error, which produce false positive (inactive compound classified as active) and false negative (active compound classified as inactive) results during analysis. False positive results are less of a problem because these compounds are retested at subsequent stages and then weeded out. In contrast, false negative results represent potential lead compounds that are lost to the development process because screening corporate files in replicate is prohibitively costly. Therefore, pooling of compounds could represent a solution to this challenge.

There is no unique way of pooling compounds together; the mathematical theory of pooling has undergone tremendous advances and has found applications in diverse areas [43]. An excellent introduction to the subject can be obtained from the book written by Du and Hwang [44]. There are several pooling schemes, depending on the aim and size of the screening.

In the late 1990s, as the corporate file was increasing significantly, Pfizer has implemented its own compression scheme. This was a combination of adaptive and orthogonal screening. As described in Figure 15.7, 7×96 – well blocks are used in duplicates to make one compressed 96-wells plate. Each compound is distributed into two different wells with a unique plate signature. Each well therefore contains a unique mixture of 14 different compounds. Four 96-well compressed plates are then consolidated into one final 384-well compressed plate containing 2520 unique compounds in duplicate. Because each compound has its own wells signature, the two wells containing the same compound have to be active for the compound to be confirmed as hit. The benefit of this strategy, which is still in use currently, is undeniable. Indeed, 14 singleton 384-well plates have to be screened to obtain the same result representing a 93% gain in productivity (Figure 15.8). Pooling at the level of individual compounds using this sevenfold compression scheme is a labor-intensive and costly process, but the return on investment is immediate

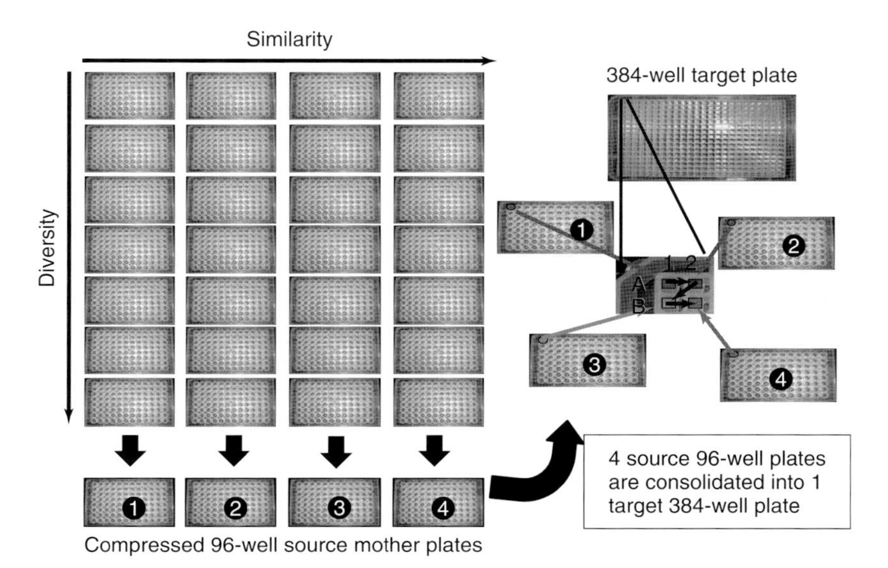

Figure 15.7 Schematic of the Pfizer sevenfold compression scheme in which 2520 unique compounds are pooled in duplicates into one 384-well plate.

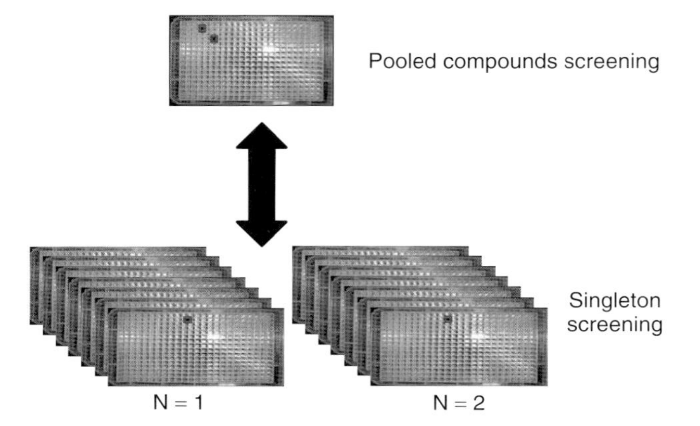

Figure 15.8 One 384-well compressed plate contains 5040 compounds, 2520 unique compounds in duplicates which is the equivalent of 14 384-well singleton plates representing a 93% cost-saving benefit. One true hit (black) with unique signature is obtained in duplicates in the same plate whereas 14 plates have to be screened for the same results.

after only few HTS campaigns. Furthermore, for some key targets, this is the only format whereby the full corporate file can be screened in a timely manner.

The recent decline in success of the application of the 'one compound, one well' strategy to high-throughput drug screening suggests that the strategy could benefit from further optimization. Pooling of compounds in HTS offers several potential gains, not limited to resource savings and error tolerance. Pooling also offers to address some of the critical issues experienced by HTS, such as the rapid increase in compounds and targets to be screened, the ever-present possibility of false negative testing errors, and the search for new classes of multi-compound, multi-target therapeutics. There are several challenges to the success of pooling in HTS, including the creation of optimal pooling design and analysis strategies, physical implementation of these pooling strategies, and management of synergistic or antagonistic behavior of compounds. Some of these challenges are also opportunities to extend the current capabilities of HTS in drug discovery maybe toward peptide or antibody screening.

In addition to time savings in primary screening, other factors favoring the pooling format include reagent conservation up to 80% (radioligands, purified proteins, plates, transfected cells, media, etc.), smaller storage requirements for the screening library, and reduced need for human resources. Since the cost of biological assays can vary from a few cents to dollars per well, with the potential cost of a moderately priced screen exceeding US$200 000, the cost savings from pooling are significant. Finally, target types expressed at low levels, or which are difficult to purify, become accessible by reducing the number of test points through compound library pooling.

Therefore, Pfizer and others took the initiative of pooling their corporate file in the 1990s. There are numerous factors motivating the effort to increase the density of compounds within the screening well in order to decrease the number of test points performed per screen.

There are, however, some caveats to the pooling approach:

1) A number of factors limit the number of compounds per screening well, including compound reactivities, ionic state, solubility, and biological effects. Approaches for strategic pooling of compounds to reduce reactivity, solubility, and ionic strength have been presented before [41].
2) Because compounds can enter a screening file in a non-random manner, compounds with similar structural features can reside within the same source, and, ultimately, assay plate. On the other hand, as compound libraries often contain clusters of structurally similar compounds, strategic pooling to increase compound diversity within screening plates can reduce the rate of false positives (i.e., compounds identified as hits in the HTS that are inherently inactive against the target).

In cell-based screening, the ultimate readout of the assay often results from a cascade of biological activities in the cell that are not related to the target of interest. Compounds which are cytotoxic, or which possess unknown activity against host proteins, may obscure the engagement of the target of interest. For example, an HTS designed to identify an agonist could be compromised by the presence of compounds in the same well that nonspecifically obliterate the effect on the target. Recent reports on high density (1536 and beyond) format implementations would suggest that screening in simplex is a possible choice for efficient screening of targets to obviate issues arising from compound pooling. However, it must be understood that high-density screening technology and assay formats are limited, leaving multiplexing of the library as the most practical alternative choice for the foreseeable future.

High organic load for a long period of time affects the viability of cells and most of the time triggers stress-related events which induce proteins secretion including read-out-dependent reporter enzyme causing a high level of assay interference (data not shown). By all means, compression screening should be part of the screening strategy options but certainly not the default. It is therefore important to assess at the screen validation stage whether the assay format is suitable for singleton or compressed screening. Most of the time, except for the example mentioned above, pool screening is the best option, the only caveat being that because of the amount of compounds per well, the compressed source plate has a lower concentration per compound than the equivalent singleton and for some targets such as protein–protein interactions where a concentration higher than 20 μM is required, compressed format could be limiting.

15.8.2
Re-designing the Future File

The primary purpose of the Pfizer screening file is to find multiple hit series for projects seeking leads for oral small-molecule medicines. The definition of largest tracts of SAR around each hit should not be the primary focus of HTS but the first stage post-HTS as the project moves into lead optimization. It has also been recognized that there was a need for chemical space flexibility around specific targets as we expand the druggable target space or at least the lead-like space toward a difficult target (i.e., non-oral delivery, protein–protein interaction, Epigenetics targets, tool-seeking exercise for deep knowledge target and pathways (DKTP) work, fragment screening, ⋯). All of these should be project-driven decisions and not default options.

Medicinal chemistry is based largely on the concept that similar chemical structures are likely to share similar biological activities [45]. This concept has been more rigorously developed recently to the point where chemical similarity can be numerically defined, and probability values can be assigned to the likelihood that a compound will share similar biological activity to a known active with related structure. Based on these hypotheses, there are clearly portions of the screening collection that are over-represented for the sole purpose of finding hit series. The aim is then to develop methodology to allow us to determine how many compounds we need in a hit series neighborhood to have circa 95% chance of finding a single active compound that is representative of that series, so we could then remove the unnecessary molecular redundancy. A combination of several methodologies was examined. To start with, medicinal chemist experts have compared multiple probe structures and have assessed the integrity of retention of features of the probe in neighbors as the Tanimoto similarity scores decrease. They also assessed the ability to identify hit series within the neighborhood of an active probe once the redundancy has been eliminated based on the belief theory which has been used successfully in lead hopping [46]. After repeating this comparison with multiple probes using multiple Tanimoto scores, decisions around file redundancy reduction could be implemented. Based on these computational methods, the file has been re-structured with a Tanimoto cut-off of 0.6 to eliminate similarity fail compounds and with a maximum cluster size of 40 compounds based on the belief theory using the hypothesis where probability of finding an active is between 5 and 10% (Figure 15.9). The other options such as using Tanimoto cut-offs <0.6 are also feasible, but lead to clusters of compounds with varying structural type where the concept of a chemical series is lost and would carve out a lot more compounds from the file. On the same note, using Tanimoto cut-offs >0.6 is feasible but too conservative, and would leave more residual redundancy in screening file.

During the similarity search, the fingerprint of the input structure is built up and compared with the fingerprints of the database entries using the Tanimoto coefficient described before and used for GDRS compound selection. A molecule with a Tanimoto coefficient ≥ 0.85 to an active compound is assumed to be

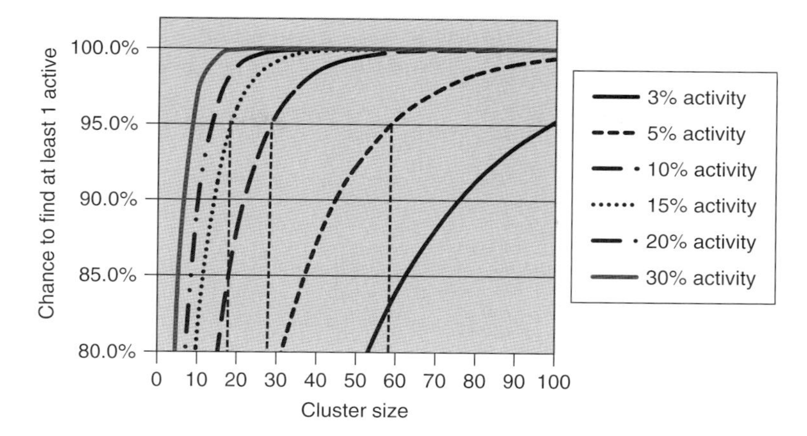

Figure 15.9 Graph showing the compound series cluster size determination based on the belief theory. The theory estimates the cluster size needed for >95% chance of finding at least one active by screening clusters in which the chance of finding any active is known. For clusters where probability of finding an active is 15, 10, or 5%, clusters of about 20, 30, and 60 molecules respectively are required.

biologically active itself [45]. The Tanimoto calculations are performed using Pipeline Pilot tool software (SciTegic).

Defining the target size for library is based on the belief theory [46]. For example, a cluster of 80 with a target size of 40 (redundancy $= 1 - (40/80) = 0.5$), each compound has 50% chance of being selected into the file, and so on. Exceptions arise when some compounds need to be kept in the file because they are part of subset or have a pharmacological activity (Keep list), the calculation can be biased as follows: cluster of 80 with a target size of 40 but with 20 in Keep list, we now need to select 20 from remaining 60 to give cluster of 40, redundancy corrected will be $1 - [(40-20)/(80/40)] = 0.67$. Each compound has now only a 33% chance of being selected.

Together with the Tanimoto scores, the ECFP4 similarity coefficient has been used to create the Tier 1 similarity filters. Less stringent than the ECFP6 similarity coefficient described earlier for the selection of the Plate-based diversity set (PBDS), the ECFP4 looks at four atoms radius. All the extended connectivity descriptors have been gathered and compared recently [47]. As an example, an ECFP4 scale has been calculated based on the PDE5 inhibitor sildenafil, where a value of 1 is for the sildenafil probe and 0 for the most dissimilar compound. Figure 15.10 describes a range of other PDE5 inhibitors, valium and how a randomness from the file would score on that scale. It is noticeable that minor change within a molecule could have significant impact on ECFP4. On the other hand, although similar compounds are likely to have the same activity, Viagra® and Cialis® are at the extreme of the ECFP4 scale, but they still inhibit the same target with similar potency, fortunately for the pharmaceutical industry.

The impact of reshaping the file into Tier 1 and 2 has been assessed by looking retroactively at the read-out of a diverse panel of 13 targets (T1 to T13). The series

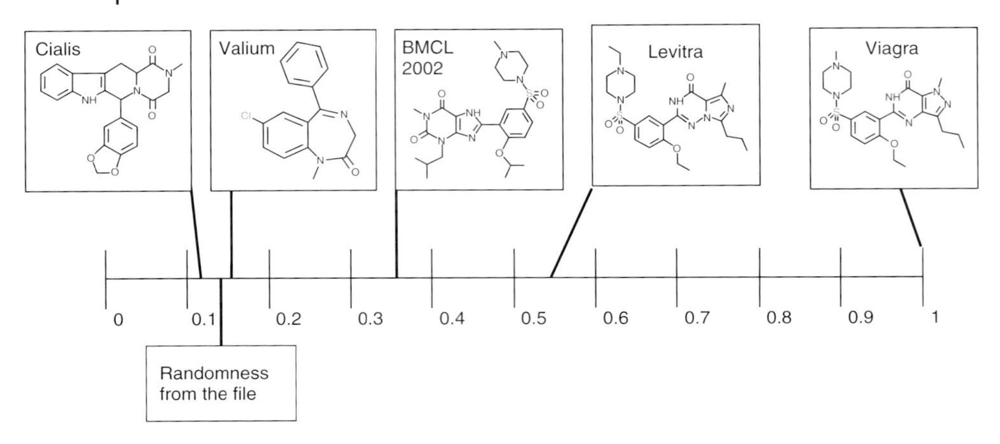

Figure 15.10 Example of ECFP4 similarity scale using Sildenafil (Viagra™) as probe. ECFP4 scores were calculated for other PDE5 inhibitors (Levitra™, Cialis™, and BMCL2002) and against Valium as unrelated drug.

obtained from these screening campaigns have been clustered and numbered. The analysis looks at how many of these series were removed by the new file entry filters, for example, how many would have been missed if only the Tier 1 and/or 2 were screened. Figure 15.11 shows that on all 13 targets the impact of removing similar compounds has been minimal and that the vast majority of the key active series have been selected in the new file.

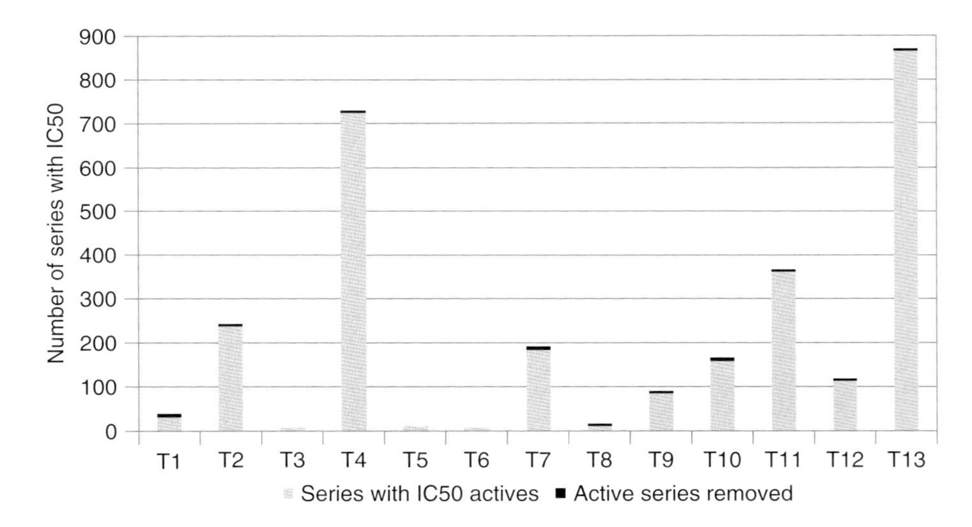

Figure 15.11 *In silico* analysis of impact of redundancy reduction on series retrieval using IC50 level data on 13 passed HTS targets (T1 to T13). The impact of the similarity filters on the active series removal is minimal.

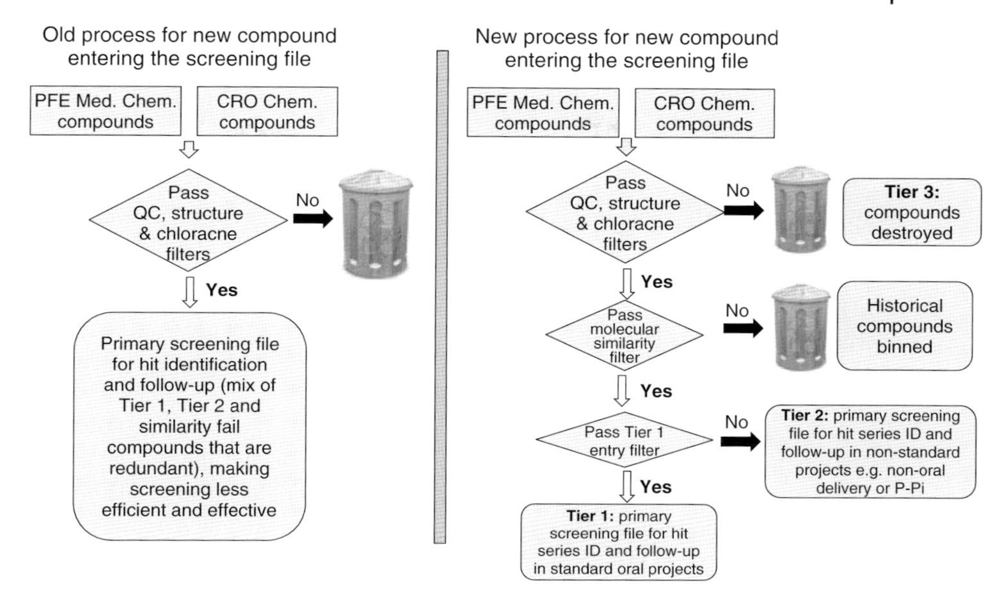

Figure 15.12 Current and old selection process for compounds entering the corporate file. New screening file focuses on ID of novel chemical series for TA RU projects, not SAR generation from HTS. It separates compounds more suitable for non-standard projects, for example, protein–protein interactions, into Tier 2.

Thus, the new screening file entry process is significantly different from the earlier one. As shown in Figure 15.12, the old process was only applying basic chemical alert and Ro5 filters, but all the large combinatorial libraries were still making their way into the file, creating an unproductive level of redundancy in the corporate file. By re-submitting the whole file to the newly implemented similarity filters, not only are a large numbers of compounds deselected as redundant, but also each library is well represented, with a maximum number of representatives of 40 compounds, and the size of the file is becoming more manageable for HTS. This new process also allows for the file to be segmented according to the type of target to screen with a Tier1 main file to be used for 'standard' targets and a Tier2 more dedicated to historically intractable targets and protein–protein interactions.

15.9 Future Routes for Hit Identification

There are key questions to reflect before jumping into a hit identification campaign with always the same aim: The discovery of new series with confidence that most hits will be attractive.

Reshaping the file brings new ways of identifying hits from HTS. The file needs to be constantly evolving in order to keep pace with the target space. Its diversity needs to include the newly patented molecules against legacy intractable families such as ion channels, protein–protein interaction, or lipid kinases.

Biologics are not yet part of the file except some Tier2 peptides with relatively low MW. Peptoids are also a new class of potential drugs worth considering [48]. SiRNA, aptamers, and other oligonucleotides are now part of pharmaceutical pipelines, but because of their specific handling they cannot be part of the small-molecule hit ID process and are incorporated in the biologics process [49, 50].

However, at the start of the project there is nothing preventing the two approaches from being pursued in parallel as they could end up being complementary, specific to an indication, or tailored to the target localization (e.g., monoclonal antibodies for peripheral targets or CNS for small molecules).

Screening technologies have also evolved considerably in parallel with the target space. High-content technologies, with some exceptions like Wyeth, which have heavily invested in high-throughput high-content screening, are still a medium-throughput-screening approach. High-content screening is looking at cellular physiological processes as a whole or could be targeting specific pathways using bioimaging analysis, as has been described for oncology programs [51].

Thus, depending on the main objectives of the project, different screening strategies could be adopted. The corporate file composition is very well linked to the future screening perspectives and target landscape, which are well described by Mayr and Bojanic [52].

Figure 15.13 describes three typical scenarios. Target 1 is the simple screening cascade for a ligand-gated ion channel as an example; calcium-flux is usually compatible with this target type. Because of the short exposure to the compounds, the compressed file is used as screening set and provides a quick exposure to the whole corporate chemical space; it is therefore unnecessary to proceed with enrichment interactions as most of the SAR should be obtained by the primary screening. On the other hand, example 2 describes a protein–protein interaction project where the Tier 1 primary file might not be the most appropriate approach. Also, in general, small molecules binding to hot spots discovered by fragment screening are a weak starting point, and it is therefore preferable to screen at a high concentration if possible. The Tier 2 file, which includes 'unwanted' molecules or is outside of the Ro5, is well adapted to this target. The diversity subset also offers the option of screening at higher concentration, which could be beneficial if one is looking for weak binders. But because those subsets may not provide enough SAR, it is recommended that hits are enriched by an iterative post-primary screening process, which could extend project timelines but is fairly necessary if the project is looking for primary series. The last example is described for completeness of the early drug discovery process. High-content screening could be used as a phenotypic screen to validate a target, and using a subset approach with this technology is perfectly suitable if you are looking for a tool. Because a tool does not have to be a lead-like molecule, there might be no need for further hit expansion. The hit identification process can stop immediately after the primary screening of the

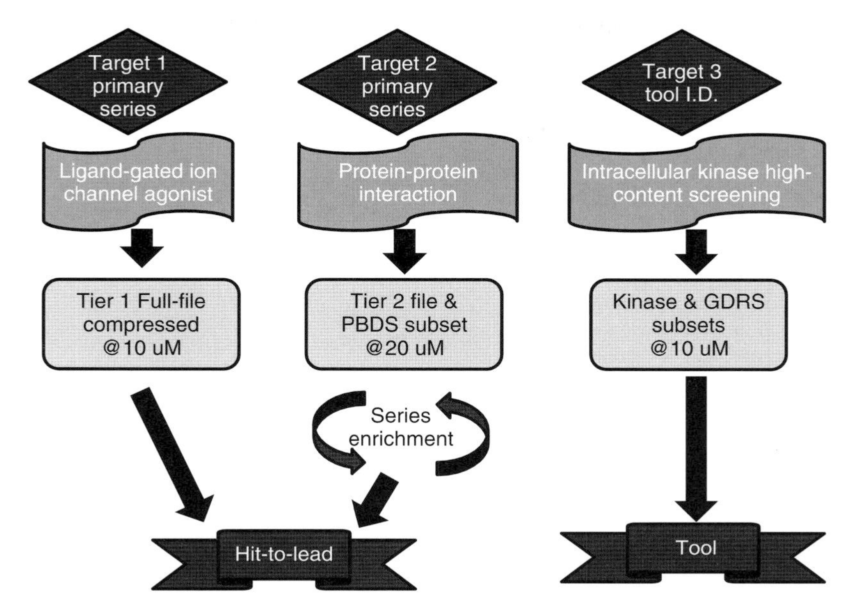

Figure 15.13 Future hit identification strategy; three target examples are described. T1 and T2 aim for primary series identification whereas T3 is looking for a tool compound.

GDRS for diversity or the kinase subset if the target is part of the intracellular pathway.

The diverse subsets are therefore ideal tools to investigate these new horizons such as high-content and label-free screening, surface plasmon resonance, high-throughput electrophysiology (IonFlux (Fluxion) or Barracuda (MDC) platforms) or Mass Spectrometry (RapidFire, Agilent). Also, bioinformatics has become increasingly popular with multinationals, speeding up the drug discovery process by predicting the biological properties of a particular lead compound and its viability as a potential drug candidate. When used in conjunction with bioinformatics, HTS can rely on quality lead compounds to generate considerable time and cost savings. At the end of the day there is no 'HTS file of the future.' There is a flexible substrate for multiple screening strategies, and each project should be able to assess a portion of it whether it is in combination with bioinformatics (knowledge-based screening) or is purely testing the diversity (random screening).

References

1. Finlay, A.C., Hobby, G.L., P'an, S.Y., Regna, P.P., Routien, J.B., Seeley, D.B., Shull, G.M., Sobin, B.A., Solomons, I.A., Vinson, J.W., and Kane, J.H. (1950) Terramycin, a new antibiotic. *Science*, **111** (2874), 85.
2. Watson, J.D. and Crick, F.H.C. (1953) A structure for deoxyribose nucleic acid. *Nature*, **171**, 737–738.
3. Tjio, J.H. and Puck, T.T. (1958) Genetics of somatic mammalian cells. II. Chromosomal constitution of cells

in tissue culture. *J. Exp Med.*, **108** (2), 259–268.
4. Merrifield, R.B. (1963) Solid phase peptide synthesis. I. The synthesis of a tetrapeptide. *J. Am. Chem. Soc.*, **85** (14), 2149–2154.
5. Mitchell, A.R.K., Kent, S.B.H., Engelhard, M., and Merrifield, R.B. (1978) A new synthetic route to tert-butyloxycarbonylaminoacyl-4-(oxymethyl)-phenylacetamidomethyl-resin, an improved support for solid-phase peptide synthesis. *J. Org. Chem.*, **43** (13), 2845–2852.
6. Ogata, K., Isomura, T., Yamashita, H., and Kubodera, H. (2006) A quantitative approach to the estimation of chemical space from a given geometry by the combination of atomic species. *QSAR Comb. Sci.*, **26** (5), 596–607.
7. Balakin, K.V., Savchuk, N.P., and Tetko, I.V. (2006) In silico approaches to prediction of aqueous and DMSO solubility of drug-like compounds: trends, problems and solutions. *Curr. Med. Chem.*, **13** (2), 223–241.
8. Lepp, Z., Huang, C., and Takashi, O.T. (2009) Finding key members in compound libraries by analyzing networks of molecules assembled by structural similarity. *J. Chem. Inf. Model.*, **49** (11), 2429–2443.
9. Bender, A., Jenkins, J.L., Scheiber, J., Sukuru, S.C.K., Glick, M., and Davies, J.W. (2009) How similar are similarity searching methods? A principal component analysis of molecular descriptor space. *J. Chem. Inf. Model.*, **49**, 108–119.
10. Muegge, I. (2003) Selection criteria for drug-like compounds. *Med. Res. Rev.*, **23**, 302–321.
11. Lipinski, C.A., Lombardo, F., Dominy, B.W., and Feeney, P.J. (1997) Experimental and computational approaches to estimate solubility and permeability in drug discovery and development settings. *Adv. Drug Del. Rev.*, **23**, 3–25.
12. Lipinski, C.A. (2004) Lead- and drug-like compounds: the rule-of-five revolution. *Drug Discov. Today: Technol.*, **1** (4), 337–341.
13. Rishton, G.M. (2003) Non leadlikeness and leadlikeness in biochemical screening. *Drug Discov. Today*, **8**, 86–96.
14. Lu, J.J. *et al.* (2004) Influence of molecular flexibility and polar surface area metrics on oral bioavailability in the rat. *J. Med. Chem.*, **47**, 6104–6107.
15. van de Waterbeemd, H., Smith, D.A., Beaumont, K., and Walker, D.K. (2001) Property-based design: optimization of drug absorption and pharmacokinetics. *J. Med. Chem.*, **44**, 1313–1333.
16. Proudfoot, J.R. (2005) The evolution of synthetic oral drug properties. *Bioorg. Med. Chem. Lett.*, **15**, 1087–1090.
17. Lemm, J.A., O'Boyle, D., Liu, M., Nower, P.T., Colonno, R., Deshpande, M.S., Snyder, L.B., Martin, S.W., St. Laurent, D.R., Serrano-Wu, M.H., Romine, J.L., Meanwell, N.A., and Gao, M. II (2010) Identification of hepatitis C virus NS5A inhibitors. *J. Virol.*, **84**, 482–491.
18. Leeson, P.D. and Brian Springthorpe, B. (2007) The influence of drug-like concepts on decision-making in medicinal chemistry. *Nat. Rev. Drug Discov.*, **6**, 881–890.
19. GVK Bio (2007) GVK Bio Databases. GVK Bio web site *www.gvkbio.com/informatics.html* [online].
20. Bleicher, K.H., Böhm, H.J., Müller, K., and Alanine, A.I. (2003) Hit and lead generation: beyond high-throughput-screening. *Nat. Rev. Drug Discov.*, **2**, 369–378.
21. Fergus, S., Bender, A., and Spring, D.R. (2005) Assessment of structural diversity in combinatorial synthesis. *Curr. Opin. Chem. Biol.*, **9**, 304–309.
22. Jadhav, A., Ferreira, R.S., Klumpp, C., Mott, B.T., Austin, C.P., Inglese, J., Thomas, C.J., Maloney, D.J., Shoichet, B.K., and Simeonov, A. (2010) Quantitative analyses of aggregation, autofluorescence, and reactivity artifacts in a screen for inhibitors of a thiol protease. *J. Med. Chem.*, **53**, 37–51.
23. Rishton, G.M. (1997) Reactive compounds and in vitro false positives in HTS. *Drug Discov. Today*, **2**, 382–384.
24. Mayer, S.C., Butera, J.A., Diller, D.J., Dunlop, J., Ellingboe, J., Fan, K.Y., Kaftan, E., Mekonnen, B., Mobilio, D.,

Paslay, J., Tawa, G., Vasilyev, D., and Bowlby, M.R. (2010) Ion channel screening plates: design, construction, and maintenance. *ASSAY Drug Dev. Technol.*, 8 (4), 504–511.

25. Walters, W.P. and Murcko, M.A. (2000) Library filtering systems and prediction of drug-like properties. *Meth. Principles Med. Chem.*, **10** (15), 15–30.
26. Wager, T.T., Hou, X., Verhoest, P.R., and Villalobos, A. (2010) Moving beyond rules: the development of a central nervous system multiparameter optimization (CNS MPO) approach to enable alignment of druglike properties. *ACS Chem. Neurosci.*, **1** (6), 435–449.
27. Xie, X.Q. and Chen, J.Z. (2008) Data mining a small molecule drug screening representative subset from NIH PubChem. *J. Chem. Inf. Model.*, **48**, 465–475.
28. McAlpine, J., Romano, A., and Ecker, D. (2009) Natural products, both small and large, and a focus on cancer treatment. *Curr. Opin. Drug Discov. Dev.*, **12**, 186–188.
29. Twede, V.D., Miljanich, G., Olivera, B.M., and Bulaj, G. (2009) Neuroprotective and cardioprotective conopeptides: an emerging class of drug leads. *Curr. Opin. Drug Discov. Dev.*, **12**, 231–239.
30. Schulz, M.N. and Hubbard, R.E. (2009) Recent progress in fragment-based lead discovery. *Curr. Opin. Pharmacol.*, **9**, 615–621.
31. Prakesch, M., Denisov, A.Y., Naim, M., Gehring, K., and Arya, P. (2008) The discovery of small molecule chemical probes of Bcl-xL and Mc11. *Bioorg. Med. Chem.*, **16**, 7443–7449.
32. Yeap, S.K., Walley, R.J., Snarey, M., van Hoorn, W.P., and Mason, J.S. (2007) Designing compound subsets: comparison of random and rational approaches using statistical simulation. *J. Chem. Inf. Model.*, **47**, 2149–2158.
33. Godden, J.W., Xue, L., and Bajorah, J. (2000) Combinatorial preferences affect molecular similarity/diversity calculations using binary fingerprints and Tanimoto coefficients. *J. Chem. Inf. Comput. Sci.*, **40**, 163–166.
34. Valler, M.J. and Green, D. (2000) Diversity screening versus focussed screening in drug discovery. *Drug Discov. Today*, **5** (7), 286–293.
35. Bemis, G.W. and Murcko, M.A. (1996) The properties of known drugs. I. Molecular frameworks. *J. Med. Chem.*, **39** (15), 2887–2893.
36. Snowden, M. and Green, D.V. (2008) The impact of diversity-based, high-throughput screening on drug discovery: 'chance favours the prepared mind'. *Curr. Opin. Drug Discov. Dev.*, **11** (4), 553–558.
37. Crisman, T.J., Jenkins, J.L., Parker, C.N., Hill, W.A., Bender, A., Deng, Z. *et al.* (2007) 'Plate cherry picking': a novel semi-sequential screening paradigm for cheaper, faster, information-rich compound selection. *J. Biomol. Screening*, **12**, 320–327.
38. Sukuru, S.C.K., Jenkins, J.L., Beckwith, R.E.J., Schreiber, J., Bender, A., Mikhailov, D., Davies, J.W., and Glick, M. (2009) Plate-based diversity selection based on empirical HTS data to enhance the number of hits and their chemical diversity. *J. Biomol. Screening*, **14**, 690–699.
39. Rogers, D. and Hahn, M. (2010) Extended-connectivity fingerprints. *J. Chem. Inf. Model.*, **50**, 742–754.
40. Rogers, D., Brown, R.D., and Hahn, M. (2005) Using extended-connectivity fingerprints with Laplacian-modified Bayesian analysis in high-throughput screening follow-up. *J. Biomol. Screening*, **10**, 682–686.
41. Dorfmann, R. (1943) The detection of defective members of a large population. *Ann. Math. Stat.*, **14**, 436–440.
42. Kainkaryam, R.M. and Woolf, P.J. (2009) Pooling in high-throughput drug screening. *Curr. Opin. Drug Discov. Dev.*, **12** (3), 339–350.
43. Hann, M., Hudson, B., Lewell, X., Lifely, R., Miller, L., and Ramsden, N. (1999) Strategic pooling of compounds for high-throughput screening. *J. Chem. Inf. Comput. Sci.*, **39**, 897–902.
44. Du D.Z. and Hwang F.K. (2006) Pooling designs and nonadaptive group testing. *Ser. Appl. Math.*, **18**.
45. Martin, Y.C., Kofron, J.L., and Traphagen, L.M. (2002) Do structurally

similar molecules have similar biological activity. *J. Med. Chem.*, **45** (19), 4350–4358.

46. Muchmore, S.W., Debe, D.A., Metz, J.T., Brown, S.P., Martin, Y.C., and Hajduk, P.J. (2008) Application of belief theory to similarity data fusion for use in analog searching and lead hopping. *J. Chem. Inf. Model.*, **48** (5), 941–948.
47. Haranczyk, M. and John Holliday, J. (2008) Comparison of similarity coefficients for clustering and compound selection. *J. Chem. Inf. Model.*, **48** (3), 498–508 (ECFP-4).
48. Heizmann, G., Hildebrand, P., Tanner, H., Ketterer, S., Pansky, A., Froidevaux, S., Beglinger, C., and Eberle, A.N. (1999) A combinatorial peptoid library for the identification of novel MSH and GRP/Bombesin receptor ligands. *J. Recept. Signal Transduction*, **19** (1-4), 449–466.
49. Zhao, Y. and Ding, S. (2007) A high-throughput siRNA library screen identifies osteogenic suppressors in human mesenchymal stem cells. *Proc. Natl. Acad. Sci. U.S.A.*, **104** (23), 9673–9678.
50. Nonaka Y., Sode K., and Ikebukuro K. (2010) Screening and Improvement of an Anti-VEGF DNA Aptamer. *Molecules*, **15**, 215–225.
51. Murray, B.W., Guo, C., Piraino, J., Westwick, J.K., Zhang, C., Lamerdin, J., Dagostino, E., Knighton, D., Loi, C., Zager, M., Kraynov, E., Popoff, I., Christensen, J.G., Martinex, R., Kephart, S.E., Marakovits, J., Karlicek, S., Bergqvist, S., and Smeal, T. (2010) Small-molecule p21-activated kinase inhibitor PF-3758309 is a potent inhibitor of oncogenic signalling and tumor growth. *Proc. Natl. Acad. Sci. U.S.A.*, **107** (26), 9446–9451.
52. Mayr, L.M. and Bojanic, D. (2009) Novel trends in high-throughput-screening. *Curr. Opin. Pharmacol.*, **9**, 580–588.

16
New Enabling Technology

Neil Hardy, Ji Yi Khoo, Shoufeng Yang, Holger Eickhoff, Joe Olechno, and Richard Ellson

16.1
Introduction

The introduction of new technology into any area is very much dependent upon the openness and willingness of the existing population to embrace and actively look for new solutions and ways of working.

History is full of examples of serendipity, or being in the right place at the right time, where a chance meeting or discovery has taken an industry down a direction that would have been impossible to predict five years earlier. However, further to being in the right place at the right time the potential application has also to be recognized.

The early pioneers within science and engineering were often great polymaths able to not only have the vision to theorize but also the practical skill to turn theory into the reality of a physical product or process. In today's world there tends to be greater, and earlier, career specialization making it easier to lose sight of the broader picture and perhaps be less influenced by developments in other areas. It is more important therefore to actively seek out new technology and innovation from other industries and areas of study. There should be no shame in taking an idea from elsewhere if it's going to benefit your particular specialism. Somebody else's development can save you a lot of time and money and give you a commercial advantage quicker than starting from scratch. While taking an evolutionary approach to development can work, it is typically driven by one of the following questions:

- How can I do this faster?
- How can I do this more cheaply?
- How can I do this with less material?

Finding answers to these questions will certainly result in a better process, but if the process itself remains essentially the same then any advantage is only likely to be slight, whereas answering the question 'how can I do this differently?' can often result in a quantum leap in performance or capability.

Management of Chemical and Biological Samples for Screening Applications, First Edition.
Edited by Mark Wigglesworth and Terry Wood.

As an example, acoustic droplet ejection (ADE) technology has been around since the 1920s [1] (see Chapter 11 for details). With its recent application to compound addition, other areas of the sample management process are also changing in order to work better with the new technology and take advantage of the new capabilities it opens up, as described below in Section 16.4.

Acoustic dispensing is a good example of a new technology that has relatively quickly changed the way many sample management processes operate. While this has no doubt enhanced what we do, we should not lose sight of other potential new technologies, such as the HP noncontact inkjet technology as presented at SBS 2011 [2]. It is also important that the systems we deploy in our routine operations remain flexible enough to cope with these new tools as seamlessly as possible such that we avoid generating the 'monolithic silos of excellence' described in Chapter 7.

More often today a solution will only be turned into an established product or process if it is marketed correctly and implemented in such a way that it is easily used by its target audience and adapted into their existing process. The adoption of VHS over Betamax in the now superseded video recorder market is a good example. We have seen this within the laboratory automation area. As the reliability and robustness of the physical automation has improved, the decision to buy one robot rather than another is now often made according to how easily that robot can be programmed in the language of the end user. Early robots were often built from scratch resulting in unreliable machines which people tolerated because using them still provided an advantage. Modern sample management robotics that followed is far superior and has benefited from advances in the semiconductor industry and other manufacturing areas. Today software plays an increasingly important role in whether a technology will be adopted. A technology which is scientifically elegant will not be widely taken up if it is not packaged in such a way that it is easy to use.

Competition is important to drive technology forward, be it the race into space of the 1960s or the recent explosion in mobile phone technologies. When a new technology or technique is introduced, if it is available from more than one source then users have the power to choose their supplier. The potential is then for the technique to be adapted more to the end-user's needs, with the adaptations being more market led than when a single supplier introduces an innovation. There is nothing stopping a 'good' single-source supplier providing a user-driven product, but certainly healthy competition drives implementation as well as affecting the commercial aspects of a new technology.

Different companies and organizations will be more or less able to embrace and exploit new technologies. It is generally accepted that although large, multi-national companies may have more resources (financial and individuals) they may be less flexible and able to adapt to new processes due to high inertia and bureaucracy within the organization. Many companies are now aware of these issues and are changing the way they work to make them more adaptable and receptive to change.

Risk sometimes associated with adopting new technologies can often be reduced by collaborations across the industry. Often there is advantage to be gained by competing organizations working together to exploit a new technology. By working with the vendors of a new device or instrument many end-users can help develop a more rugged and stable technology from which they can all benefit. Vendors can organize the pooling and sharing of user data in such a way that customers can apply the technology to their own specific area of interest without having to share proprietary information with commercial rivals.

One area of need within liquid transfer automation that is still to be met is actually knowing how much liquid has been transferred rather than relying on the fact that the technology has been QC'd sufficiently to be trusted that a correct transfer has taken place. With the traditional syringe based dispensing it is possible to go for long periods operating with a blocked or damaged tip. There are systems which can monitor back pressure or displacement to monitor performance during use, or vision systems that look for liquid being transferred, but the increased cost means they are not in general use. Techniques used with some acoustic dispensers mean that more is known about the solution being transferred (volume, %DMSO (dimethylsulfoxide), etc.) so the transfer mechanism can be optimized, increasing the quality of dispense. There still remains, however, the potential for the ejected droplet to fail to arrive at the target.

The past is no predictor of the future, and it is not the intention of this chapter to predict what technologies will be used in the future, but rather to present areas presently in development that may influence the way sample management operates. It is very easy to make a prediction and end up looking very foolish five years down the line when the technology is either still in development or has itself been superseded by another. The topics included in this chapter are very diverse and can only be put forward as possible solutions to presently perceived needs within the sample management arena. The items presented examine storage and dispense technology for solution samples, dry powder dispense technology, and possible future applications of acoustic dispensing technology. These cover a range of potential applications within sample management and elsewhere, and at the time of writing cover different positions in the technology lifecycle.

16.2
A Drop-On-Demand Printer for Dry Powder Dispensing

There is a great need to automate the process of metering and dispensing small quantities of dry powders in a wide range of processes and industries [3]. In pharmaceutical industries as with many laboratory situations, powders in small yet precise quantities are weighed out and dispensed for conventional solid-phase organic synthesis and routine analysis for large compound libraries [4]. In combinatorial chemistry of drug development, powders are often metered in interim dose forms, with each dose as low as hundreds of micrograms and up to dozens of milligrams [5]. Weighing the required amount of solid samples manually by skilled operators is tedious, and the accuracy is limited to amounts of 2 mg or more. In existing high-capital-cost laboratory dispensing stations, the powders are either vacuum aspirated into a small tip which is then transferred to the destination vial, or they are fed by screw/auger/vibration directly into the destination vial. The vial is constantly or periodically weighed by a built-in balance and the dosing step is repeated until the target mass is reached. These methods have limitations when dispensing cohesive and adhesive micron-sized inhalation powders; for example, they can be time-consuming, requiring high levels of operational complexity, and they can provide a lower fill yield due to loss of fine powders at the filter. The mass content uniformity in each dosage is especially crucial as it assures consistent therapeutic benefits in the patient [6].

Recently a promising method was developed to 'print' dry powders at high accuracy and precision, analogous to a desktop drop-on-demand inkjet printer. By application of an acoustic energy driven system, with glass capillary as funnel and delicate computer control system, the vibrations from a piezoelectric disc can precisely initiate and halt the flow of powder from a fine nozzle [7]. Once the ultrasonic vibration is switched off, the powder flow arrest is brought about by the formation of domes in the capillary due to wall–particle and particle–particle frictions. The powder flow rate can be adjusted by varying the frequency and amplitude of the vibration [8].

16.2.1 Dispensing Device Setup

The experimental facility, shown in Figure 16.1, comprises a computer, an analog waveform generator (National Instruments Corporation Ltd., Berkshire, UK), a power amplifier (PB58A, Apex Co., USA), a piezoelectric ring (SPZT-4 A3544C-W, MPI Co., Switzerland), a glass nozzle (made from a capillary tube), a purpose-built water vessel, and a microbalance (2100 mg $\pm$ 0.1 μg, Sartorius AG, Germany). The system generates a voltage signal, which can be varied at different waveforms (e.g., square, sine, triangle, and sawtooth), frequencies, and amplitudes [7, 9, 10].

The piezoelectric transducer excited by the high-frequency signal (>20 kHz) transmits the vibration to the capillary through water. The inner diameter of the water tank is 40 mm, made of a 10 mm (inner diameter) glass tube drawn to different nozzle sizes, varying between 0.1 mm and 3 mm. The upper section functions as a hopper for the powder sample. The piezoelectric ceramic ring was

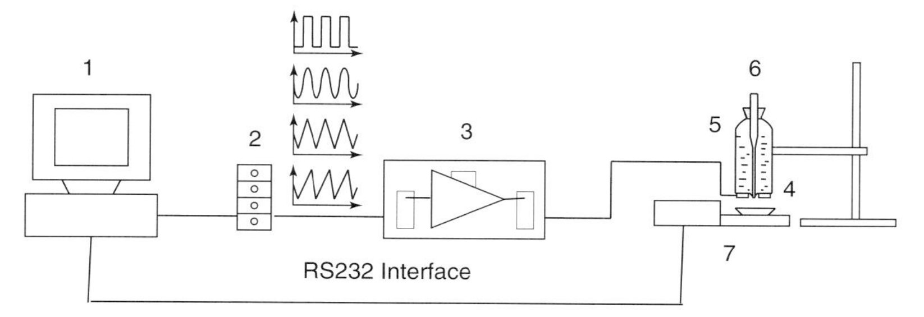

Figure 16.1 Experimental arrangement of the ultrasonic vibration controlled micro-feeding system [7].

Table 16.1 List of materials tested with the ultrasonic vibration controlled micro-feeding system.

Materials	Particle density ($kg\,m^{-3}$)	Average particle size (μm)	Angle repose (°)	Manufacturing company
SorboLac® 400[a]	355–780	<32(≥90%)	55	Meggle GmbH Wasserburg, Germany
InhaLac® 70[a]	590–660	200	31	
SpheroLac® 100[a]	690–840	<200(≥75%)	38	
Lactohale® LH-100[a]	590	125–145	34	Friesland Foods Domo, The Netherlands
Lactohale® LH-200[a]	390–650	50–100	42	
Microcrystalline cellulose	280	50	45	FMC Biopolymer, Brussels, Belgium
Avicel® PH-101	–	–	–	–
Tungsten carbide (WC)	15 500	12	36	Sandvik Coventry, UK
H13 tool steel powder	7 800	<22 (80%)	54	Osprey Metals Ltd., UK
TiO_2	4 150	0.18	38	Tioxide UK Ltd.
MgO	3 580	0.10	53	PI-KEM Ltd, UK
Starch	1 500	10	49	Merck KGaA, Germany
Glass beads	2 300	42	25	Whitehouse Scientific Ltd., UK
Carbonyl iron	7 500	1.1	–	BASF GmbH, Germany

[a]Commercial α-lactose monohydrate.

attached to the bottom of the glass tank with an adhesive commonly used in ultrasonic cleaning tank construction (9340 GRAY Hysol Epoxi-Patch Structural Adhesive, DEXTER Co., Seabrook, USA). The microbalance is employed to verify and record the dose mass.

A wide range of powders with different flow characteristics (Table 16.1) have been tested based on numerous parameters: density, particle size distribution, particle shape, and processing 'history.' Angle of repose[1] is used as a relative measure of friction and cohesiveness of the powders.

16.2.2 Effect of Powder Dispensing Parameters on Micro-feeding

Flow properties have a major impact on the dispensing performance of the system, and successful micro-feeding is determined by the powder structure in the capillary tube. Depending on the type of powder structure: arching, plugging, and blocking, the powder falls out of the nozzle as discrete particles, partially compacted

1) The angle of repose is the steepest angle of descent of the slope relative to the horizontal plane when powder on the slope face is on the verge of sliding. This angle is given by the number (0–90°).

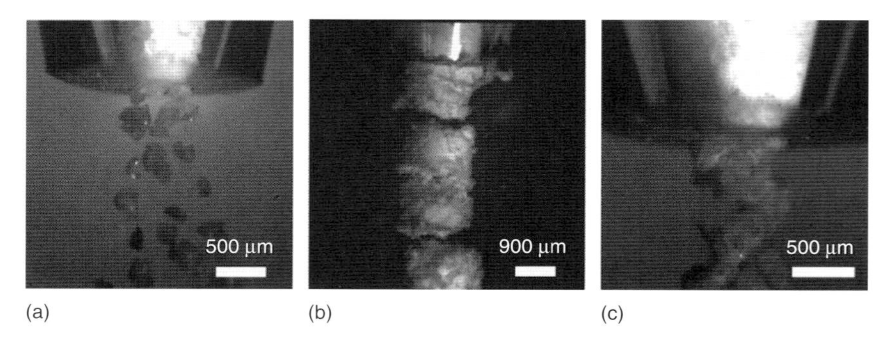

Figure 16.2 High speed images of (a) InhaLac®70 coarse powder, (b) SorboLac®400 fine powder, and (c) interactive blend of InhaLac®70 and 20 wt% SorboLac®400 powders.

columnar rods, clots, or clusters of particles [10, 11]. The relatively less cohesive and denser powders cause the preferred arching phenomenon which allows controlled dispensing, as demonstrated in Figure 16.2a. Cohesive and low-density powders, however, are more prone to induce blockage at the capillary tube as they descend from the hopper, generating plugs of short rods, as shown in Figure 16.2b. The structure of powder in the capillary tube is strongly related to the powder cohesion, diameter of capillary tube, and intensity of vibration. The system works well with interactive mixtures of free-flowing and cohesive powders (Figure 16.2c), with no particle segregation being observed during the micro-feeding process. Problematic fine powders often require higher voltage amplitude to initiate the powder extrusion process, which can be improved by adjusting other key process parameters.

The nozzle diameter, water depth, waveform, voltage amplitude, frequency, and oscillation duration influence the mean dose mass of dispensed powder. A range of different nozzle diameters is required to accommodate each powder, and that nozzle in turn determines the dose mass range. For example, in Figure 16.3a, α-lactose monohydrate: InhaLac®70 was alternately compacted and extruded from the nozzle tip under the influence of vibration, at a time interval of 6 s. The dispensed mass increased steadily as in a 'ladder' curve, demonstrating a uniform dispensing process with discrete doses. Generally, when a larger nozzle is used, the dose range is greater. An equation has been derived for mono-dispersed systems to estimate the upper and lower boundaries of nozzle diameter for different powders [7].

Water depth in the transmission tank is another key component to achieve the intended dose mass range because it changes the resonance frequency, which affects the resulting acoustic pressure levels, axially and radially [11]. Therefore, at a given frequency, there is an ideal water depth for inducing the maximum feed. Under the same conditions of frequency, amplitude, and oscillation duration, square wave excitation provides stronger vibration response and offers more stable dosing capability, compared to other types of waveform.

Experimental results have shown that the dose mass increases monotonically with excitation voltage amplitude and oscillation periods and in some cases almost linearly, as exemplified in Figure 16.3. Typical oscillation periods are from approximately 0.01–10 s, more usually, from 0.1 to 1 s. The burst period controls the time between dispensation of each dose, and the time interval chosen will therefore depend on the particular application, for example, the geometry of the production or other environment in which the device is operated. In order to obtain

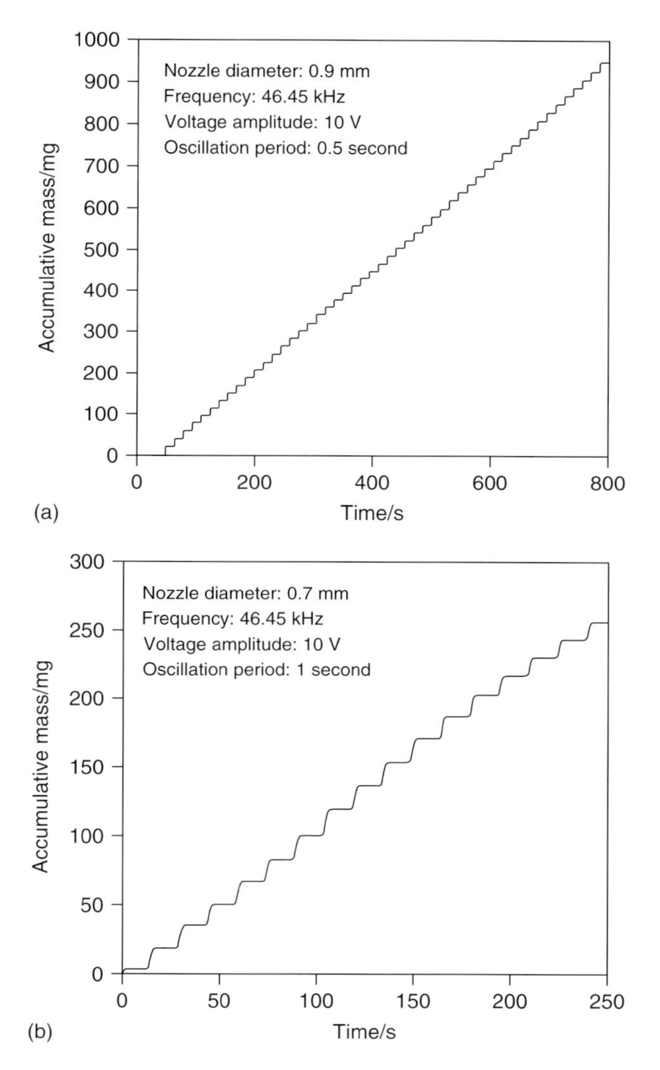

Figure 16.3 Accumulated amount of (a) InhaLac® 70 and (b) SpheroLac® 100 dispensed uniformly at different nozzle diameters and oscillation periods.

the optimum uniformity of dose mass, all the aforementioned key parameters should be considered to moderate the signal voltage.

Based on the minimum mean dose mass investigated, the system is capable of and extremely effective in dispensing dry powders at as low as 0.1222 mg (±0.0053 mg) and at low relative standard deviation (RSD < 7%). It is evident that the system is particularly suitable for powders with wide particle size distribution, cohesive powders, and even interactive mixtures, that is, dry powder inhaled formulations. This acoustic energy-controlled technology provides significant advantages of increased speed and reduced complexity compared with currently available powder dispensing systems, in addition to its adaptability to scale-up for high speed production-line applications.

16.3
Piezo Dispense Pens: Integrated Storage and Dispensing Devices and their Potential in Secondary Screening and Diagnostic Manufacturing

The piezo dispense pen (PDP) is a resealable ultra-low-volume liquid handling dispensing device with an integrated reservoir equipped with a read and write radio frequency identification (RFID) tag. The PDP can be used multiple times for dispensing, with intermediate storage of the whole device. PDPs allow for dispensing from picoliter to microliter volumes with outstanding dispensing accuracies. The reservoir size ranges from 20 μL to 1 mL. Every dispensing step is logged on the integrated RFID tag with date, time, user, and volume dispensed. The PDP technology can be integrated into many standard automation applications. The PDP dispensing pen is an innovative integrated storage and dispensing device and designed for applications in high-throughput screening (HTS) and high-content screening (HCS) as well as for diagnostic production purposes.

16.3.1
An Introduction to Piezo Dispensers

Piezo inkjet dispensers used today are ideally suited for handling ultra-small volumes of bio-molecules [12]. Nevertheless, advanced users have experienced limitations in several market segments where high probe numbers have to be handled (HTS, HCS, high-density biochip arraying, proteomics, and to some extent diagnostics) [13]. Limitations occur especially where handling of a compound library or arraying thousands of different oligonucleotides or proteins on a biochip is needed. Current systems, which require tedious pipette-tip exchanges or washing procedures and aspiration steps in between the deposition of two samples, slow down a technology which is fast in principle, and which can reach dispensing speeds of several thousand droplets a second [14].

With the development of disposable piezo inkjet cartridges and PDPs many limitations of present systems can be overcome. PDPs are the result of a logical

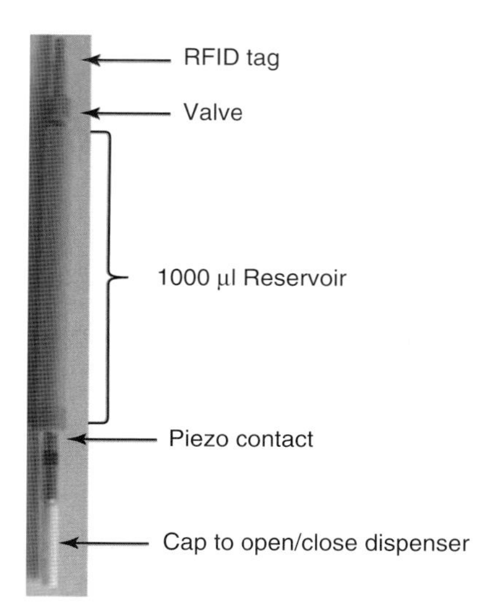

Figure 16.4 Core characteristics of a piezo dispense pPen (PDP). A 1 mL PDP has a diameter of 8 mm and a length of 100 mm.

further development of the present standard piezo dispense technology and represent universal tools, which have the potential to improve screening processes in the biotech, pharmaceutical, and agrochemical industry by combining compound storage with highly precise dispensing in a single entity (Figure 16.4). PDPs will simultaneously solve several problems currently encountered in the handling of biomolecules, like cross contaminations, dead volumes, and no access to single compounds without having to thaw complete libraries in HTS. Thereby PDPs have the potential to largely replace liquid handling robots, microplates, pipettes, and pipette tips. Fitting into existing robotic systems, these tools have another unique feature: they can be implemented into hand-held as well as robotic devices. This will provide single bench scientists as well as big screening groups with ultra-precise reagent storage and handling.

16.3.2
PDP Mode of Operation and Its Advantages

The PDP capillary, which is a piezo dispense capillary as used elsewhere, is filled with a chemical substance or biomolecules. In contrast to established systems, PDPs do not have a tubing connection to the robot or manually operated systems (Figure 16.5). In order to generate the required holding pressure a dedicated valve is directly mounted to the dispensing unit. The PDP design enables filling of empty cartridges with a predetermined solution and sliding these prefilled cartridges either in robotic systems or into handheld devices. The reservoir volume depends on the applications chosen and can be adjusted by the reservoir size. The dispensing solution can contain any kind of biomolecules or chemicals, and these can easily

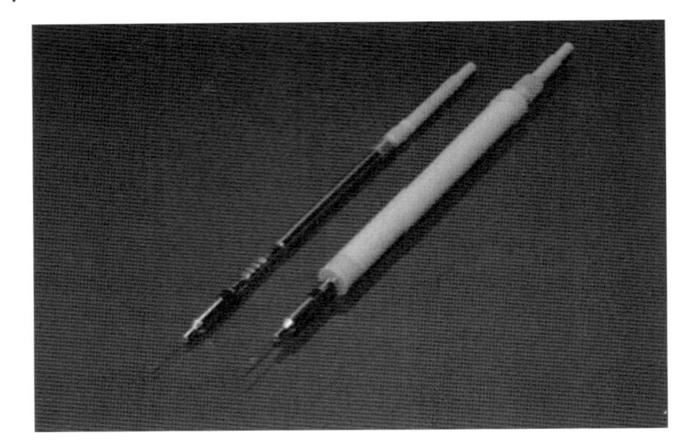

Figure 16.5 PDPs with a storage volume of 30 μL (left) 1000 μL (right). 30 μL PDPs with a 4 mm diameter allow for two-dimensional heads which are compatible with standard SBS-compatible 384-well plates while 1000 μL PDPs are compatible with 96 well plates. Both types can load SBS-compatible microplates from 96- to 3456-well formats.

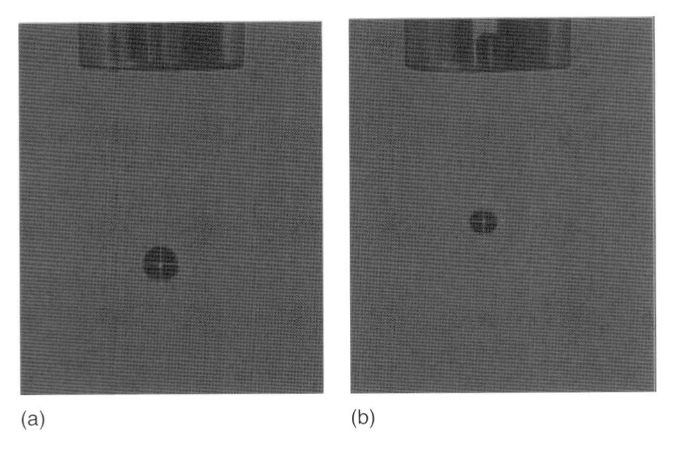

Figure 16.6 Different solvents can be easily handled with PDPs. This picture shows the orifice of the PDP (upper part) in a side view. Released droplets can be visualized by a stroboscopic setup (see, for example, *http://www.scienion.de/index.php?mid=38&vid=&lang=en* or N. J. Dovichi *et al.* [15]). Water-based buffers (a) as well as organics solutions (DMSO, b) can easily be stored and dispensed using PDPs. Dispense parameters for the pictures given above are: H_2O @ 87 V 48 μs and 242 pL/drop, 100% DMSO @ 82 V 30 μs and 132 pL/drop.

be identified by a barcode or RFID tag attached to the dispensers (Figure 16.6). The sample is filled once into the PDP and stored therein. For any type of experiment PDPs are taken from a store and put directly into the dispensing system – no additional tips, tubes, or intermediate plates are necessary. Single drops as small as

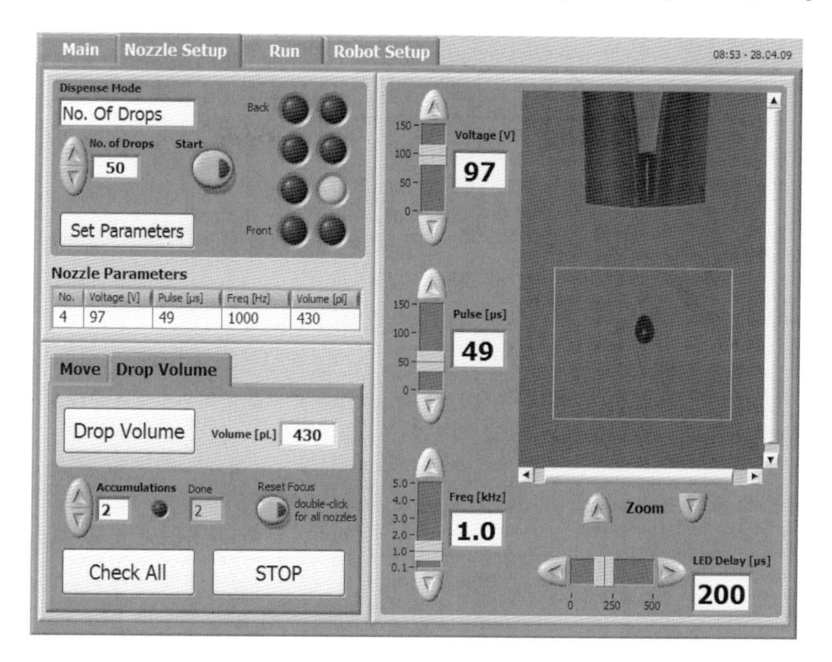

Figure 16.7 Online volume determination. The screen shot of the PDP driving software within a sciSWIFTER instrument shows the online measurement of the drop volume. An image of the drop is taken and the volume is calculated. Adjusting voltages and pulse times allows dispensing of different drop volumes for single drops. Multiple drops form the complete dispense volume.

100 pL, can be dispensed at frequencies of up to 2000 Hz resulting in accumulated volumes of up to $1\,\mu L\,s^{-1}$ (Figure 16.7). The resulting workflow is considerably simplified – initial screening, hit validation, and dose response experiments can all be done from one and the same container/sample. After usage the dispenser can be either withdrawn or put back onto a shelf, where a robotic arm can pick up the next dispenser. The PDP comprises a read and write RFID tag providing full traceability. Every use and every dispensed drop is logged and saved at the storage and dispensing cartridge itself.

Current PDPs are manufactured from glass, but due to production costs other inert materials like ceramics or plastics may replace the glass at a later stage. Their dispensing accuracy and precision is outstanding, with CVs smaller than 1.5% for volumes between 100 pL and 10 µL (Figures 16.8 and 16.9).

PDPs are technologically very similar to other actuator-driven dispensing technologies, but these cartridge-based dispensing systems offer various advantages:

- Use of PDPs is time- and cost-saving.
- PDP cartridges are autonomous, disposable, and require no washing.
- PDPs minimize contamination risks.
- PDPs are closed containers and thereby minimize precipitates caused by water condensation into DMSO solutions.

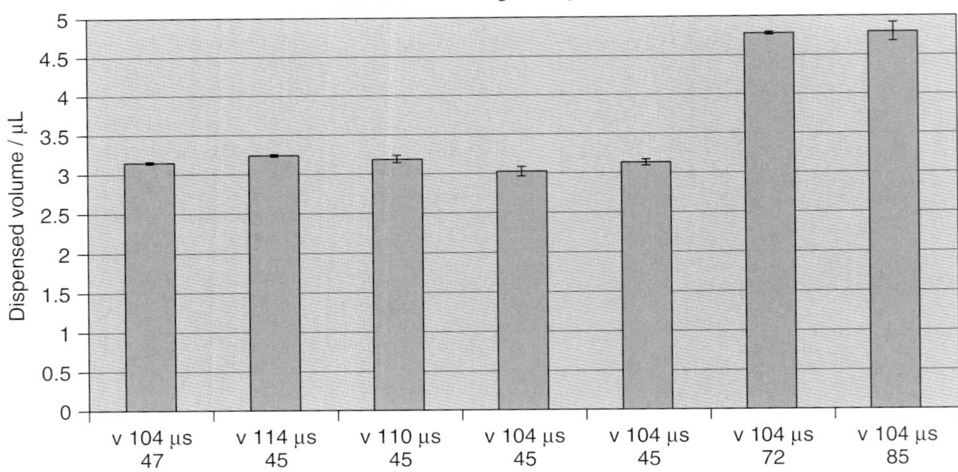

Figure 16.8 Precision of dispensing volumes. This figure shows the dispensing results of one PDP actuated with different settings in terms of pulse width and pulse height. Different settings enable different volumes to be dispensed. In each run, 10 000 drops were dispensed and gravimetrically determined. Error bars are given in the graphics; typical CVs were less than 1.5%.

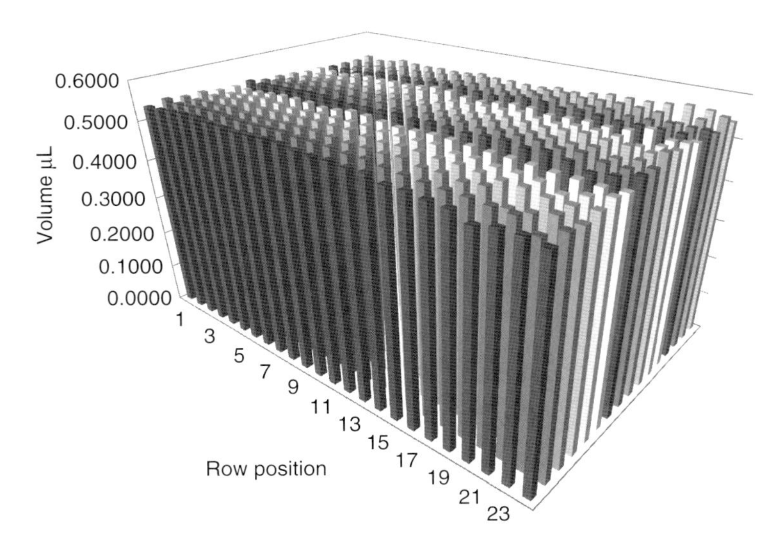

Figure 16.9 Graphical representation of PDP dispensing into a 384-well plate. A volume of 520 nL of a colored solution was dispensed into a standard 384-well screening plate. Read out with a plate reader showed CVs < 1.5%.

- PDPs are suitable for miniaturized assay formats.
- PDPs have high dispense-volume flexibility: pico- to microliters.
- PDP technology ensures high accuracy in dispensing precise volumes.
- PDPs allow for sterile long-term storage of samples and reagents.
- PDP cartridges are equipped with an RFID-tag for complete traceability.
- PDPs can be integrated into existing assay infrastructures.
- PDPs can be used in handheld devices and in robotic systems.
- The PDP technology platform is easily scalable; it can be operated by single PDP users in research as well as by large companies and laboratories.

16.3.3
PDPs in the High-Throughput Screening Environment

Pharmaceutical and biotechnology companies compete to become the first to market new drugs to capture the maximum market share and optimize revenue generation within the first year of entry. HTS is regarded as a major factor in shortening the drug discovery cycle; HTS also increases the chances of clearing clinical trials by addressing issues related to pharmacokinetics, toxicity, and side effects. In spite of the significant challenge presented by a high equipment costs and poor quality of lead compounds, HTS has the potential to yield both mechanistic insights and new drug leads with unparalleled efficiency through investment and innovation [16].

Technological advancements in HTS are fueling the development of numerous novel products and are creating demand for more accurate, higher-throughput, and cost-effective technologies. These HTS technologies are being used to develop improved drugs and discover new bio-markers, as well as for personalized medicine and biodefense. The market for novel HTS products was estimated to reach $225 million in 2004 and approximately $540 million worldwide in 2009, these products consisting of emerging labeling and detection methods (42%), novel microarray technologies (55%), and informatics (3%).

The drug discovery process requires constant and large-scale experimental studies necessitating large volumes of samples together with their preparation and processing. Screening large volumes of potential drug targets places a strain on capital resources – that is, money spent on research consumables and labor resources. The adoption of automated liquid handling research instrumentation cuts down experimental run time by 40% in certain instances while increasing experimental reproducibility. It also reduces the use of the consumable reagents resulting in direct savings for the company.

A significant number of pharmaceutical companies regard HTS as one of their core technologies for identifying novel compounds. Typical drawbacks existing in screening units of the pharmaceutical industry can be overcome with PDP technology, which will ultimately enable pharma companies to screen better, faster, and more safely with smaller amounts of reagents. Typically, in libraries for HTS several hundred thousand compounds are stored in large cooled units equipped with highly sophisticated logistics for compound handling. After solubilization these compounds are usually stored in microplates in organic solution. Several copies of these grandmother plates have to be made by several cycles of thawing and freezing to generate mother and daughter plates. For accessing a single probe, a complete microplate has to be thawed, affecting all other compounds of the library, thereby blurring future screening values. During this thawing operation, moisture from the air may condense into the sample substances, which can lead to their impairment, and during processing a high percentage of the compounds is lost because of the complex handling procedure and the large number of individual steps which may lead to contamination, cross-contamination, dead volumes, delays, dilution errors, or dilution due to condensing water. Other reasons for the loss of compounds are precipitation of the stock solution and chemical degradation.

However, PDPs offer a unique solution to most of these problems and therefore possess the potential to largely replace microplates in this area. They are particularly advantageous for producing so-called working copies for screening processes and multiple assays in the pharmaceutical industry [17].

Here, the sample substance is filled directly from the source container into the microdispenser, which is then used to produce the screening plates, for example. There is no need for the mother and daughter plates which would otherwise be necessary. Individual samples can be dispensed without having to thaw the entire microtiter plate and are stable in the container for months (Figure 16.10).

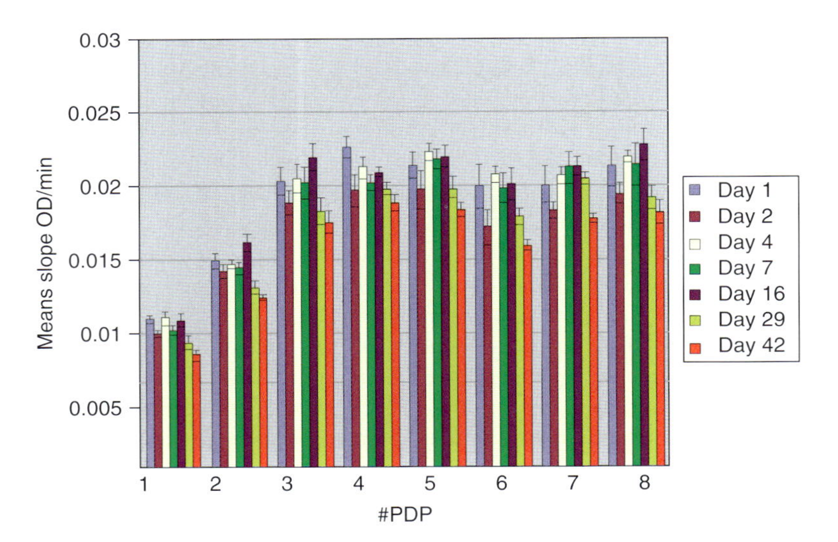

Figure 16.10 Stability of polymerases and phosphatases in PDPs. Polymerases and phosphatases were stored for six weeks in PDPs. Assays with a colorimetric detection were done at several time points and recorded. As a control, conventional pipetting was done, and this is named in the graphics as PDP#1. All reagents showed high stability after six weeks. No influence of the PDP on reagents stability could be measured.

Compelling advantages of PDPs for HTS include:

- Drastic reduction in the number of mother daughter replications, since PDPs can directly dispense any volume above 100 pL needed in an assay directly into a screening plate or into a screening assay.
- Cherry picking can be done on a single-access basis for each compound WITHOUT having to move or touch any other compounds or the whole library.
- The PDP itself is inert and, due to its design, has no condensing effects.
- Efficient mixing and miniaturization of assay components, thereby saving millions of Euros per year in each single screening unit.

16.3.4
The Instrument to Operate PDPs in a Pharmaceutical Laboratory: sciSWIFTER

For pharmaceutical research the sciSWIFTER is a system that operates PDPs and is a time- and reagent-saving drop-on-demand dispenser. This unit dispense up to 16 different substances in parallel into various multi-well plate formats. Dispense range is 100 pL to 100 μL with dead volumes close to zero. Each nozzle has online drop control and optical volume determination (Figure 16.11).

The sciSWIFTER can operate as a stand-alone device and includes integrated touch-screen controls, robotics shuttle, and external software application programming interfaces (APIs) for integration into automated robotics platforms, including AutoMAP of Matrical Bioscience. The sciSWIFTER allows for quick and easy change of setup for different assays and minimizes cross contamination. The

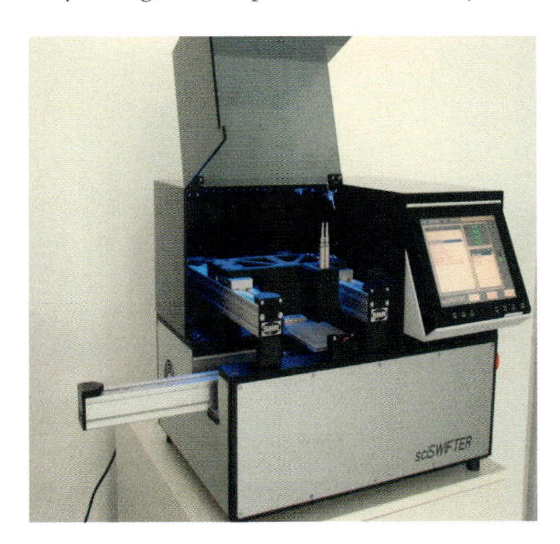

Figure 16.11 Operational system for PDPs: sciSWIFTER. The sciSWIFTER holds up to 16 different PDPs/substances in parallel. Each well of a destination or screening plate can be filled individually (substances and amount of each substance) for up to 16 compounds in parallel. The system provides online volume calibration, a metal-free liquid path, and an integrated shaker. The system is touch screen operated and has a real-time display of the dispensing process.

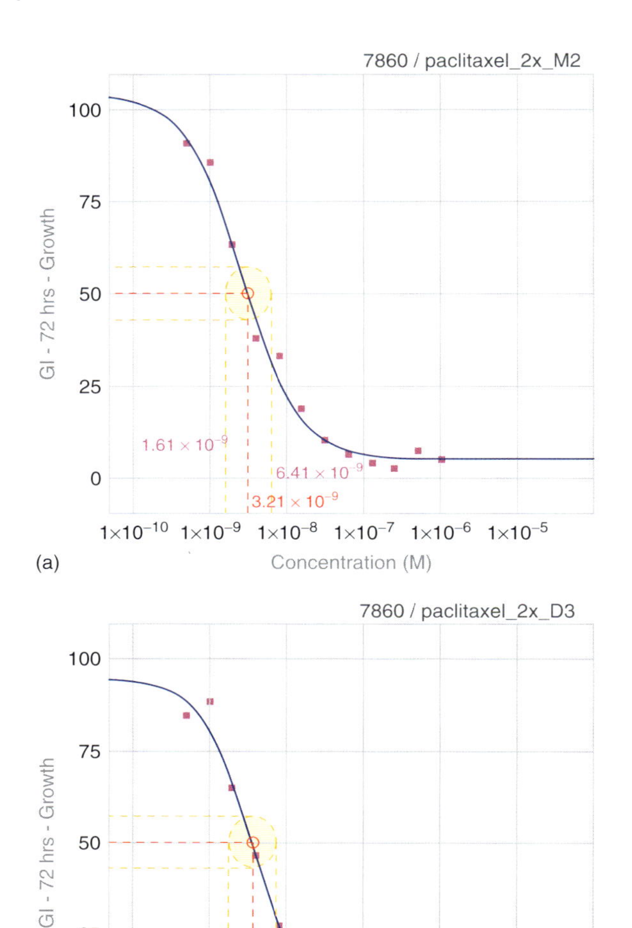

Figure 16.12 PDP results in cellular screening in a model system for combination therapy. For a cellular screen in a growth inhibition assay with cancer cell lines, a manual pipetting scheme (a) was compared to PDP usage in a sciSWIFTER (b) in a combination study. Within the error bars, identical performance could be measured, although no dilution series of the compound had to be prepared when PDPs were used. PDP values were obtained by direct pipetting of concentrated paclitaxel/compound solution into the assay well. By doing so, less compound (factor 10) and time (factor 10) were required to get assay results. (Data are courtesy of Dr. Igor Ivanov, Oncolead GmbH, Munich.)

requirement for intermediate dilution plates is eliminated by the dynamic range of volumes dispensed (6 orders of magnitude), and the system can be used effectively in cellular screenings, for example, in studies for combination therapies, where different amounts of various drugs are applied directly into a cell culture in a microplate (Figure 16.12).

16.3.5
PDPs for the Sterile and Contamination-Free Production of *In Vitro* Diagnostics

Diagnostics and other related industries face formidable challenges today [16]. While costs have to decrease, an increasing number of compounds need to be stored, managed, and delivered in even smaller volumes of expensive compounds and bio-reagents to perform biochemical reactions. From a practical perspective, numerous storage and retrieval steps, dilutions, pipetting, and reformatting steps have to be performed (Figure 16.13). These processes are erratic, time consuming, and expensive due to high investment costs and the high amount of dead volumes and disposables like pipetting tips, tubes, microtiter plates, and so on [18].

For multiparameter assay research, development, and manufacturing, several drop-on-demand technologies are used today. These technologies permit the production of microarrays, loading of biosensors, or just mixing of nanoliter volumes on a variety of surfaces and reaction vessels, including microtiter and nanotiter plates.

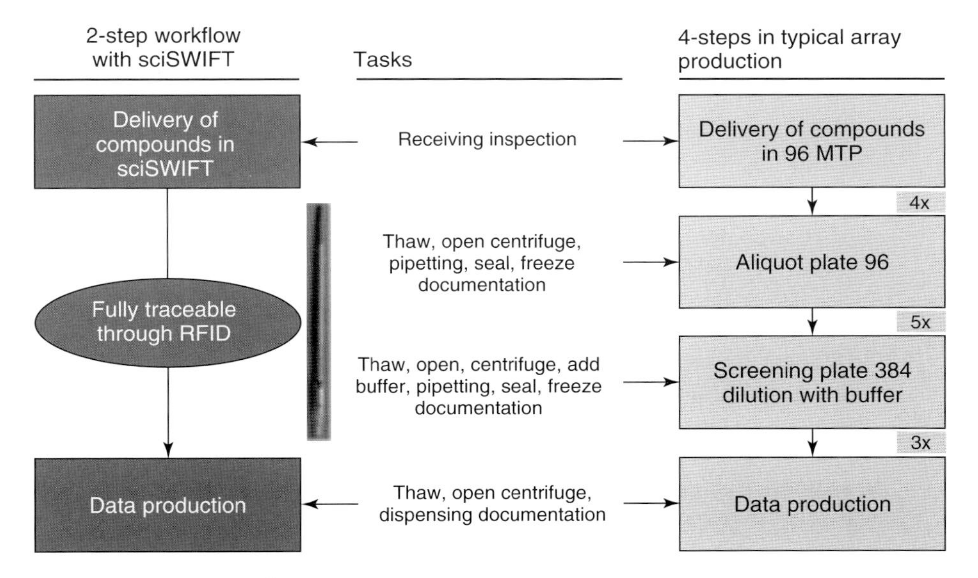

Figure 16.13 Working scheme using PDPs versus conventional probe preparation schemes in microarray manufacturing. Instead of multiple steps prior to printing microarrays in a conventional setup, PDPs simplify and shorten the process significantly.

Drawbacks of existing low-level liquid handling systems include possible cross contamination, biofilm formulation in dispense lines, tedious washing procedures, and aspiration steps in between spotting two samples, and these slow down a technology which is fast in principle, and which can reach dispensing speeds of several thousand droplets a second. Moreover, liquid handling procedures in the production of *in vitro* diagnostics face one more challenge, and that is a crucial one: the FDA (or other regulatory authorities) require conformity of the production process such that every single production step has to be documented and approved. As mentioned above, PDPs avoid various kinds of problems related to tubing, washing, and cleaning steps and other risks of contamination, thus offering new possibilities in the approval of regulated diagnostic manufacturing.

In the diagnostics sector, the PDP technology therefore offers a great advantage over today's alternative technologies, since the dispensers can be filled sterile and contamination free. In current approaches, most systems contain tubing and connectors or metal pieces. These surfaces are prone to contamination or biofilm formation when used in a production environment, while the PDP surface is inert and allows for precise dispensing onto any diagnostic substrate. In addition, diagnostic companies benefit from time savings experienced with PDP systems, since no time-consuming washing or conditioning steps will be required. As there is also no tubing or single use of the PDPs, no cleaning is required, either.

Major advantages of PDPs for the diagnostics industry include:

- Sterile and contamination-free handling of markers and reagents in a production environment.
- No time-consuming washing or conditioning steps will be required.
- PDPs avoid any cleaning procedure involved in today's production processes.
- PDP can easily be integrated into existing piezo systems in diagnostic manufacturing.

The whole dispensing process when PDPs are used is monitored via online drop control and a highly precise optical volume determination. Equipped with software – as well as hardware – interfaces for integration into robotic systems, the PDPs also enable easy access to secondary screenings as well as dose response tests.

Different molecules can be dispensed with different dispense settings (pulse width or pulse shape of the actuator). The reservoir can have different sizes, depending on the needed throughput per day or production shift.

PDPs are able to deliver droplet volumes between 100 and 1000 pL over a distance of 10 mm with high positional accuracy, allowing filling of microplates from above without having to move the PDP or pipette tip into the reaction vessel.

16.3.6 Summary and Outlook

The PDP technology represents a next-generation arraying technology. PDPs, integrated storage, and dispensing devices are brand new and highly innovative tools in their respective market segments and provide users with unique key features. The PDP technology has reached an advanced development status and is already available to early-access customers. As well as dispenser labeling and sealing, automated filling and alternative materials to glass are judged to be key components to broaden the user base of this technology in pharma and diagnostics significantly.

16.4 Future Directions in Acoustic Droplet Ejection Technology

16.4.1 Introduction

This section covers possibilities and potential of ADE technology. The currently accepted benefits and uses are described in Chapter 11. For much of the information below, feasibility has been demonstrated in the laboratory, but use is yet to become widespread. Therefore, what will actually be commercially available will depend upon the needs of the growing user base and the results of efforts on the part of the R&D department of the manufacturer.

16.4.2 Stretching the Boundaries of Current ADE Uses

We have seen from Chapter 11 how ADE is commonly used to transfer solutions to and from microtiter plates. However, few people will realize the flexibility of ADE in that it can be used to move solutions from a wide variety of containers or surfaces to many different destinations. As source volumes are reduced, the need for containing walls is also reduced. At the extreme, ADE can transfer droplets of fluid directly from a standing drop of liquid, as shown in Figure 16.14.

Acoustic energy not only provides the motive force for generating drops, but can be applied to understand the aspects of the container, surface, and fluid that impact the generation of the droplet. The use of acoustics for this type of measurement is commonly referred to as *acoustic auditing*. In particular, the ability to measure acoustic impedance in each individual well of a multi-well plate, and from that determination extract the acoustic velocity of the medium, allows the system to auto-focus the energy at the meniscus of each individual well by automatically moving the transducer vertically. This automatic focus adjusts for changes in the acoustic velocity and fluid depth over time and for each well. Each well of a microplate is audited individually in milliseconds for both acoustic and fluid depth. Auditing of the source microplate is usually performed while the destination plate is loading, in order to improve throughput.

After auditing, the transducer is positioned to deliver the properly focused acoustic energy to achieve optimal droplet ejection in the transfer process. Auditing is essential for precise and accurate generation of droplets. Acoustics are further

Figure 16.14 ADE is not limited to transferring fluids from multi-welled plates. In this instance, the larger reservoir droplet consists of 20 nL of water sitting on a glass microscope slide. The smaller ejected droplet has a volume of 20 pL with a diameter of 34 μm.

used to audit on a real-time basis while a well is being depleted. Real-time auditing allows the instrument to automatically adjust the position of the transducer to allow for changes in volume due to sample transfer as well as changes due to evaporation or hydration. For HTS libraries in DMSO, the range of well composition is often narrow, and typically consists of mixtures with 30% water or less and small molecular compounds (under 1000 Da) in concentrations under 30 mM. Advances in the technology of acoustic systems have improved auditing capabilities, enabling systems to monitor and eject fluids of much greater variability in composition. For example, by expansion of the range of measurable acoustic impedance to include more viscous materials and to enable proper focusing, the fluids in commercial protein crystallography sets are easily transferred. Another major advance was a rapid technique for determining the acoustic power required to eject a droplet from a well, this determination being rapid enough to be performed just prior to ejection in every well, enabling compensation for wide variations in viscosity as well as surface tension. This enables a diverse set of materials to be placed in a single microplate without any *a priori* knowledge of the materials or their locations in the plate. Details of these advances and their implications on life science applications are described below.

16.4.2.1 High-Viscosity Fluids

The transfer of solutions containing cells has been described in Chapter 11, demonstrating that such solutions can be transferred acoustically without damage to the cells. Many biological fluids are, however, very viscous and do not easily lend themselves to transfer by any liquid management device. For example, the viscosity of cellular cytoplasm has been measured between 139 and 78 mP · s (1 milliPascal-second (SI units) = 1 centipoise (cP, cgs units)), depending upon the position in the cell and the direction of the test [19]. Other researchers have shown that the viscosity of cytoplasm can vary by more than an order of magnitude depending upon position in the cell [20]. Furthermore, many fluids used in biological experiments are also very viscous. Among these solutions are those of polymers (e.g., polyethylene glycol, Jeffamine, dextran), glycerol, sugars, proteins, and nucleic acids. Experimentation with these fluids may be limited by the high viscosity encountered in concentrated solutions.

ADE offers the potential to transfer many viscous fluids. Recent research has shown that glycerol solutions up to 60% can be transferred acoustically with excellent precision and accuracy (Table 16.2). Efforts are in progress to optimize the precision and accuracy of the transfer of even higher concentrations of glycerol.

Of particular interest is the use of glycerol as an additive to fluids. McCloud and his coworkers [22] showed that binary (DMSO and glycerol) and ternary (DMSO, glycerol, and water) mixtures often dissolved library compounds better than pure DMSO. Additionally, at least some of the solutions tested resisted further hydration or dehydration from atmospheric humidity which could eliminate the need for plate sealing. Furthermore, the DMSO-glycerol mixtures are less toxic to cells and less damaging to enzymes.

Recent work at Labcyte suggests that glycerol–water solutions retard evaporation. This observation, coupled with the bacteriostatic properties of glycerol [23], suggests that these solutions may be a better choice for storing many enzymes and oligonucleotides. Furthermore, the glycerol protects against enzyme degradation and, at least in the case of quantitative polymerase chain reaction (qPCR), appears to improve assay results.

Some materials may resist acoustic transfer not because they are extremely viscous but because the length of the molecules themselves may be greater than the distance between the source and the destination. While viscous solutions of sheared genomic DNA are easily transferred acoustically, solutions of the unsheared nucleic acid stretch in the acoustic environment forming a long thin neck connecting the main droplet and the source plate.

Table 16.2 Highly viscous solutions of glycerol are transferred with high precision and accuracy at low nanoliter volumes.

Glycerol composition	0%	10%	20%	30%	40%	50%	60%
Viscosity [21] 20 °C	1.005	1.31	1.76	2.50	3.72	6.00	10.8
Volume CV (%)	1.4	1.9	1.7	1.3	2.3	2.4	2.2
Volume error	4.0%	NM	0.4%	NM	2.9%	NM	0.5%

NM = Not Measured.

Table 16.3 Surface tension of some commonly used fluids that have been transferred acoustically.

Liquid	Surface tension (dyne/cm)
Acetonitrile	19.1
Dimethyl formamide (DMF)	36.8
Dimethyl sulfoxide (DMSO)	43
Dow200 silicone oil	19.8
Ethanol	22.3
Ethanol (40%) + water (25 °C)	29.6
Ethylene glycol	48
Glycerol	63
Isopropanol	21.7
Methanol	22.6
1-Methyl-2-pyrrolidinone	40.3
Mineral oil	~30
Sodium chloride 6.0 M aqueous solution	82.6
0.48% Sodium dodecylsulfate (SDS)	32
Sucrose (55%) + water	76.5
Tetrahydrofuran	26.4
Water (25 °C)	72.0
0.1% Triton X-100 (25 °C)	30

16.4.2.2 Low-Surface-Tension Fluids

Low-surface-tension fluids, including many organic solvents, are not easily transferred by pipette-based devices because the fluid tends to drain from the pipette before it can be moved to the appropriate receptacle. Pressure-based liquid transfer systems tend to require that all liquids being transferred must have similar surface tension and viscosity to ensure that the volume transferred remains constant. ADE has been used to transfer a number of fluids from a single microplate that contains both high- and low-surface-tension fluids including methanol, ethanol, and aqueous solutions of surfactants, as outlined in Table 16.3.

Protein Crystallography X-ray crystallography can solve the structure of proteins, allowing researchers to better understand how a drug or substrate interacts with both active and allosteric sites and, with that knowledge, develop better drugs. Crystallization of proteins, often a difficult process, is based upon slowly precipitating proteins out of solutions so that they pack into a regular lattice. Rapid precipitation leads not to crystals but to disordered, sometimes denatured, solids which cannot be used for structure determination. A wide range of materials are used to encourage the protein to slowly leave solution and form a crystal. These solutions typically contain a mixture of salt(s), buffer, and one or more precipitants, (solvents, polymers and 'other'), as outlined in Table 16.4.

Solutions typically used to crystallize soluble proteins have been successfully transferred by ADE [24]. These solutions cover a wide range of viscosity (roughly

Table 16.4 These compounds and concentrations are components of commercially available protein crystallization solutions.

Family	Compound	Typical concentrations
Buffers	Acetate, bicine, bis-TRIS, bis-TRIS propane, citrate, cacodylate, HEPES, imidazole, malonate, MES, phosphate, succinate, tartrate, TRIS	100 mM
Metal cations	Cadmium, calcium, caesium, cobalt, iron, lithium, magnesium, nickel, potassium, sodium, zinc	5–3000 mM
Organic solvents	*t*-Butanol, 1,4-dioxane, ethanol, ethylene glycol, glycerol, 1,6-hexanediol, isopropanol, 2-methyl-2,4-pentanediol, propanediol	2–70%
Polymer	Jeffamine ED-2001, Jeffamine M-600, polyacrylic acid 5100, polyethylene glycol (PEG) 200 through 20 000, Polyethylene glycol monomethyl ether 550 through 5000, polypropylene glycol P400, polyvinylpyrrolidone K15	0.5–50%
Other	Ammonium, ethylene imine polymer, hexadecyltrimethylammonium, pentaerythritol ethoxylate, pentaerythritol propoxylate, trimethylamine N-oxide	0.2–35%
pH	3.5–10.5	–

0.5–50 cP) and surface tension (roughly 20–80 dynes cm^{-1}) yet are transferred by an Echo liquid handler using a single calibration. The viscosities of Hampton Crystal Screen 2 fluid #26 (0.2 M ammonium acetate, 0.1 M sodium citrate, 30% methyl-2,4-pentanediol, pH 5.6) and #35 (0.1 M HEPES sodium, 0.8 M monosodium phosphate, 0.8 M monopotassium phosphate, pH 7.5) were measured at 23.5 and 16.7 cP, respectively. Droplet placement of these liquids was very good and allows the use of the two most common methods used in protein crystallization, the sitting drop, and hanging drop techniques.

While many crystallization formulations are available, many researchers would prefer to have a mechanism to make their own mixtures. Theoretically, this can be done with existing technology. In practice, the volumes transferred by existing technology are so great that new combinations are obtained only with larger volumes, which must then be sub-divided to obtain the small volumes required to mix with the droplet of protein solution. With ADE small droplets can be easily combined allowing the user greater control of the final concentration of all components in the mixture without requiring the large volumes to be produced.

Typically, it is difficult to form large crystals. All too frequently, small or malformed crystals are generated. Malformed crystals can be isolated and broken into fragments. These fragments or microcrystals obtained from solutions can be used to seed other protein solutions. These seed crystals act as crystallization centers creating the larger crystals required for some X-ray units. Traditionally, seeding has been done by hand with the researcher using a filament (often a cat's whisker or a fine wire) to transfer microcrystals to droplets of crystallization fluid and protein. ADE has been shown to be a viable alternative to add microcrystals to solutions. Similarly, researchers from Brookhaven National Laboratory [25] have shown that they were able to use ADE to transfer small crystals directly from a crystallization solution to the loops used to place crystals in the X-ray beam line. Normally this transfer is done manually. While this is relatively easy with large crystals grown in nonviscous solutions, as the crystals get smaller and the solution more viscous, it can take minutes to mount any crystal. Improvements in beam line intensity have reduced the need for large crystals and have shown that ADE is nondestructive to the microcrystals and is a rapid mechanism for mounting.

When crystallographers have investigated reasons for the lack of reproducibility of protein crystallization, they have often found that variations in solution composition are to blame. As the solutions age, they may concentrate due to evaporation of water or volatile components such as dioxane or alcohols. This affects the reproducibility of crystal generation. Other solutions change as they age due to the oxidation or precipitation of components. Some researchers have developed dye-based techniques to monitor the quality of the crystallization solution libraries [26]. However, this requires the addition of colored dyes to the solutions, which may have an impact on the proteins being crystallized. Efforts are in progress to determine whether tracking changes in acoustic impedance as well as solution volume can determine when concentration shifts in library solutions occur that could impact the reproducibility of crystallization experiments.

Membrane proteins are an extremely important target for many drugs, but crystallization of membrane proteins has been difficult. Lipidic cubic phases (LCPs) are mixtures of a lipid (commonly monoolein) with a buffer. Under certain conditions, the monoolein will form a three-dimensional structure, the LCP (as shown in Figure 16.15), which is conducive to the crystallization of membrane proteins.

Typically, protein solutions are mixed with monoolein and then passed through a narrow aperture many times to generate the LCP. The LCP, which is extremely viscous and not amenable to standard liquid handlers (the viscosity is usually compared to that of toothpaste), is covered with one of the crystallization fluids, and the protein slowly crystallizes in the LCP. Alternatively, LCP may be prepared that is free of protein. This is placed in a well and covered with a solution containing both the membrane protein and the crystallization fluid. In this case, the protein migrates from the solution into the LCP, where it crystallizes. Recent work at Labcyte has suggested that the LCP does not need to be mechanically manipulated but will form spontaneously. Monoolein was dissolved in methanol in a 2 : 1 ratio (wt:wt) to form a clear solution. Fifty nanoliters of water (or buffer) was transferred acoustically to a crystallography plate. Two hundred and twenty five nanoliters of monoolein solution was transferred directly onto the existing droplet of water. The two solutions mixed thoroughly, driven by surface tension [27], to form a single pool of clear and highly viscous material that matches the physical properties of LCP prepared in bulk. Confirmation of the identity of the material as an LCP will be confirmed by future experiments.

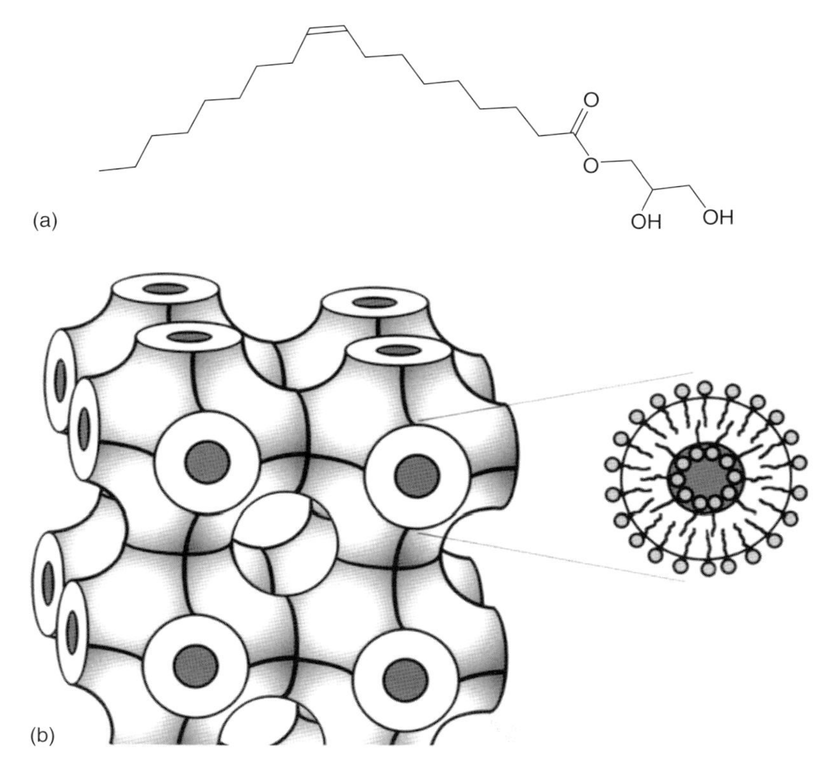

Figure 16.15 (a) Structure of monoolein. (b) Representation of a lipidic cubic phase. The monoolein forms a bi-layer that self-organizes into a three-dimensional structure with water both inside and outside the structure. Hydrophobic proteins move from the aqueous phase into the monoolein phase and crystallize.

16.4.2.3 **Layered, Bi-Phasic Fluids**

Mutz *et al.* [28] first showed that when ADE is performed on a layered, bi-phasic solution, it is possible to eject a droplet of the lower fluid that is coated with a layer of the upper fluid, as shown in Figure 16.16.

If the upper fluid (Fluid A) is a nonvolatile fluid such as mineral oil, it can protect a volatile lower liquid, such as water, from evaporation. Not only is the lower liquid protected from evaporation in the source well but the bi-phasic droplet is protected from evaporation upon landing, as the upper phase stays on the surface. Mutz showed that this technique allowed the miniaturization of peptide binding assays to volumes as low as 10 nL while maintaining accuracy.

The present algorithms on commercial instruments do not allow a determination of the depth of the upper fluid. However, if the thickness of the fluid layers and the difference in acoustic impedance between the liquids enables the reflection at the interface to be distinct from other signals and above the background noise,

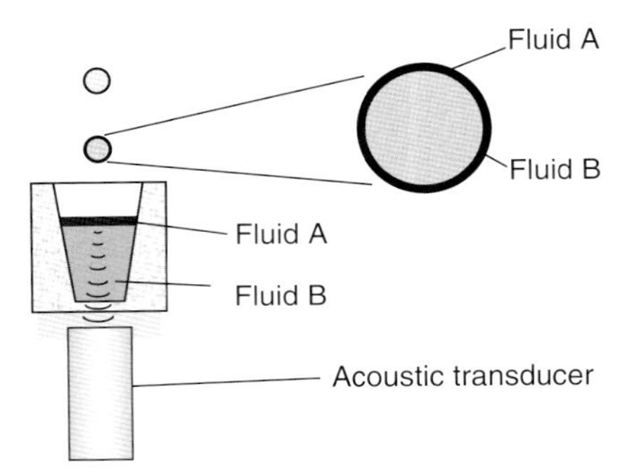

Figure 16.16 When acoustic energy is focused at the interface between two immiscible liquids, the droplets that are formed are pellicular. The thickness of the coating is affected by the depth of the upper fluid, as well as by the viscosity and surface tensions of the two fluids.

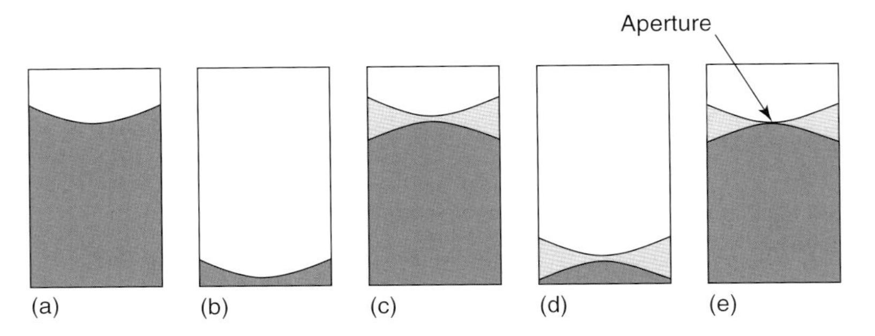

Figure 16.17 A denser, volatile fluid (e.g., buffer) shown in dark gray is covered with a nonvolatile fluid (e.g., mineral oil) shown in light gray.

information for the depth measurement exists within the acoustic signal to allow this to be determined for both fluid layers in the well surface.

In Figure 16.17, Frame C shows how the oil can cover the buffer in a well to reduce or eliminate evaporation. The curved shapes of the fluids provide another potential use. Frame B shows how liquids that wet the plastic surface of the well (such as water or DMSO) cling to the walls of the well and reduce the volume of fluid at the center of the well. Since ADE ejects from the center of the well and requires a depth of fluid approximately three times the diameter of the ejected droplet, a limit to the ejectable volume is reached while there is still significant fluid in the well. Coating the fluid with a fluid like mineral oil causes the aqueous layer to recede from the wall and to collect at the center of the well. This may allow users to increase the working volume and decrease the dead volume of the well. Furthermore, the relatively high viscosity of the oil may damp surface waves, allowing higher ejection rates.

If users would like to reduce evaporation of a volatile fluid without transferring oil into their assays, they may be able to take advantage of the fact that as the amount of mineral oil is decreased, an aperture is made in the center of the well as shown in Frame E of Figure 16.17. For a single well in a 384-polypropylene ECHO qualified microplate, the amount of oil needed to coat the entire surface is approximately 30 μL. As the volume of the oil is reduced below that value, the resulting aperture becomes larger. The user may be able to adjust the oil coating to optimize evaporation control while minimizing the amount of oil transferred.

16.4.2.4 **Combinatorial Chemistry**

Acoustic liquid handling has been used to miniaturize combinatorial one-pot syntheses and multi-component reactions (Thomas Baiga, manuscript in preparation. Waiting for more information). Since acoustic transfer uses a single transducer to facilitate the transfer of any source well to any destination well, it is ideally suited for combinatorial work with complex patterns and mixtures of materials. Work done at the Salk Institute showed that the spot-on-spot capability of transfer provided synthetic reactions in volumes less than 100 nL. Of significance was that the researchers found that DMSO, contrary to expectations, was an extremely good solvent for a variety of conditions and reagents including those used in the Groebke, Ugi, Azido Ugi, Petasis, Pictet-Spengler, and Povarov reactions. These researchers showed that yields were very good, and, in the case of some reaction types, extraordinary (Groebke >80% yield; Azido Ugi >90% yield). Not only were yields excellent but in many cases the purity of the reaction was greater than 90%.

These researchers also measured changes in yield based upon additives to the DMSO solvent, including trifluoroethanol, methanol, dioxane, acetone, hexane, chloroform, and others. Surprisingly, they showed that small amounts of water, rather than reducing yield as predicted, actually enhanced the yields of most reactions. (Water did have a negative impact on yields for the Groebke reaction, apparently due to one component, the amine, coming out of solution.) New work showing the transfer of ionic liquids promises to increase the range of useful solvents for organic synthesis.

Using ADE as part of a combinatorial chemistry approach means that hundreds or thousands of different compounds can be synthesized in a single plate. Baiga has shown data describing the discovery and development of new solvent and catalyst systems for combinatorial chemistry (including both one-pot syntheses and multi-component reactions). He also demonstrated the scalability from microliter to nanoliter, and micromolar to nanomolar reactions. The high yields and purity he found in his miniaturized reactions suggest that initial screens can be performed without requiring significant (or any) purification and could speed drug development. The ability to make relatively pure compounds and immediately screen them provides a virtual library that can be developed 'on demand' in quantities appropriate for the screening laboratory. The synthesis laboratory typically required the production of many milligrams. By eliminating the need to turn to the synthetic chemist even when creating analogs, the drug discovery process may be accelerated creating compounds 'on demand' and in quantities as low as that required for a single assay. Of particular interest is the use of the technology to rapidly make analogs of compounds identified in an initial large library screening.

16.4.2.5 Particle Formation

Droplets ejected by acoustic technology can be converted to particles in a number of ways. The resultant particles differ from particles developed by other techniques in that they are completely spherical and have a very small size distribution. Typical mono-dispersed particles using traditional techniques exhibit a variation in diameter of plus or minus 10% or more. With particles generated via ADE, the variation in diameter drops to less than 2%.

Particles generated via ADE may be homogeneous or, if they are generated from a biphasic solution as previously described, coated with a liquid or solid layer. If the outer layer is solid, the inner material could remain a liquid.

Droplets can be converted into particles in various ways: solutions of activated monomers may polymerize in transit; the solvents of droplets can evaporate during transit leaving only the solute; or droplets of materials can be captured in fluids with which they react, for example, solutions of alginic acid can be captured in solutions high in calcium ion to yield solid alginates. The size of the particles can be regulated by choice of the size of droplets, the solidification process, and the composition of the liquid. An example acrylamide bead is shown in Figure 16.18.

Particles generated via ADE could be used in a number of different applications including the generation of mono-dispersed particles containing drugs for pulmonary delivery, packing materials for ultra-HPLC, cells protected by semi-solid gels of alginate as preparation for tissue engineering, and the preparation of finely divided catalysts. ADE-generated particles can be made with diameters ranging from 10 μm or less to 1 mm or more.

16.4.2.6 Precision Coating

ADE holds the promise of coating surfaces with materials with a high degree of positional accuracy. For example, stents are typically coated by spray, vapor deposition, or dipping into a solution of compound. All of these methods have drawbacks. When a stent is expanded, certain areas of the surface twist and deform.

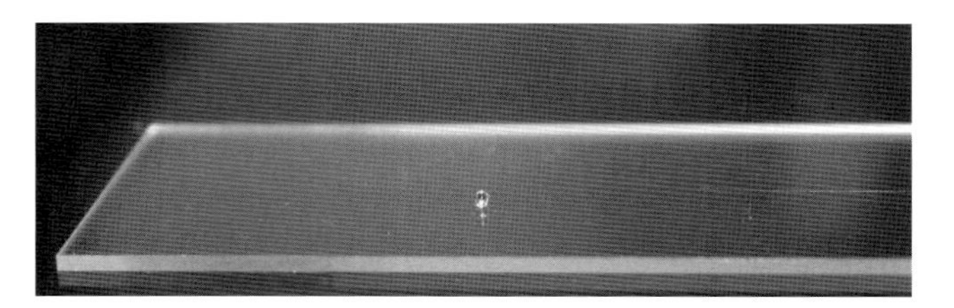

Figure 16.18 This 800 μM polyacrylamide bead was generated by acoustically transferring a solution of 30% acrylamide in water (v/v) to which 2.5% ammonium peroxydisulfate (by weight) was added. The droplets were captured in a reservoir of mineral oil containing 25% tetramethylethylenediamine (TEMED) by weight.

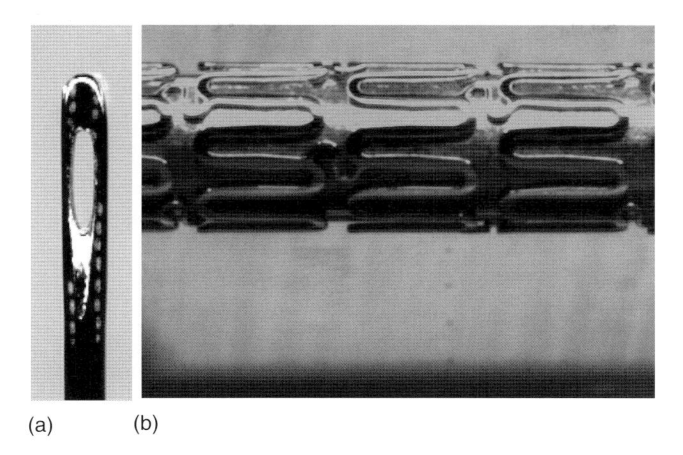

Figure 16.19 (a) A needle is coated with polylactic acid exactly where the researcher chooses without blow-by or clumping. (b) Droplets of polylactic acid are deposited directly onto a stent, one droplet at a time. The mandrel that holds the stent is moved horizontally as it rotates allowing the droplets to be directed to any specific spots and, more importantly, to leave bare those portions of the stent that flex when inserted and expanded.

The stress associated with this deformation is enough to cause material deposited on the surface to flake off. This places the drug not on the stent but in a bolus in the artery or duct. Using ADE it is possible to 'paint' spot by spot the precise areas of the stent that need to be coated and to disregard the sections that undergo surface stress.

Spot-by-spot coating (Figure 16.19) also eliminates the problem of blow-by, whereby expensive drugs are lost to surfaces of the coating instrument. Also eliminated is the formation of clumps of drug that often aggregate on surfaces when sprayed as a fine mist. These clumps tend to fall off during handling.

Using ADE, it is possible to add a second coating of compounds in one of two ways – either directly on the previously laid spots or interspersed with the previous spots. In the first case, reactions between the coatings can be encouraged. In the second case, compounds that could react with each other if in physical contact are held separate from each other until inserted and expanded *in vivo*.

16.4.2.7 **Touchless Transfer of Dangerous Materials**

US patent 7,405,072 demonstrates the unique advantages of using ADE to transfer fluids within a sealed container, free from any physical contact. This process can be applied to the testing of pathogens and other dangerous fluids. For example, a pathogen-containing liquid can be introduced into a vessel and permanently sealed. Acoustic energy can then move that liquid with high precision and accuracy within the vessel without ever physically touching the liquid. This could be used to transfer a droplet of a liquid, including blood, urine, cerebrospinal fluid, toxic chemical waste, and so on, from a sample pool within the sealed vessel onto a receiving surface that contains specific test chemistry and where the result can be then read. The use of ADE eliminates the generation of contaminated pipette tips and labware as it eliminates wash solutions. Transferring liquids within a sealed container minimizes exposure risk for technicians who handle the samples, while enabling easy miniaturization of the assay.

16.4.2.8 Assay Miniaturization

The small volumes associated with acoustic transfer translate directly into assay miniaturization. Material presented in Chapter 11 points out that acoustic liquid transfer is already used in 384-well microtiter plates. The present section refers to miniaturization beyond that well-accepted level. Since the most significant cost of many assays is due to the reagents, a 10-fold reduction in volume can lead to appreciable savings. Higher-density formats provided by assay miniaturization also lead to greater throughput, allowing researchers to spend less time at the bench. Finally, many assays can be automated simultaneously with their miniaturization, freeing researchers from drudgery.

Multi-well Plates Multiwell microtiter plates have been used within HTS with significant cost saving benefit for some time [29, 30]. While common in drug screening, the 1536-well format is not yet widely used in biological research. This is due, in part, to the wide-spread use of manual pipetting and to the lack of automation in many laboratories. As automation costs drop, there will be a tendency for smaller laboratories to move to simple robotics to free researchers to pursue more intellectual matters. The ultra-high throughput associated with many genomic studies (DNA chips) has also changed the mind-set of researchers who now have, or can more easily obtain, data analysis programs that can work with large data sets. Many existing absorbance and fluorescence detectors are compatible with the 1536-well forma, although sensitivities may be lower because of reduced amounts of light-absorbing material or fluorescence. Researchers have recently shown that real-time polymerase chain reaction (RT-PCR) or qPCR can be done in the 1536-well format at total assay volumes of 250 nL, a 20-fold drop from traditional volumes.

3456-well Format Researchers have shown that standard HTS assays can be reduced to the 3456-well format while maintaining the sensitivity, precision, and accuracy associated with 384-well assays [31, 32]. Typical 3456-well assays use 1/30th the volume of reagent per assay compared to the 96-well format. These assays are especially useful with chemiluminescence detection where the reagents are expensive but the light output is significant over a wide dynamic range.

Higher-Density Formats ADE has been shown to be compatible with the 6144-well format. Moving toward the limits, Mutz has shown that assays can be run in volumes as low as 100 nL with the distance between assays being less than 1 mm. Researchers at Trinity College, Dublin, showed the transfer of live cells at these high-density formats (Figure 16.20).

This is equivalent to a standard SBS plate with 7420 individual wells. Arrays have been made with acoustic transfers in imaging mass spectrometry where the

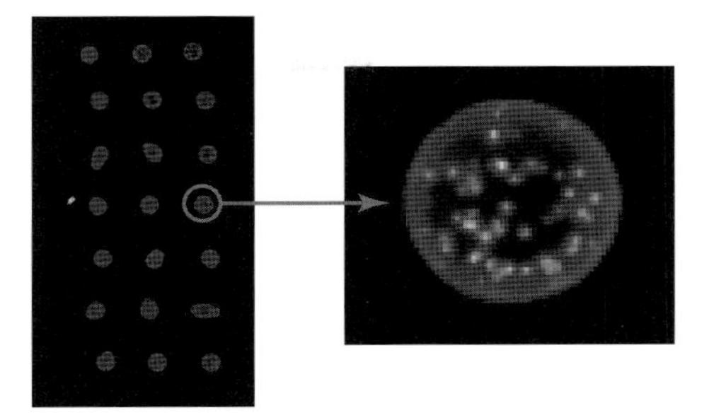

Figure 16.20 Using an Echo liquid handler, researchers at Trinity College, Dublin transferred 10 nL of THP1 cells (from a starting density of 8×10^6 cells/mL) into an array format on a microscope slide with 1000 μm spacing. The magnification on the right shows the live cells in a single spot of the array.

space between discrete droplets is 200 μm, which translates to over 100 000 spots in a surface defined by SBS/ANSI standards [33–36]. This opens the possibility of ultra-miniaturized assays based on small droplets or spots of materials. This ultra-high density application may have ramifications in high-throughput gene expression analyses, DNA enrichment processes, and miniaturized ELISA tests. These formats begin to approach the densities associated with DNA chip-based analyses.

One of the major impediments to miniaturization has been that traditional liquid handlers provide poor accuracy and precision at low volume [37]. As the surface-to-volume ratio increases with smaller volumes, the bottleneck now moves to the evaporation of fluids from the reagent sources and from the final assay destinations. Acoustic liquid handling provides a robust solution for this problem. As already described, ADE allows the transfer of fluids that have low volatility, including glycerol. Addition of glycerol to reagent solutions can not only help stabilize enzymes present but can also reduce the impact of microorganisms while it reduces volatility. ADE is also compatible with transferring fluids through a layer of oil, as previously discussed. The protective oil coating can reduce evaporation even under extreme conditions.

One final impediment to miniaturization has been standardized protocols that have been developed based upon the limitations of previous liquid handlers. In many cases, workflows have been designed specifically to work around the deficiencies of existing laboratory equipment: volumes have been increased in order to maintain an acceptable precision, large volumes of dilute, labile materials have been used rather than small volumes of more stable, concentrated reagents, and low-density plates have been used to overcome detector insensitivity. The precision and accuracy of acoustic liquid handling coupled with enhanced sensitivity of modern detectors have eliminated the reasons behind many protocols. As this technology becomes more widespread it will enable researchers to redesign such limited protocols and improve workflows.

16.4.2.9 Transfection via Sonoporation

Transfection is the nonviral process by which nucleic acids are moved into cells. Transfection can be mediated by chemical processes (calcium phosphate, cationic polymers, cyclodextrin, and liposomes) where the nucleic acid to be moved into the cell is noncovalently bound to a reagent. This reagent is taken up by the cell via various mechanisms and the nucleic acid is released. Many of these transfection reagents show a degree of cytotoxicity.

Physical processes for transfection include electroporation, laser-based optical transfection, and impalefection. These techniques create holes in the membranes of the cells through which the nucleic acid may pass. There are also solid particle-based techniques such as ballistic transfection where nucleic acids that are bound to gold microparticles are shot into cells with a gun, as well as magnetofection where nucleic acid-coated microbeads of iron oxide are concentrated onto cells which then take up the beads via endocytosis or pinocytosis.

Sonoporation has also been explored as a transfection mechanism. Until recently, the belief was that cavitation caused by acoustic energy generated holes in the cell membranes through which nucleic acids could enter. Cavitation is a violent procedure where acoustic energy produces small bubbles that rapidly collapse. During collapse, the surface temperature of the bubbles can reach thousands of degrees and the pulse caused by the collapse of the bubble is powerful enough to strip the metal from the propellers of ocean-travelling ships. While sonoporation has been used for transfection, many people felt that it was too damaging to the cells to be generally useful.

Recent work [38–40] shows that cavitation is not required for sonoporation and that the effectiveness of sonoporation drops when the energy is high enough to lead to cavitation. Instead, sonoporation appears to be driven by a much gentler process called '*microstreaming*,' where areas of localized flow drive nucleic acids to the surface of cells, where they are taken up. The energy required to generate microstreaming is very similar to the energy required for acoustic mixing, and experiments are in progress to determine whether an instrument designed to transfer fluids can also be used to transfect cells. The advantages of sonoporation are that the cells suffer less damage than they do by physical means that poke holes in the membrane, as they do not expose the cell to toxic carriers associated with the chemical transfection mechanisms.

16.4.2.10 Expanded Reporting Capabilities

Transfer volumes are controlled and exact when fluid composition is within a predefined range on a factory-calibrated instrument. For situations where the fluids being transferred have not been calibrated, the user may choose to generate a standard curve correlating actual volume transferred with reported volume. Even when transferring a fluid for which there is no specific calibration, the reproducibility will be maintained by a combination of autofocus, impedance measurement, and measurement of droplet ejection power. Labcyte may modify their user interface so that researchers can choose to have results reported either as nanoliters (as with the instruments today) or as number of droplets (which can be converted to volume specific to that fluid with a fluid-dependent factor).

16.4.2.11 Transfer of Droplets of Different Volume – Smaller Droplets

The volume of a droplet transferred by acoustic processes is determined primarily by the frequency of the sound energy used to form the droplet – higher frequencies (shorter wavelengths) produce smaller droplets. Volume is proportional to the inverse of the cube of frequency; alternatively, droplet diameter is proportional to acoustic wavelength. Other parameters, including the surface tension and viscosity of the fluid have smaller but noticeable effects.

Smaller droplets can be useful in the production of arrays, especially in the area of proteomics with the production of micro-ELISA tests using extremely small spots of antibodies packed at high density. These tests would be useful to determine the presence of numerous components in extremely small samples, as might be encountered when samples are limited, for example, diagnostics involving neonatal or geriatric blood, forensic samples, and primary cells.

Smaller droplets may also be of use in the miniaturization of assays. Researchers have shown that assay volumes can be reduced significantly. However, a mismatch between assay volume and the volume of a solvent used to carry a sample may exist. That is, while it is very possible to reduce the volume of an assay to 25 nL if the minimum droplet volume is 2.5 nL, adding the smallest amount of sample dissolved in DMSO will yield an assay solution that is approximately 10% DMSO. Numerous researchers have shown that concentrations of DMSO greater than 1% can have serious negative effects on assays due to loss of helical structure in proteins as well as to increased enzyme degradation and aggregation [41, 42]. Using a smaller droplet to carry the sample reduces this problem.

Likewise, in miniaturized organic syntheses, the ratio of the size of the smallest droplet that can be added to a reaction compared to the total reaction volume may be too great to allow optimization. In this case smaller droplets may also be of use. Finally, smaller droplets of sample can be deposited into assay-ready plates and dried down for easy distribution. The samples are more readily resolubilized, possibly due to the smaller size of crystals in the precipitate, and the problems seen when larger volumes are taken to dryness are eliminated.

While it is possible to acoustically transfer droplets as small as 1 pL, the distance that the droplet is able to travel becomes extremely small. Smaller particles have a greater surface-to-volume ratio than larger droplets. The small droplets are slowed by drag from the air through which they travel. While droplets of 2.5 nL can be easily transferred 3 cm, droplets of 1 pL travel approximately 1–2 mm before coming to a halt. Because of the high surface-to-volume ratio, small droplets also have the tendency to rapidly evaporate. While a 2.5 nL droplet of water may evaporate in 15 s, a droplet of 2.5 pL will evaporate in less than 2 s. The rapid evaporation of small volumes needs to be considered when designing small volume assays.

Researchers have recently shown that qPCR reactions can be reduced to 250 nL total reaction volume while retaining excellent analytical results [43]. Recent investigations have suggested that PCR can be achieved in droplets of 2.5 nL or smaller, providing an acoustic approach to digital gene expression via PCR or isothermal amplification techniques.

16.4.2.12 Transfer of Droplets of Different Volume – Larger Droplets

As previously described, the volume of a droplet ejected is strongly related to the frequency of the acoustic energy used for ejection. Lower frequencies produce larger droplets. Ellson and his coworkers have used transducers at lowered frequencies to transfer droplets as large as 1 μL [44]. Larger droplets have multiple potential uses including the formation of particles, as previously discussed, and they enhance the utility of acoustic transfer for bulk reagents. Increasing the size of the droplet for bulk reagents reduces the amount of time required to completely assemble the assay plate. The parameters required for transfer of droplets larger than 2.5 nL have not yet been optimized for accuracy, precision, and throughput as they have for smaller droplets.

The same considerations associated with making smaller droplets are also associated with larger droplets – a transducer different than the one currently employed in the system will need to be installed.

16.4.3 Expanded Auditing Capabilities

The use of ADE for auditing water content in DMSO solutions has been detailed by Tang *et al.* in Chapter 11. Additional capabilities that may be realized over the next few years as acoustic instruments continue to increase their capability to audit intelligently, and within seconds, provide information on the contents of each well of a multi-well plate. Included in this information could be the depth of fluid, the acoustic impedance, and information about the concentration of analytes. The measurement of concentration coupled with a measurement of fluid depth (which can be converted to volume) could provide researchers with the knowledge needed to replenish the solvent, be it water or an organic solvent that has evaporated, in order to restore the initial concentration.

When a well is audited, a pulse of acoustic energy insufficient to cause ejection is transmitted by the transducer. The sound propagates through the coupling fluid (water) to the bottom surface of the microplates. At the interface between the water and the plastic plate, some energy is reflected and some continues through the plastic. The energy that is reflected returns to the transducer where it is collected and analyzed. When the acoustic energy that entered the plate reaches the next interface, that between the plastic of the plate and the fluid in the well, some of the energy is reflected while the remainder travels into the fluid in the well. As before, the reflected sound is received by the transducer. Finally, the acoustic energy reaches the surface of the fluid – the interface between the fluid and the air above it. At the air interface almost all of the energy is reflected back through the liquid and received by the transducer. The three reflections, or echoes, are analyzed by the instrument for two specific characteristics, the amplitude of the reflection at the plastic–fluid interface and the time of flight measured for the reflection at the fluid surface.

The amplitude of the reflection at the plastic–fluid interface provides a measurement of the acoustic impedance of the fluid in the well by the equation[2]:

$$R_{\text{plastic - fluid}} = \frac{(Z_{\text{plastic}} - Z_{\text{fluid}})^2}{(Z_{\text{plastic}} - Z_{\text{fluid}})^2} \tag{16.1}$$

where, $R_{\text{plastic–fluid}}$, the reflection at the plastic–fluid interface, is equal to the square of the difference between the acoustic impedance of the plastic, Z_{plastic}, and the fluid, Z_{fluid}, over the square of the sum of those two values (see Figure 16.21). The reflection is the measured amplitude of the echo from the interface; the acoustic impedance of the plastic is known, and therefore the acoustic impedance of the fluid is easily determined. Once the acoustic impedance is determined, it is possible to correct the positioning of the transducer to ensure that when ADE is performed, the focus is at the meniscus.

16.4.3.1 **Auditing for Volume**

The auditing process determines the depth of fluid in order to position the focus of the acoustic energy at the meniscus. For many users, the ability to know how much liquid is in each well is of significant importance. Hayward (personal communication) reported that 1–3% of sample wells tested prior to liquid transfer actually were dry, with either the sample having been sampled until nothing remained or the solvent having evaporated, leaving a thin, solid coating at the bottom of the well. Auditing eliminates the false negatives that result from attempting to test from an empty sample well. Furthermore, knowing the volume of a well can provide the user with important information concerning the usage rate of sample. Some Echo liquid handlers convert the depth value into a more user-friendly volume value. Conversion from depth to volume requires an understanding not only of the geometry of the well but an understanding of how different liquids are drawn to the edges of the well. Figure 16.22 shows that pure DMSO has a much greater curve to the meniscus than does 70% DMSO. In this case, the same depth of fluid can yield a 2.5 μL difference in volume, which amounts to 10% of the minimum working volume.

Papen *et al.* [45] reported an innovative use of the volume-auditing capacity of an ADE system. They showed that acoustic auditing could be used as a quality control mechanism for assembled assay plates. They used the volume monitoring function to check that all wells of a 1536-well plate were filled equally when fluids were transferred by a 384-channel head and a peristaltic line filler. Errors due to

2) Equation (16.1) is based upon a plane wave of energy. We use this simple equation for the sake of clarity. Since focused waves are used in the Echo system, the relation is actually more complicated. However, the instrument automatically converts the information gained from the echoes to adjust the position of the transducer in order to maintain focus at the fluid meniscus.

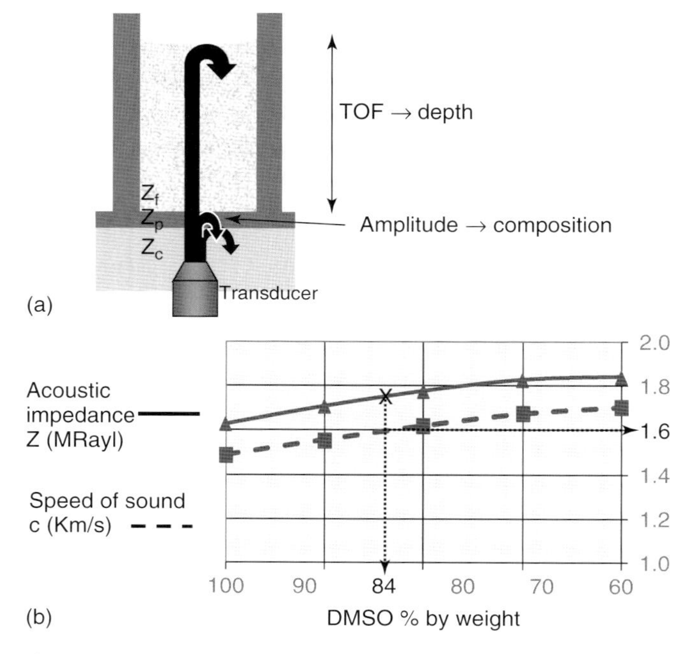

Figure 16.21 (a) Acoustic energy emitted by the transducer travels until it meets an interface between materials of different acoustic impedance (Z), in this case the fluid in the well (Z_f), the bottom of the plate (Z_p), and the fluid coupling the transducer to the plate (Z_c). At the interface, some energy is reflected and some continues on. Almost 100% of any remaining energy is reflected at the meniscus due to the large difference in acoustic impedance between the solution and air. The energy that is reflected back to the transducer is measured and used to determine the acoustic impedance of the liquid from Equation 16.1 and the reference Z_p plate. (b) Once that impedance, Z_f, is known, (marked with 'X' on the solid line) it is converted to composition by dropping a perpendicular line (dotted) to the horizontal axis (84% DMSO) and to the speed of sound by determining where the perpendicular crosses the dashed line (1.6 km s^{-1}). See Figure 16.23 for an example of the sonogram produced by the interrogating acoustic pulse.

blocked lines or loose tips were easily identified by the reduced volume in a well, as were 'transfers' from empty source wells. The auditing capacity of an acoustic liquid handler can easily identify potential filling problems associated with other fillers.

16.4.3.2 Auditing for Restoration

As water evaporates from an aqueous solution, the concentration of salts and buffers increases. This change can be exploited to allow the user to determine exactly how much water has evaporated from a source well. The user has multiple options including:

1) Modify the volume transferred in order to keep the amount of solute constant in the transfer.

Figure 16.22 Pure DMSO has a much more pronounced curve on its meniscus than 70% DMSO. Two wells that have identical depth of fluid can vary by as much as 2.5 μL depending upon the level of hydration. Meniscus shape is also affected by surfactants, detergents, proteins, salts, and solvents. In a well-defined situation, as with binary mixtures of water and DMSO, it is possible to convert depth to volume.

2) Add water to the source plate in the appropriate amount to re-establish the original concentrations.
3) Continue with more concentrated solutions, converting results, as appropriate, in the analysis step.

Replenishment of water to re-establish the original concentration is especially interesting. The existing Echo system can easily add water back to the source plates in incremental steps of 2.5 nL. Not only would this maintain the original concentration (something that is not maintained with any current liquid handler) but it extends the lifetime of the source well.

An interesting modification of this application is to use the fluid in the source well until the minimum working volume (dead volume) is reached, whereupon water is added to the well to create a diluted source. The user can transfer larger aliquots of this diluted solution to maintain the correct amount of solute transferred. This approach can significantly reduce the effective dead volume of the source well.

16.4.3.3 **Auditing for Solute Information**

Currently being investigated is the use of acoustic auditing as a noninvasive mechanism to determine the concentration of DNA in a solution. This technique could potentially simplify procedures in both sequencing and gene expression analysis and, in particular, could enable a rough re-normalization of DNA concentration. Preliminary investigations suggest that as the concentration of sheared DNA increases in solution, the acoustic impedance of the solution increases. This simple, noninvasive and nondestructive measurement could be useful for measuring over the range $1–10\,mg\,mL^{-1}$.

In a similar way, libraries of protein crystallization fluids can be monitored to ensure that their concentrations remain constant over time. The fluids used to encourage proteins to crystallize from solution are many and extremely varied. They include solutions of buffers with heavy metal salts, solvents including ethanol, isopropanol, methylpentanediol, hexanediol, and dioxane, and polymers including polyethylene glycol 30 000, polyvinyl pyrrolidone, and Jeffamine (Table 16.4). Newman [26] has used wide-range pH-sensitive dyes to monitor changes in crystallization fluids, but the dye itself may bind to the protein of interest. Acoustic auditing offers the possibility of monitoring these fluids rapidly and noninvasively to check the acoustic impedance of the solutions at the time of use and compare to their original state.

16.4.3.4 **Auditing Bi-Phasic Solutions**

ADE instruments have been designed to transfer single-phase solutions. However, researchers have shown that biphasic solutions can be ejected [27]. In this case, a nonmiscible fluid covers a denser fluid. When audited, a new echo is recorded from the interface between the two immiscible fluids as long as the acoustic impedances of the two liquids are not equal (see Equation 16.1). Monitoring the thickness of the upper layer may prove essential in preparing coated droplets (see below) or in developing evaporation-retarding barriers.

16.4.3.5 **Auditing for Sample Quality**

When solutes precipitate from solutions in the form of microplates, these are often found on the bottom of the well, where they can alter the magnitude of the sound reflected in the acoustic audit. In present Echo systems, wells with anomalous echo signals are denoted in the data output as wells where the transfer of fluid is not attempted. The ability to see changes associated with solute precipitation can be useful to determine the health of a compound library by comparing the acoustic reflection of the fresh solution with the reflection from an aged sample. Figure 16.23 shows the impact on the acoustic reflection from the interface between the plastic at the bottom of the well with the fluid immediately above it. As crystals precipitate from the solution, a change is noted. More work is being done to determine the limits of crystal detection, including the impact on acoustic monitoring of crystal structure (needles vs prisms), size (microcrystals vs visible crystals), positioning, and density.

It has been shown experimentally that acoustic velocity and attenuation vary with the concentration of suspended cells [46]. It may, therefore, be possible to determine the density of cells in suspension by monitoring the delay and amplitude of echoes generated. It may also be possible to acoustically disturb the surface of the cell suspension and use frequency analysis to gain information on the number of cells in suspension. Backscatter measurements of acoustic energy have been used to measure aggregation of red blood cells [47].

These techniques could be achieved with an Echo liquid handler with increased capabilities and could be applicable to blood, as well as to suspensions of bacteria, yeast, or plant and animal cells. As mentioned in Chapter 11, cells can be safely ejected without negative biological impact because the amount of energy required for a transfer is very low.

While it has already been shown that it is possible to monitor cell density acoustically, it has not yet been determined whether it is possible to monitor changes

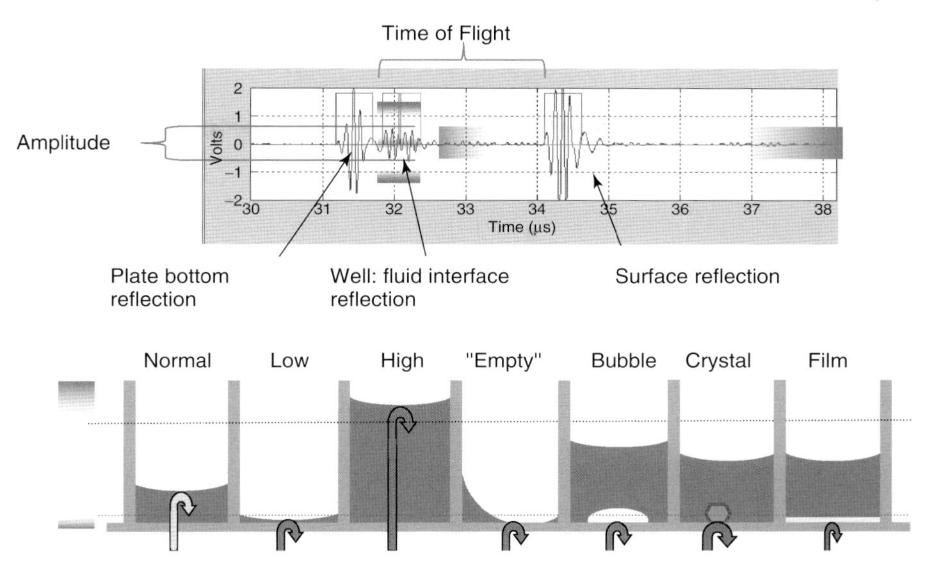

Figure 16.23 The sonogram of a standard audit has three sets of echoes – one from the bottom of the plate, one from the interface between the plastic of the plate and the fluid in the well, and one from the meniscus. The amplitude of the second echo, that from the plastic–fluid interface, provides information about the acoustic impedance and speed of sound in the solution. If the signal is too small (as with the film coating) or too large (as with a bubble or crystal) and falls into the colored areas of the sonogram, the instrument will flag the well for visual inspection and will not attempt to transfer fluid. If the time of flight is too short (as with low fills or empty wells) or too long (as with an over-filled well) and falls into the orange-colored areas, the instrument will signal that the fluid is not within the working range and will not attempt to transfer the fluid. The inset shows a large crystal in the bottom of a well in a 384-well microplate. These crystals will occlude the acoustic energy and eliminate or greatly reduce the reflections from the meniscus.

in the cells as they change size or as they sequester materials (e.g., oils or starches) in vacuoles. Research into the analysis of cell changes is currently in progress. This type of information could be used by researchers in a number of areas, including the fields of food genetics, bio-fuel production, and cell differentiation.

16.4.3.6 **Frequency-Domain Analysis**

Real-time surface monitoring refers to a newly expanded auditing process to track the energy required for droplet ejection. In this procedure, energy sufficient to deform the surface but not enough to cause ejection (a sub-threshold pulse) is transmitted into the fluid. As the mound of fluid forms and relaxes at the meniscus, a separate acoustic pulse is sent to the meniscus, and its reflection is collected for analysis in the frequency domain (a procedure similar to frequency analysis performed in Doppler shifts for velocity measurement). The information obtained can be essential when ejecting samples with varying amounts of surfactants, allowing the instrument to change the power directed to a well in real time to

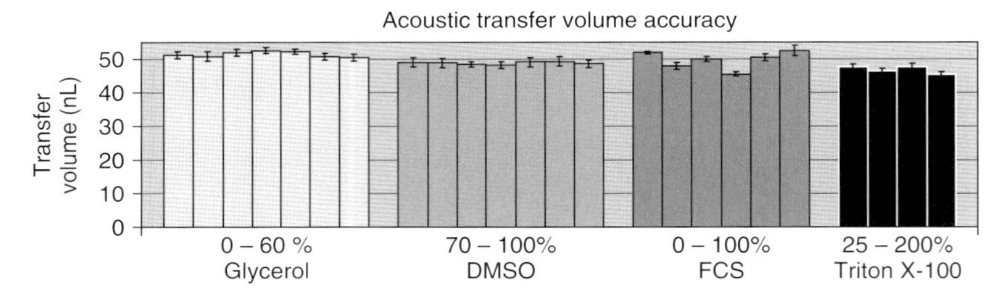

Figure 16.24 Using real-time surface monitoring, transferring 50 nL of fluids from different classes (viscous: glycerol, compound libraries: DMSO, blood/serum: FCS (fetal calf serum), surfactant: Triton X-100) with different viscosities and surface tensions provides superior precision (CV of all transfers <3%) and excellent accuracy (less than 8% deviation from nominal.

ensure optimal droplet precision and accuracy (Figure 16.24). This information can augment information obtained by measuring the acoustic impedance of the fluid used for focusing through the fluid to allow the instrument to compensate for surface tension or other factors that influence the amount of acoustic energy required for droplet formation.

16.4.4 New Software Advances

The introduction of applications software packages including those for dose-response experiment setup, plate reformatting, and auditing have relieved users of the burden of step-by-step instruction to the ADE instrument. Other software packages, modifications to the algorithms used in analysis, or changes to the user interface may be developed as user needs become apparent.

16.4.4.1 Improved Meniscus Scan

Existing software adjusts the volume of the well to the depth depending upon the tendency of the fluid to be pulled up the walls of the well by surface tension (Figure 16.25). Fluid may also be pushed toward a particular edge of a well by centrifugation (Figure 16.26). While existing Echo liquid handlers are able to measure depth correctly even when a meniscus has been tilted by centrifugation, they do not adjust the reported volume because the degree of tilt depends upon the centrifuge and is not an intrinsic property of the fluid and the well. By measuring the depth of the fluid at more points in each well, it is possible to map out the surface shape and provide a better evaluation of total well volume.

16.4.5 ADE Summary

In the few years since the commercial introduction of acoustic liquid handling, the technology has become a workhorse to transfer samples in compound management

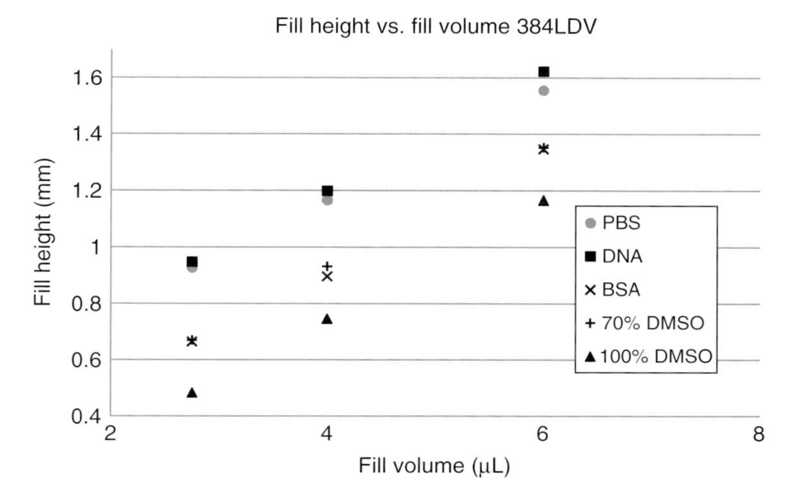

Figure 16.25 Equal volumes of fluids show different fluid depths in identical wells (in this case, a 384 low-dead-volume plate). 100% DMSO has a greater curve in the meniscus than any of the other fluids, while both phosphate-buffered saline (PBS) and DNA in a Tris-EDTA buffer have the flattest surfaces and, therefore, the greatest depth. 2.75 μL of PBS has the same fluid depth as 4 μL of 70% DMSO and, by interpolation, 5 μL of 100% DMSO. Almost half of a 5 μL fill of 100% DMSO is captured in the fluid at the edges of the well. The Echo system automatically adjusts the depth of the fluid to a volume by first determining the DMSO concentration.

and HTS laboratories in the pharmaceutical industry. Advances in the technology have already broadened the variety of fluids with which the technology is compatible. Researchers using the existing technologies have adapted it for uses for which the instrument was not originally intended. These new applications include the monitoring of DMSO hydration in library plates, the transfer of cells for miniaturized cell-based assays, the setup miniaturized combinatorial organic syntheses, and the mixing of components in source wells.

Interactions between users and manufacturer will guarantee the development of new features to provide advantages across the laboratories of current users as they introduce the technology to a new range of researchers in a variety of scientific and technological areas.

ADE has progressed from a cutting-edge idea to a state-of-the-art necessity in many laboratories worldwide. The precision and accuracy of transfer, coupled with pinpoint positionality of the transferred fluid, have expanded the technology from that used solely to replicate multi-welled plates into a broad new technology with hundreds of applications. As this technology diffuses to new researchers, and as existing users increase their usage, we are confident that the use of the technology will escalate as the number of applications proliferate. Already the technology is being explored for use in areas far afield from drug discovery and is being tested for applications in the electronics industry, diagnostics, and the flavor and fragrance

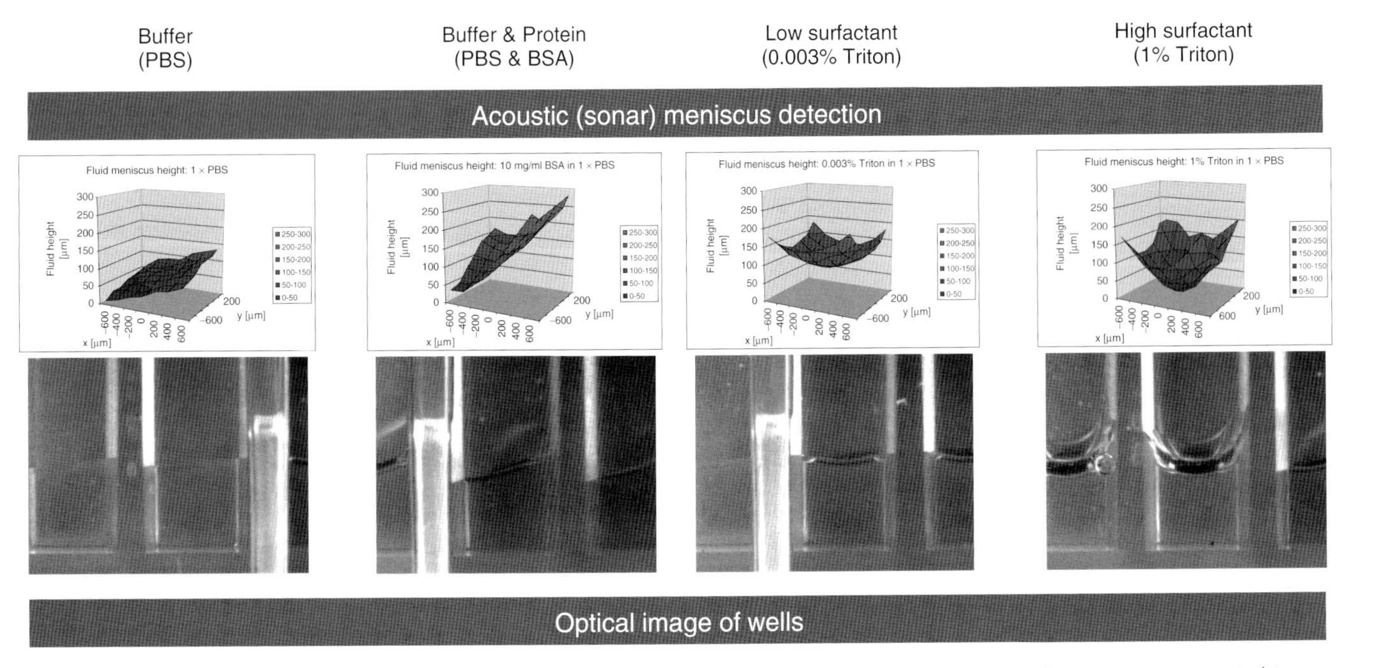

Figure 16.26 When a plate is centrifuged, the fluid in the wells at the edges of the plate have a significant tilt in the meniscus (two optical images on left). The degree of slope is affected by the components of the fluid. The shape of the meniscus changes as the surface tension of the solution changes (compare two images at right). The Echo liquid handler correctly determines fluid depth at the position of the meniscus from which the sample will be ejected and reports volumes adjusted for meniscus shape.

industry. Undoubtedly, new ideas will develop as more people are exposed to ADE, and we expect that this technology will soon be as common as spectrophotometers and centrifuges.

16.5 Closing Remarks

We have covered technologies at different stages of development within this chapter. How fully exploited these ideas will be within the sample management arena will more than ever depend upon both the commercial viability and the technical worth of the process, with each novel technology or application being rigorously assessed on a cost–benefit analysis. This is particularly well exemplified by the ADE discussion. Here there is already a well-established use of this technology, with many further potential applications. Yet with only a finite resource pool available to develop them, and a varying level of alternative technologies for each of the applications, only time will tell which ones are the most beneficial and will become routinely utilized. Hence, it is the industry which will decide how these technologies evolve in the future.

The ability to automatically dispense powders has long been perceived as a need within sample management, but no single technology has yet been able to become accepted within the industry for this to become a reality. We have presented one good example of a technology that has potential to address this need.

Adoption of some of these ideas is also dependent on how sample management itself develops as a process. We have seen in Chapter 10 the classical store, solubilize, and dispense as separate technology-driven workstreams. However, it can be seen that, with the sciSWIFTER device, technology offers the possibility that the core activities of sample storage, retrieval, and dispense can become closely coupled to the downstream process. This type of process innovation can offer significant efficiencies to the pharmaceutical business. Hence, we encourage the reader not to look at the technologies we present here in isolation but to think about how they are likely to benefit the sample management process and how these or similar technologies can be applied to your process, or, if these are not the technologies you require, to look to engage with equipment vendors to encourage them to develop new technology, or as has happened many times in the past, to take your ideas and launch your own technology company. The right ideas will usually be the ones that become standard practice in the laboratory, and the vendors that innovate with their customers are usually the most successful.

References

1. Wood, R.W. and Loomis, A.L. (1927) The physical and biological effects of high frequency sound waves of great intensity. *Philos. Mag.*, **4** (22), 417–436.
2. Ward, K., Day, M., Dudenhoefer, C., Nielsen, J., Paris, H., Peters, K.F., Thomas, D., and Yu, J. (2011) Improved Dose-Response Analyses

Using Direct Titration with Picoliter Dispense, Published online *http://www.labautopedia.com/mw/index.php/W560:Ward:Improved_Dose-Response_Analyses_Using_Direct_Titration_with_Picoliter_Dispense* (accessed 11 November 2011).
3. Yang, S. and Evans, J.R.G. (2007) Metering and dispensing of powder; the quest for new solid freeforming techniques. *Powder Technol.*, **178** (1), 56–72.
4. Islam, N. and Gladki, E. (2008) Dry powder inhalers (DPIs) – a review of device reliability and innovation. *Int. J. Pharm.*, **360** (1–2), 1–11.
5. Morissette, S.L. *et al.* (2004) High-throughput crystallization: polymorphs, salts, co-crystals and solvates of pharmaceutical solids. *Adv. Drug Deliv. Rev.*, **56** (3), 275–300.
6. Bell, J.H., Hartley, P.S., and Cox, J.S.G. (1971) Dry powder aerosols I: a new powder inhalation device. *J. Pharm. Sci.*, **60** (10), 1559–1564.
7. Yang, S. and Evans, J.R.G. (2004) Acoustic control of powder dispending in open tubes. *Powder Technol.*, **139**, 55–60.
8. Yang, S. and Evans, J.R.G. (2005) On the rate of descent of powder in a vibrating tube. *Philos. Mag.*, **85** (10), 1089–1109.
9. Yang, S. and Evans, J.R.G. (2004) A dry powder jet printer for dispensing and combinatorial research. *Powder Technol.*, **142** (2–3), 219–222.
10. Lu, X.S., Yang, S.F., and Evans, J.R.G. (2006) Studies on ultrasonic microfeeding of fine powders. *J. Phys. D: Appl. Phys.*, **39** (11), 2444–2453.
11. Lu, X.S., Yang, S.F., and Evans, J.R.G. (2007) Dose uniformity of fine powders in ultrasonic microfeeding. *Powder Technol.*, **175** (2), 63–72.
12. Vetter, D., Thamm, A., Schlingloff, G., and Schober, A. (2000) Single bead parallel synthesis and screening. *Mol. Divers.*, **5** (3), 111–116.
13. Eickhoff H., Ivanov I., and Lehrach H. (2001) Technical system management in *Microsystem Technology: A Powerful Tool for Biomolecular Studies* (ed. H.P. Saluz), Birkhäuser Verlag, pp. 16–29.
14. Seitz, H., Hutschenreiter, S., Hultschig, C., Zeilinger, C., Zimmermann, B., Kleinjung, F., Schuchhardt, J., Eickhoff, H., and Herberg, F.W. (2006) Differential binding studies applying functional protein microarrays and surface plasmon resonance. *Proteomics*, **6** (19), 5132–5139.
15. Hager, D.B. and Dovichi, N.J.. (1994) Behavior of microscopic liquid droplets near a strong electrostatic field: droplet electrospray *Anal. Chem.*, **66** (9), 1593–1594.
16. Moeller, T.A. (2010) From in vitro to in vivo. *Drug Discov. Dev.*, **13** (5), 10–13.
17. Schaack, B., Reboud, J., Combe, S., Fouqué, B., Berger, F., Boccard, S., Filhol-Cochet, O., and Chatelain, F. (2005) A 'DropChip' cell array for DNA and siRNA transfection combined with drug screening. *NanoBiotechnology*, **1**, 183–189.
18. Citeau, H., Horn, S., Lebrun, S.J., and McIntosh, B. (2007) Setting Up And Running A Protein Microarray Core Facility, Lab Manager, January.
19. Scherp, P. and Hasenstein, K.H. (2007) Anisotropic viscosity of the *Chara* (Characeae) rhizoid cytoplasm. *Am. J. Bot.*, **94**, 1930–1934.
20. Gradl, G. and Sterrer, S. Mobility of Molecules and Particles within the Cytoplasm of a Living Cell. *http://www.zeiss.com/C12567BE00472A5C/EmbedTitelIntern/Application_Cellular_Cytoplasm/$File/CELLMEASURE2.PDF* (accessed 11 November 2011).
21. Viscosities of Aqueous Glycerine Solutions in Centipoises/ mPa s. available at *http://msdssearch.dow.com/PublishedLiteratureDOWCOM/dh_0032/0901b803800322bd.pdf?filepath=glycerine/pdfs/noreg/115_00678.pdf&fromPage=GetDoc* (accessed 11 November 2011).
22. (a) Waybright, T.J. and McCloud, T.G. (2004) Overcoming the problems associated with long-term storage of compounds in DMSO. Poster presented at the NCI-Frederick/Ft. Detrick Spring Research Festival, May 12–13, 2004; (b) Waybright, T.J., Britt, J.R., and McCloud, T.G. (2009) Overcoming problems of compound storage in DMSO: solvent and process alternatives. *J. Biomol. Screen.*, **14**, 708.

23. Litsky, W., Libbey, C.J., and Mariani, E.J. Jr. (1971) Sterility Testing and Antimicrobial Activity of Commercial Grade Glycerine. Report prepared for Glycerine Producers' Association. *http://www.aciscience.org/docs/Sterility_Testing_and_Antimicrobial_Activity_of_Commercial_Grade_Glycerine.pdf* (accessed 11 November 2011).
24. Stearns, R., Huang, R., Harris, D., Ellson, R., Olechno, J., and Datwani, S. (2011) Acoustic, touchless transfer of compounds, proteins and protein crystallography fluids for crystallographic screening. Poster presented at the Society for Biomolecular Screening 2011 Meeting, Orlando, FL.
25. Soares, A.S., Engel, M.A., Stearns, R., Datwani, S., Olechno, J., Ellson, R., Skinner, J.M., Allaire, M., and Orville, A.M. (2011) Acoustically mounted microcrystals yield high resolution X-ray structures. *Biochemistry*, **50** (21), 4399–4401.
26. Sayle, R.A., Fazio, V.J., and Newman, J. (2010) High-throughput pH measurements: seeing is believing. Poster presented at the American Crystallographic Association, July 2010, Chicago, *https://aca.conference-services.net/resources/786/2077/pdf/ACA2010_0290.pdf* (accessed 11 November 2011).
27. Mitre, E., Schulze, M., Cumme, G.A., Rößler, F., Rausch, T., and Rhode, H. (2007) Turbo-mixing in microplates. *J. Biomol. Screen.*, **12** (3), 361–369.
28. Mutz, M., Harris, D., Kannegaard, E., Pringle, J., Stearns, R., and Ellson, R. (2002) Array production by acoustic ejection of single and multilayer fluids and cell suspensions. Poster at SmallTalk Conference, July 30, 2002, San Diego, CA.
29. Allenby, G., Rawlins, P., and Swift, D. (2006) High Throughput Screening of Human Neutrophils in 1536 Well Format, Web-based Seminar, December 2006, *http://labcyte.com Webinars/Default.480.html* (accessed 11 November 2011).
30. Lou, Z. Bioassays with POD 810 at Wuxi AppTec, Presentation available at *http://www.labcyte.com/fileupload/Bioassays%20with%20POD%20810.ppt* (accessed 30 January 2011).
31. Massé, F., Guiral, S., Kargman, S.L., and Brideau, C. (2005) Compound transfer to 3456-well assay plates using the Echo 550. Poster presented at the Society for Biomolecular Screening 11th Annual Conference, September 11–15, 2005, Geneva, Switzerland, *http://labcyte.com/_fileupload/Image/Resources/compound.transfer.3456.pdf* (accessed 11 November 2011).
32. Johnson, E. (2011) Doing more with less: advances in miniaturization for uHTS. Presentation at the Sixth Annual Screening and Imaging Summit, June 2011, Philadelphia, PA.
33. ANSI/SBS 1-2004 Standard for Microplates - Footprint Dimensions. available at *http://www.slas.org/education/standards/ANSI_SBS_1-2004.pdf* (accessed 11 November 2011).
34. ANSI/SBS 2-2004 Standard for Microplates - Height Dimensions. available at *http://www.slas.org/education/standards/ANSI_SBS_2-2004.pdf* (accessed 11 November 2011).
35. ANSI/SBS 3-2004 Standard for Microplates - Bottom Outside Flange Dimensions. available at *http://www.slas.org/education/standards/ANSI_SBS_3-2004.pdf* (accessed 11 November 2011).
36. ANSI/SBS 4-2004 Standard for Microplates - Well Positions. available at *http://www.slas.org/education/standards/ANSI_SBS_4-2004.pdf* (accessed 11 November 2011).
37. Comley, J. (2004) Continued miniaturisation of assay technologies drives market for nanoliter dispensing, *Drug Discov. World*, Summer, *http://www.ddw_online.com/* 43–54.
38. Forbes, M.M., Steinberg, R.L., and O'Brien, W.D. Jr. (2008) Examination of inertial cavitation of optison in producing sonoporation of Chinese hamster ovary cells. *Ultrasound Med. Biol.*, **34** (12), 2009–2018 [Epub 2008 August 9].
39. Forbes, M.M. (2009) The role of ultrasound contrast agents in producing sonoporation. PhD Thesis. Department

of Bioengineering, University of Illinois at Urbana-Champaign, 2009.
40. Forbes, M.M., Steinberg, R.L., and O'Brien, W.D. Jr. (2011) Frequency-dependent evaluation of the role of definity in producing sonoporation of Chinese hamster ovary cells. *J. Ultrasound Med.*, **30** (1), 61–69.
41. Tjernberg, A., Markova, N., and Hallén, D. (2005) Is DMSO a friend – or a troublemaker? Poster presented at the 11th Anniversary Meeting of the Society for Biomolecular Screening, September 2005, Geneva, Switzerland.
42. Tjernberg, A., Markova, N., Griffiths, W.J., and Hallén, D. (2006) DMSO-related effects in protein characterization. *J. Biomol. Screen.*, **11** (2), 131–137.
43. Glazer, C., Sonntag, M., Lee, H., Jarman, C., Pickett, S., and Datwani, S. (2010) Miniaturized Quantitative PCR in 1536-Well Plate Format Using the Echo Liquid Handler, Labcyte Applications Note 410, *http://www.labcyte.com/_fileupload/qPCR-Echo-LabcyteAppNote410.pdf* (accessed 11 November 2011).
44. Ellson, R. (2002) Picoliter: enabling precise transfer of nanoliter and picoliter volumes. *Drug Discov. Technol.*, **7** (5, Suppl.), S32–S34.
45. Papen, R., Travis, M., Qureshi, S., and Brown, C. (2005) Use of focused acoustics for auditing assembled assays – a QC validation in HTS. Poster presented at the Lab Automation 2005.
46. Treeby, B.E., Zhang, E.Z., Thomas, A.S., and Cox, B.T. (2011) Measurement of the ultrasound attenuation and dispersion in whole human blood and its components from 0–70 MHz. *Ultrasound Med. Biol.*, **37** (2), 289–300.
47. Franceschini, E., Yu, F.T.H., and Cloutier, G. (2008) Simultaneous estimation of attenuation and structure parameters of aggregated red blood cells from backscatter measurements. *J. Acoust. Soc. Am.*, **123** (4), EL85–EL91.

17
The Impact of Future Technologies within Biobanking

Manuel M. Morente, Laura Cereceda, and María J. Artiga

17.1
Introduction

To talk about applied technology means to talk about the future – about something unknown and almost impossible to foresee. The current rate of technological development implies that the most modern platforms will become obsolete within a very brief period of time. We can take as an example the case of the capability and accessibility of sequencing systems which allowed us to decrease the cost of sequencing 1 million bases pairs from 10 000 USD in 2000, using automated Sanger sequencing, to 1 USD in 2010 by third-generation sequencers [1] with a processing speed and accuracy unimaginable when the Human Genome Project was published. Furthermore, nothing indicates that this development is about to stop. What would be the sequencing gold standard in 2025? What new type of goals would biotechnology allow us to deal with in the next 25 years? And more importantly, what will be the scientific goals and approaches, concepts, and perspectives?

The present chapter focuses on this challenge from the perspective of biobankers and it is framed by the awareness of the prevailing public service vocation of biobanks. The reader should understand the difficulty to act as a prophet in this extremely changing society with respect to biotechnology and information technologies, and should not look for concrete solutions, but rather for general approaches to the increasing complexity of biobanking, some intuitions about how new technologies could modify the daily activities of biobanks, goals and procedures, and some proposals for the road ahead.

17.2
The Role of Biobanks in Biomedical Research

17.2.1
Biobanking Activity Is Based on Commitments

Biobanking is an ancient activity and a young discipline, and therefore it is especially necessary to establish its position in biomedical research in these changing times,

Management of Chemical and Biological Samples for Screening Applications, First Edition.
Edited by Mark Wigglesworth and Terry Wood.

and this position should only be understood in terms of public service. Biobanks are not a goal, but rather a scientific tool for accomplishing the real objective of gaining knowledge by performing research of excellence, especially health care-driven projects, following the concept of Biological Resource Centers (BRCs) proposed by the Organisation for Economic Co-operation and Development [2, 3] summarized as service providers in terms of public service.

Biobanking activity is mediated by a fourfold commitment: social, ethical, technical, and scientific. Social commitment refers to the promotion of knowledge (basic science) and health (translational/applied research). The ethical commitment refers to the guarantee of donors' rights and the chain of custody of samples and personal data to be protected.

Because of these two commitments, biobanks are obliged to use the most suitable methodology in order to ensure the highest quality of service, this being their technical commitment. Finally, biobanking should be directly linked to scientific challenges, interests, and advances. For these reasons, new technologies used or developed in the scientific scenario should be of the greatest interest for biobanking professionals, who are compelled to collect, preserve, and provide samples not only for current studies but also for those to be promoted in the coming 5 or 10 years.

This is one of the more challenging dimensions of biobanking and, at the same time, one of its most important limitations. In 2005, the US National Cancer Institute promoted a huge initiative to sequence the whole genome of a number of cancer cases, This was called The Cancer Genome Atlas (TCGA), and is a comprehensive and coordinated effort to accelerate our understanding of the genetics of cancer using innovative genome analysis technologies. Technology was available, funds were obtained, and personnel were perfectly trained. However, most of the samples corresponding to some tumor types were inappropriate for ethical and technical reasons [4]. These samples were, for sure, collected following the gold standards of the time of collection, but were inappropriate at the time of use for a whole genome analysis project. Can we foresee what type of samples and data, and their collecting conditions, will be necessary 5 or 10 years ahead? Can we improve our methods to ensure the best service today and in the medium to long term?

17.2.2
Scientific Commitment: Biobanks Must Be Open to Updating and Redefinition According to the Ever-Changing Scientific Requirements

Progress in translational research does not depend only on suitable technology platforms for genomics, epigenetics, transcriptomics, proteomics, and metabolomics. It is also highly dependent on large series of high-quality samples and data corresponding to affected and unaffected individuals. Integration of these resources with high-throughput molecular and '-omics' approaches and integrated bioinformatics tools holds the promise of furthering the advance of our knowledge about the development of diseases, leading to better prevention and treatment strategies.

Although it is impossible to know the future in science, it is very important for biobank professionals to study current trends in order to better prepare for this future. Looking at the last five years, we can detect a clear trend to reconsider tissues from a more global perspective and not merely based on isolated components. Looking at oncology research, in the last two decades many studies focused on the recognition of altered processes within neoplastic cells but without paying attention to the stroma surrounding them; many molecular studies have even been carried out using tissue homogenates, which are by definition heterogeneous. These approaches raised expectations that on many occasions have not been reflected in robust results and/or results applicable to the clinical practice. On the other hand, the so-called systems biology and systems pathology propose not only an integrated approach to the different '-omics,' but also retrieve the concept of the tissue as a whole.

This approach has impact on the activity of the biobanks. For years, there has been a friendly scientific debate between proponents of the extracted and ready-to-use products as the basis for biobanks and those who found it best to preserve the primary sample to be able to extract the necessary derivatives for specific projects later on. The postulates of systems pathology seem to agree with the proponents of the preservation of the primary sample, because we cannot know what will be the final products on which the research project using these samples will focus, and therefore it seems best, based on scientific commitment, to preserve the intact primary sample in multiple preservation formats, if possible, so that we will be able to provide the researchers with the products they really need. For instance, if we only store DNA, we are losing the possibility for other studies. The epigenetic revolution, based on studies including methylation, acetylation, and mRNA, which has come about in the field of biology during the last few decades, has challenged the established traditional view that the genetic code is the only key determinant of cellular gene function and that its alteration is the major cause of human diseases [5].

This comprehensive view not only affects sample storing and manipulation, but also the tissue to collect. Talking, once again, about cancer, the most common standard was to collect representative pieces from the center of the tumor; today it is necessary to modify this routine to try to collect not only a representative piece from the center of the whole tumor, but also from the tumor border and from the neighboring mesenchymal area as well.

17.2.3
Personalized Medicine on the Horizon: Biobanks for Better Health

During the past two decades, we have witnessed an explosive growth of the amount of genomic and proteomic data, major advances in unraveling the molecular mechanisms of human diseases, and a rapid development of new technologies for molecular diagnostics and therapy. This has ushered in a new era of molecular medicine in which disease detection, diagnosis, and treatment are tailored to each individual's molecular profile [6].

Molecular, phenotypic, and patient tracing of disease initiation and progression before, during, and after therapeutic intervention is paramount for the successful treatment of (complex) diseases. The study of molecular interactions has been a research focus for many years and has provided much insight into biology, but now the new age has come for integrative network biology. The aim of integrative network biology is to provide models of cellular networks based on the integration of large and heterogeneous datasets originating from proteomics and high-throughput functional genomics studies. As we are moving toward multi-dimensional and integrative network models of cell behavior, we expect that such models will be useful not just for determining network structures that can be targeted, but also as predictive markers of disease emergence or progression. This will require robust quantitative network assays to become more widespread and user-friendly [7].

Translational cancer research is highly dependent on large series of cases, including high quality samples and their associated clinical and phenotypic data. Research is needed for prevention as well as for better care of those who have already acquired a disease, and biobanks are a cornerstone for this better health perspective. The generation of trial databases and/or biobanks originating in large randomized clinical trials has successfully increased the knowledge obtained from those trials [8, 9].

17.3
The Increasing Complexity of Biobanking

17.3.1
Biobanks versus Sample Collections

The current definition of a biobank refers not just to an assortment of samples collected, handled, and preserved for a concrete project, but it is mainly to be seen as a long-term repository for future studies that are undefined at the time the samples are collected.

This concept of biobanking is different from, but complementary to, banking collections specifically assembled for a project or specific research in which it is possible to know *a priori* the handling procedures to be followed for these samples in accordance with the applicable technology and the scientific objectives of the project. This distinction is particularly important with respect to information associated with the sample, and specifically in those studies in which it is necessary to know at the time of the analysis the progression of the disease or biological process under study; thus, for instance, the use of retrospective samples with 10 or 15 years of follow-up to study the natural evolution of superficial bladder carcinoma will always be more useful than collecting a prospective series during this lengthy period before starting the project.

On the other hand, this approximation to biobanking activities should have regard to the complexity of models, depending on the research activity that is being developed. Biobanking is, at one and the same time, a unitary and a multiple concept because under a common global definition it should bring a service to very different activities which may modify its design, methods, and sample types.

17.3.2 Biobank Diversity

It appears to be necessary to discriminate and achieve a deeper knowledge and recognition of differences and specific characteristics of the various types of biobank and their future challenges. Since easy access to these samples for the scientific community is considered as the main bottleneck for health research, those who are involved in the development of concepts and strategic management of biorepositories are the most appropriate to try to resolve this issue [10, 11].

Biobanks are often developed in relation to a research field having its own strategy and specific demands on quality and annotation of the collected samples. In a brief approach to such diversity, three major human tissue-related biobank types can be described: population-based biobanks, disease-oriented biorepositories for epidemiology-driven collections, and clinic-based and specific disease-oriented biobanks [12]. Whereas the first category usually is government promoted and funded, the second usually is promoted and coordinated by scientific consortia, while the third is hospital-based and managed.

The ultimate goal of population-based biorepositories is to enable researchers to obtain information on determinants of susceptibility and population identity. Usually, their operational substrate is germline DNA from a huge number of healthy donors, representative of a specific country/region or ethnic cohort [13]. Advances in genomics, IT, and bioinformatics are essential for this type of study, and for biobanks *per se*.

The activity of disease-oriented biorepositories for epidemiology-driven collections is focused on determinants of exposure (including environmental or occupational aspects), biological monitoring of mostly huge numbers of samples following a healthy exposed cohort/case–control design, and studying germline DNA, serum and/or urine markers, or environmental probes (i.e., soil, water, air, or nutrients). The vast majority of these collections are designed with focus on very specific questions requiring detailed collected data or protocols for the collection and annotation of probes/samples. Epigenetics are seen as crucial in new epidemiological approaches.

Biorepositories with inherent collections of samples and data related to specific diseases including cancer (the latter often referred to as '*tumor banks*' – as a relic of our former limited understanding of biological collections) encompass prospective and/or retrospective collections of disease and non-disease samples and their derivates (DNA/RNA/proteins). The goal of such biorepositories is the search for determinants of a given disease including its biological behavior through associated clinical data. In some cases, these collections are associated with clinical trials. However, the majority of collections are not brought together for a specific research project, and thus measures of quality control vary significantly, since at the time of collection specific conditions are not as elaborated as they might have been if the collection was for a certain project in the future. Especially with regard to the follow-up of certain conditions, these collections may often only rely on healthcare records, with all the implications regarding completeness and quality. This type of banking, not related with a specific research project, is especially useful in oncology research because it allows promoting and developing retrospective projects with a suitable follow-up of patients and healthy donors. To maintain the collection of an informative follow-up over a long period is especially important for some chronic diseases and some cancer types like superficial low grade bladder cancer or lymph node-negative infiltrating carcinoma of the breast, where a minimum of 10 years follow-up is needed. All '-omics' are at the basis of the future development of projects using those biobanks, and IT could be of great value for suitable annotation.

17.3.3 Biobanking, a Young Discipline

Biobanking should not be considered a static activity. On the contrary, biobanking is a new concept with solid bases in traditional pathology, the development of which is mandatory to allow effective translational research to flourish. Biobank activity is developing as a progressively complex young discipline playing a central role in biomedical research, and it needs to evolve continuously according to the constant development of new technologies and new scientific goals; this continuous updating also affects basic biobanking protocols [14].

17.4 Future Technologies and Biobanking: How Could New Technologies Affect the Daily Activities of Biobanks?

Science, technology, and service platforms are intimately interconnected (Figure 17.1). Biobanks are platforms at the service of biomedical research and should be open to research requirements and technology developments. Therefore, new technologies should modify biobanking methodologies from a

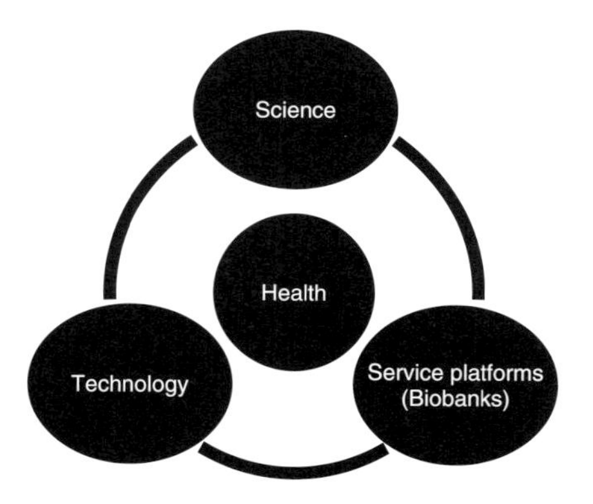

Figure 17.1 Scientific interconnections in health promotion.

double perspective: improving its own activity and providing the type of products that researchers need.

17.4.1 IT Solutions and Challenges

The real value of banked samples lies not only in tissues and derivates, but also in associated data. IT development is the hope, and the challenge is to optimize the balance between health-care users, health-care providers, policies, procedures, and technologies, applying the principles of user-centered design to support the best practice of medicine by creating tools that are easy to use, intuitive, protective, and empowering. Appropriate data mining methods and interoperability of diverse informatics platforms may combine patient demographic and clinical health information, such as medical history. It may also have the capacity to provide clinical decision support, to capture and query information relevant to health-care quality, and to facilitate exchange of electronic health information while integrating such information from other sources [15].

The data (clinical, epidemiologic, demographic, biospecimen quality, etc.) associated with biospecimens are critical for their overall value and utility for research. Best practices include assuring that data are collected according to recognized standards, in particular for identifying the pathologic status of the biospecimen, determining the common data elements that need to be collected, establishing quality control for data collection, assuring that data are collected according to privacy rules, and data protection rules and regulations are complied with.

Whenever appropriate, informatics systems within biospecimen resources should be integrated with the institution's data systems (in particular in a clinical context) and other systems, and also include data applicable to the study design (e.g., large databases of population studies). One of the major obstacles to the efficiency and accuracy of data exchange among biospecimen resources is the lack of interoperability of informatics systems. Best practice dictates that information systems should be interoperable to both integrate clinical and research data and to enhance the ability to establish biospecimen resource networks that can readily exchange information [16].

Data mining methods offer us the possibility, at least in theory, of capturing data automatically from electronic clinical files. This is a new and highly promising landscape in the accessibility to basic clinical information, essential for scientific projects focused on clinical factors and, especially, for retrospective studies.

However, different systems of handling and storage are not exempt from problems, as they can constitute a risk to the right to privacy of patients. It is possible to develop appropriate methods and accreditations to ensure perfect traceability of data while protecting the right to privacy of patients and donors. However there are limits to this capability that are not so easily reconcilable with success.

The main premise of information technologies is that, ultimately, the quality of the output data depends directly on the quality of the entered data. The limitation here is not about errors in the introduction of the data, but in technology or reference values that determine the data which are not usually reflected in the data captured by data mining. This information is valuable when the data arise and are registered, but takes on special value and can cause dispersion of information when it is used or captured some time afterwards.

When information is entered in a prospective way and it relates to a specific project, its degree of uniformity is appropriate because it is based on consistent methodologies, procedures, and benchmarks. But it should not be forgotten that the activity of a biobank also refers (in many cases with special priority) to cases and samples taken from clinical activity and incorporated into a biobank without previous knowledge of the future destiny of the samples. In this case the reference values and changing methodologies are 'frozen' with the sample, and the recovery 5, 10, or 15 years later may give rise to distortion and variability of results.

The second major challenge refers not to data quality, but to data transfer security. With the extension of Internet use and the increasing need for information exchange, a rapidly growing amount of data is transferred over the World Wide Web. Solutions to the problem exist and should be activated when the transferred information refers to health and genetic data. Legal requirements should be fulfilled, including encryption and certification. Encryption guarantees that no unauthorized access during the transmission occurs, while a certificate is the adequate support for data integrity and identification of the partners of the communication. Certification can also serve as a basis for maintaining a log of the transfers carried out. In addition, it is necessary to continuously monitor and update these safeguards to prevent sophisticated attacks [17].

What are the advances in the field of information technology that most urgently must be developed in the coming years in the service of biobanks? We suggest three: one by biobanking professionals, another on the part of engineers, and the last by scientific societies. The first relates to the introduction of specific codes to monitor issues relating to the sample, its quality, and its traceability. In this respect, the proposal of a working group from the International Society for Biological and Environmental Repositories (ISBER) is especially valuable and concerns preanalytical features of the samples [18].

The second development responds to the need to interconnect management platforms and samples from different biobanks to promote their integration into cooperative networks of biobanks. These developments have to be settled on a level above the platforms used by each of them, in terms of functionality rather than uniformity. Only in this way may extensive national and international networks develop mechanisms for fair access to large series of cases, adequately collected and stored according to standardized and harmonized working procedures.

In these networks, language is a real difficulty, for which the most appropriate solution is the use of internationally accepted and endorsed coding systems. One of these coding systems used worldwide is the International Classification of Disease (ICD) proposed by the World Health Organization [19]. Current ICD versions are published in only a few languages. The lack of a reference ICD version 10 database is a severe pitfall for the widespread usage of this classification. Hence, software publishers have to develop their own databases, but they have difficulties in validating these and so the final quality is impaired. Updates are not easily available, and any question of interpretation or ambiguity remains unresolved. The more differences there are between a specific national version and a reference source version, the lower is the quality of encoding and the fewer possibilities there are of comparison and sharing [20].

New scientific challenges require new approaches for basic problems. A clearer definition and codification of entities is an obvious example, but not the only one. A wonderful oncology catalog could be absolutely useless if the basic discriminative data are not based on scientific criteria. A clear example is the ethnic catalog to be included in genetic and clinical databases. The use of terms and categories like Hispanic or Afro-American may be useful for social purposes but not for scientific goals. How can we maintain the category 'African' when we know that the genetic diversity in African populations is higher than that in the rest of humanity? How can we maintain the category 'Hispanic' for the entire Latin-American population including in the same group people from the Sierra Madre region, people from the Amazonia area, and people from Buenos Aires?

17.4.2
Storage Mechanization

There is a growing interest in automated biobanking for large-scale automated repositories of biological samples such as blood, tissue, plasma, and DNA. These types of initiatives typically involve a library containing hundreds of thousands up to tens of millions of individual sample aliquots requiring storage at $-80\ ^{\circ}C$ or colder. The samples require careful management to guarantee sample conditions over their lifetime, while also providing a high integrity of sample tracking to ensure reliable retrieval [21]. Robotized storage and sample tracking provides more security and better preservation and space optimization, but currently is only designed for small aliquots and standardized containers and not for medium-sized tubes or cryomolds for solid samples. The advantage is evident in the case of large biobanks, but cost–benefit studies should be done for small- and medium-sized biobanks before implementation.

Classical cryotubes can be used for small pieces of solid samples, and this is a good method for biosamples to be homogenized for product extraction. But it is inappropriate for studying the whole range of biological components (in keeping with current trends of systems pathology) and for histopathological study including immunohistochemistry, microdissection, and other *in situ* techniques. The use of cryotubes for solid samples can be easily mechanized, but the specimen is usually fragmented when it is retrieved. Specific cryomolds seem to be the best option for solid samples. New cryomold designs and new storage mechanization systems are necessary to provide this facility.

17.4.3
Virtual Microscopy

Telepathology is the practice of 'pathology at a distance,' using telecommunication technology to facilitate transfer of images concerning pathological data between remote locations for the purposes of diagnosis, education, and research. Uses of telepathology include remote discussion between pathologists and between clinicians and pathologists, remote quality assurance involving pathologists and referral laboratories, and collaboration between research teams.

A relevant application of virtual slide technology is the documentation of tissue samples in tissue banks. This technology in the tissue bank can enhance traditional sample annotation, assist in ensuring that remote tissue bank customers receive appropriate tissue for their research, and form the basis of quality assurance systems. Users of this resource can view any part of an entire specimen at any magnification using a standard web browser.

The acquisition of high-quality virtual slides is still a time-consuming task, and the cost of storing the image electronically is still significantly higher than the cost of storing the information on a glass slide. This is why the application of virtual microscopy in routine diagnostics cannot be recommended at this time. With the arrival of fast and easy-to-use virtual slide systems, we expect a significant change in telepathology. Whether this will be in the near or more distant future will be determined by the costs of digital storage and the scanning time. As with images from biomedical technologies, histology images may be integrated into the digital information systems for medical diagnostics, making these data available as part of the patient's electronic file. In the future, the glass slide will be archived immediately following digitization. The work, diagnosis, teaching, and research will be done with the virtual slide [22].

17.4.4
Nanotechnology and Quality Control

The establishment and maintenance of a BRC is a knowledge- and skill-intensive activity that in particular requires careful attention to the implementation of reliable preservation technologies and appropriate quality assurance to ensure that recovered cultures and other biological materials perform in the same way as the originally isolated culture or material [23]. To improve our quality control armamentarium is a great challenge for biobanks with the common goal of accurate, reproducible, cheap, and ready-to-use methods. This goal has not been achieved with conventional methods, which are time-consuming and sometimes have poor precision. Recently, much attention has been paid to the use of electrochemical immunosensors and other nanotechnology-based platforms for the detection of markers, because of their high sensitivity, easy miniaturization, and automation.

Immunosensors combine the advantages of sensors with high sensitivity and immunoreactions with high specificity. The electrochemical immunoassay has the advantages of a low detection limit, small analytical volume, simple instrumentation, and minimal manipulation, and the assay system can be easily miniaturized and integrated into protein chips. Electrochemical immunosensors constructed on protein chips are expected to bring marked progress to the detection of quality markers. Particularly, the miniaturization and integration of electrochemical detection systems on microarray chips may allow multianalyte detection for application in point-of-care testing, combining sample quality with high reproducibility, and requiring a very small analytical volume [24]. However, despite these advances, the absence of robust quality markers for most types of samples, especially solid ones, is still the main obstacle in the way of adequate quality control of banked samples.

17.4.5 Tissue Microarrays

Tissue-microarrays (TMAs), originally developed by J. Kononen *et al.* [25] are characterized by a high-throughput platform based on the ordered distribution of large numbers of tissue cylinders previously selected from the original paraffin-embedded block, which ensures homogeneous manipulation, staining, visualization, and thus a considerable reduction of bias in the analysis of multiple samples, as well as a considerable saving in technical costs. Despite the fact that their construction is heavily linked to fixed and paraffin-embedded tissue adequate for immunohistochemical (IHC) studies and/or fluorescence *in situ* hybridization (FISH), several methods have been developed for frozen tissue [26], with some differences in their construction [27], and in some cases allowing for the implementation of other technologies such as spectroscopic analysis [28].

TMA technology has been widely used in recent years in very diverse areas of research to demonstrate and quantify the distribution of protein expression by IHC in neoplastic tissue of clinical origin, xenografts, cell lines, and so on, in order to identify new markers and relevant genes in different diseases, confirm findings from animal models and modified cell lines, identify new prognostic factors useful in the clinical practice, identify markers predicting therapeutic response, or validate molecular profiling results (from cDNA or oligonucleotides arrays, etc.) [29]. This is the reason why this technology is considered the most useful in pharmacogenomics studies and other translational research dynamics. On the other hand, its low cost, low complexity, easy shipping, and the fact that it is fundamentally based on IHC techniques, make the TMA technique suitable for development in the clinical environment. This also promotes it as a key complement to clinical trials and retrospective studies involving large series of cases requiring adequate clinical follow-up [30].

For this reason, the building of TMAs should be considered as a basic service of any disease-oriented biobank directed at providing samples to translational researchers. However, the true value of these arrays is based on a correct selection of the areas included therein, and for this, the implication of experienced pathologists as collaborators of the biobank is highly necessary. On the other hand, new technical designs may improve the correct inclusion in the paraffin block of the area previously selected from the histopathological slide.

Something similar can be said about laser-microdissection for the selection of specific target cell populations. This service may also be provided by the biobank or it may not, but if it is, great expertise is necessary, and the assistance of a pathologist is highly recommended.

17.4.6
New Fixatives

The gold standard in the preservation of samples of surgical origin is fixation in formalin and embedding in paraffin. This method, used all around the world, allows the availability of tissue suitable for histopathological examination and diagnosis, both immunohistochemically and by *in situ* hybridization, and also allows the implementation of many histochemical techniques and the extraction of DNA of sufficient quality for genetic studies. Its main defect is, however, the degradation of large RNA molecules, but it still allows microRNAs expression analysis by quantitative PCR and similar technologies. On the other hand, the samples can be kept at room temperature without an excessive loss of tissue quality.

In recent years different means of fixation have been proposed that allow the quality of RNA molecules to be preserved without loss of the qualities of paraffin embedding.

Biobanks must be very careful in incorporating these advances as long as their effectiveness has not been validated in the long term. It is not uncommon to find fixing systems on the market that are assumed to maintain not only a good quality of RNA in paraffin-embedded tissue but also a correct antigenicity for immunohistochemistry. However, when these fixatives are evaluated, sometimes the analysis of the IHC markers more commonly used in clinical practice renders results difficult to assess, if not contradictory and/or unstable. Therefore the most appropriate approach would be not to routinely assume that these new fixatives will displace fast freezing of tissue or fixing in formalin unless there is a clear excess of tissue or it is used for research projects specially focused and designed on the basis of the evaluation of new products.

It is important to remember, once again, that new markers developed by translational research groups should not only be highly informative, but that they also should be focused on their future use in the clinical context: they must be suitable for implementation in diagnostic services of pathology departments where the gold standard is classical techniques. In addition, the cost of maintenance of frozen tissues, although it is significant, is minimal if one takes into account the total price for the implementation of new diagnostic technologies based on gene expression studies used in transcriptomics, epigenetics, sequencing, and so on. However, there are some areas where research on new fixatives should be a top priority, such as those fixatives enabling stabilization of highly unstable or phosphorylated proteins.

17.4.7
Robotized RNA/DNA/Protein Extraction

One of the key steps in structural genomics and proteomics is high-throughput expression analysis of many target proteins and nucleic acids. The newer generation of molecular diagnostic technologies offers hitherto unparalleled detection and discrimination methodologies, which are vital for the positive detection and identification of new markers. The development of novel technologies can only be successful if they are transferred and used in the field with a sustainable quality-assured application that allows the optimal detection and effective control of diseases. This means the use of very large series of cases and controls as primary samples and as secondary products. To facilitate this process, robotized methods for extraction and purification of specific molecules could be very useful [31], and this automated extraction step may also be assumed by the biobank facility.

This process would provide us with better homogeneity of the products and a great speed of extraction but, again, a cost–benefit analysis should be completed, especially for smaller biorepositories. In these cases, the employment of professional service providers may be a better option. On the other hand, it should be assumed that, in most cases, researchers prefer to do the extractions themselves.

17.5
The Future of Biobanking Does Not Depend on Technological Developments Alone

It is necessary to bear in mind that biobanking is still a young discipline. Specific technological development would be very useful and will improve its service quality, but it may be more important to first clarify and develop other aspects, in particular the basic definition and scope of biobanks, including:

1) Biobanks should move toward becoming comprehensive facilities.
2) Large-scale studies call for renewed requirements of standardization and harmonization.
3) Standardization and quality management policies.
4) High-sensitivity techniques call for high-quality samples.
5) Access rules and limits.
6) Easier access to genetic information call for a new ethical approach.
7) Molecular marker-driven drug discovery needs integration of biobanks into clinical trials.
8) Increasing biobank complexity needs professional biobank management.
9) International scientific consortia call for collaborative biobanking and biobank networking.
10) The social value of translational research calls for biobanking as a public service and for public/social recognition.

Most of these requirements have been discussed elsewhere in this volume.

17.6 Conclusions

New technologies will cause a revolution in the activity and goals of biobanking, but the first milestone should be the same as it was in the last century: a microscope and a pathologist selecting and reviewing samples, and making skilled observations [32, 33]. Biotechnologies should allow us to deal with new scientific goals.

As for biobanks, it can be wrong to rely excessively on new technologies as a unique solution to their problems. A plethora of technological improvements should be key for improving our service activity, but it will not resolve all our problems and challenges, not even the most important ones, and the human dimension of biobanking as a service that guarantees patients' and donors' rights should also not be forgotten.

The real future of biobanking will not only be based on technological development but will also be linked to better training, better resources, a more comprehensive support by institutions, and especially the transition (i) from personal/institutional collections (that are not for sharing) to real BRC with the aim of public service and (ii) from isolated institutional collections to real biobank networks.

For this evolution, an open mentality and an open attitude to sharing are necessary, and this does not depend on technology. In fact, the essential preoccupation for biobanks should not be new technologies, but rather the generation of new knowledge, concepts, and scopes in biomedicine.

Acknowledgments

This work was supported by grants from the Ministerio de Ciencia y Tecnología (RD09/0076/00028).

References

1. (2010) Human genome at ten: the sequence explosion. *Nature*, **464** (7289), 670–671.
2. OECD (ed.) (2001) *Biological Resource Centres: Underpinning the Future of Life Sciences and Biotechnology*, OECD, Paris, *http://www.oecd.org/dataoecd/55/48/2487422.pdf*, ISBN: 92-64-18690-5.
3. Caboux, E., Plymoth, A., and Hainaut, P. (2007) *Common Minimum Technical Standards and Protocols for Biological Resource Centres dedicated to Cancer Research*, IARC, Lyon, ISBN: 97-89-28322442-6, *http://www.iarc.fr/en/publications/pdfs-online/wrk/wrk2/standardsBRC-1.pdf* (accessed 18 November 2011).
4. Check, E. (2007) Cancer atlas maps out sample worries. *Nature*, **447**, 1036–1037.
5. Sharma, S., Kelly, T.K., and Jones, P.A. (2010) Epigenetics in cancer. *Carcinogenesis*, **31** (1), 27–36.
6. Phan, J.H., Moffitt, R.A., Stokes, T.H., Liu, J., Young, A.N., Nie, S., and Wang, M.D. (2009) Convergence of biomarkers, bioinformatics and nanotechnology for individualized cancer treatment. *Trends Biotechnol.*, **27** (6), 350–358.

7. Erler, J.T. and Linding, R. (2010) Network-based drugs and biomarkers. *J. Pathol.*, **220** (2), 290–296.
8. Gustafsson, F., Atar, D., Pitt, B., Zannad, F., and Pfeffer, M.A., On behalf of the participants in the 10th Cardiovascular Clinical Trialists Workshop (2010) Maximizing scientific knowledge from randomized clinical trials. *Am. Heart J.*, **159** (6), 937–943.
9. Riegman, P.H.J., Morente, M.M., Betsou, F., Blasio, P., and Geary, P., the Marble Arch International Working Group on Biobanking for Biomedical Research (2008) Biobanking for better healthcare. *Mol. Oncol.*, **2**, 213–222.
10. Ozols, R.F., Herbst, R.S., Colson, Y.L. *et al.* (2007) Clinical cancer advances 2006: major research advances in cancer treatment, prevention, and screening: a report from the American Society of Clinical Oncology. *J Clin Oncol.*, **25**, 146–162.
11. Oosterhuis, J.W., Coebergh, J.W., and van Veen, E.B. (2003) Tumour banks: well-guarded treasures in the interest of patients. *Nat. Rev. Cancer*, **3** (1), 73–77.
12. Goebell, P.J. and Morente, M.M. (2010) New concepts of biobanks – strategic chance for uro-oncology. *Urol. Oncol: Semin. Orig. Invest.*, **28**, 449–457.
13. Steinberg, K., Beck, J., Nickerson, D. *et al.* (2002) DNA banking for epidemiologic studies: a review of current practices. *Epidemiology*, **13**, 246–254.
14. Morente, M.M., Fernandez, P.L., and de Alava, E. (2008) Biobanking: old activity or young discipline? *Semin. Diagn. Pathol.*, **25**, 317–322.
15. Hesse, B.W., Hanna, C., Massett, H.A., and Hesse, N.K. (2010) Outside the box: will information technology be a viable intervention to improve the quality of cancer care? *J. Nat. Cancer Inst. Monogr.*, **40**, 81–89.
16. Vaught, J., Carboux, E., and Hainout, P. (2010) International efforts to develop biospecimen best practices. *Cancer Epidemiol. Biomarkers Prev.*, **19** (4), 912–915.
17. Spahni, S. and Weber, P. (2002) Secure exchange of medical data: requirements and solutions, in *Medical Informatics Europe 2002* (eds. G. Surjan *et al.*), IOS Press, pp. 112–117.
18. Betsou, F., Lehmann, S., Ashton, G., Barnes, M., Benson, E.E., Coppola, D., DeSouza, Y., Eliason, J., Glazer, B., Guadagni, F., Harding, K., Horsfall, D.J., Kleeberger, C., Nanni, U., Prasad, A., Shea, K., Skubitz, A., Somiari, S., and Gunter, E., International Society for Biological and Environmental Repositories (ISBER) Working Group (2010) Standard preanalytical coding for biospecimens: defining the sample preanalytical code. On biospecimen science. *Cancer Epidemiol. Biomarkers Prev.*, **19** (4), 1004–1011.
19. WHO (1993 and 1997) International Statistical Classification of Diseases and Related Health Problems, Vol. 1–3, World Health Organization, Geneva, Switzerland.
20. Baud, R.H., Lovis, C., Weber, P., and Geissbuhler, A. (2002) Multilingual approach to ICD 10: on the need for a source reference database, in *Medical Informatics Europe 2002* (eds G. Surjan *et al.*), IOS Press, pp. 406–410.
21. Elliott, P. and PeakmanT.C., on behalf of UK Biobank (2008) The UK biobank sample handling and storage protocol for the collection, processing and archiving of human blood and urine. *Int. J. Epidemiol.*, **37** (2), 234–244.
22. Teodorovic, I., Isabelle, M., Carbone, A., Passioukov, A., Lejeune, S., Jaminé, D., Therasse, P., Gloghini, A., Dinjens, W.N.M., Lam, K.H., Oomen, M.H.A., Spatz, A., Ratcliffe, C., Knox, K., Mager, R., Kerr, D., Pezzella, F., van Damme, B., van de Vijver, M., van Boven, H., Morente, M.M., Alonso, S., Kerjaschki, D., Pammer, J., Lopez-Guerrero, J.A., Llombart-Bosch, A., van Veen, E.B., Oosterhuis, J.W., and Riegman, P.H.J. (2006) TuBaFrost 6: virtual microscopy in virtual tumour banking. *Eur. J. Cancer*, **42** (18), 3110–3116.
23. Day, J.G. and Stacey, G.N. (2008) Biobanking. *Mol. Biotechnol.*, **40** (2), 202–213.
24. Chen, H., Jiang, C., Yu, C., Zhang, S., Liu, B., and Kong, J. (2009) Protein

chips and nanomaterials for application in tumor marker immunoassays. *Biosens. Bioelectron.*, **24**, 3399–3411.

25. Kononen, J., Bubendorf, L., Kallionimeni, A., Bärlund, M., Schraml, P., Leighton, S., Torhorst, J., Mihatsch, M.J., Sauter, G., and Kallionimeni, O.-P. (1998) Tissue microarrays for high-throughput molecular profiling of tumor specimens. *Nat. Med.*, **4**, 844–847.
26. Fejzo, M.S. and Slamon, D.J. (2001) Frozen tumor tissue microarray technology for analysis of tumor RNA, DNA, and proteins. *Am. J. Pathol.*, **159**, 1645–1650.
27. Hidalgo, A., Piña, P., Guerrero, G., Lazos, M., and Salcedo, M. (2003) A simple method for the construction of small format tissue arrays. *J. Clin. Pathol.*, **56**, 144–146.
28. LeBaron, M.J., Crismon, H.R., Utama, F.E., Neilson, L.M., Sultan, A.S., Johnson, K.J., Andersson, E.C., and Rui, H. (2005) Ultrahigh density microarrays of solid samples. *Nat. Methods*, **2**, 511–513.
29. Fernandez, D.C., Bhargava, R., Hewitt, S.M., and Levin, I.W. (2005) Infrared spectroscopic imaging for histopathologic recognition. *Nat. Biotechnol.*, **23**, 469–474.
30. Sherman, M.E., Howatt, W., Blows, F.M., Pharoah, P., Hewitt, S.M., and Garcia-Closas, M. (2010) Molecular pathology in epidemiologic studies: a primer on key considerations. *Cancer Epidemiol. Biomarkers Prev.*, **19**, 966–972.
31. Yokoyama, S. (2003) Protein expression systems for structural genomics and proteomics. *Curr. Opin. Chem. Biol.*, **7** (1), 39–43.
32. Hainaut, P., Caboux, E., Bevilacqua, G., Bosman, F., Dassesse, T., Hoefler, H., Janin, A., Langer, R., Larsimont, D., Morente, M., Riegman, P., Schirmacher, P., Stanta, G., and Zatloukal, K. (2010) Pathology as the cornerstone of human tissue banking: european consensus expert group report. *Biopreserv. Biobanking*, **7**, 157–160.
33. Bevilacqua, G., Bosman, F., Dassesse, T., Höfler, H., Janin, A., Langer, R., Larsimont, D., Morente, M.M., Riegman, P., Schirmacher, P., Stanta, G., Zatloukal, K., Caboux, E., and Hainaut, P. (2010) The role of the pathologist in tissue banking: european consensus expert group report. *Virchows Arch.*, **456**, 449–454.

18
Outsourcing Sample Management

Sylviane Boucharens and Amelia Wall Warner

18.1
Outsourcing in the Pharmaceutical Industry

In comparison to other industries, the pharmaceutical industry has been slow to embrace outsourcing, but the momentum in this direction is now growing as the benefits are appreciated. The pharmaceutical industry now considers outsourcing as no longer optional, but mandatory. The total pharmaceutical and biotech outsourced market is currently estimated at $33 billion and is expected to reach some $48 billion, whereas R&D outsourcing expenditure is approximately 50% of the total market [1, 2]. Outsourcing is being used more strategically as an ongoing part of a company's overall business strategy, and outsourced activities in the pharmaceutical industry now cover various fields, from drug discovery through to manufacture of the product. Pharmaceutical companies have long outsourced functions such as manufacturing, packaging, clinical trials, and sales force mobilization. The only activities outsourced in the 1970s were the marketing of products, including the areas of advertising, market survey, and so on. In the 1980s, outsourcing was broadened to include aspects of formulation, clinical trials, and registration. To date, outsourcing in pharmaceutical development is estimated at 30–35% and is continuing to grow [3]. As the pharmaceutical industry faces unprecedented cost pressures, companies have more recently considered outsourcing at earlier stages in drug discovery. In the late 1990s there was a jump to include the early stages of target identification and the production of compounds to expand collections. Hit-to-lead and lead optimization were finally included in the last decade [4, 5].

R&D and other fields in life science that previously stayed close to company headquarters are beginning to be outsourced. All parts of the process can be, and are, the subjects of regular outsourcing by the largest to the smallest of pharmaceutical companies.

Management of Chemical and Biological Samples for Screening Applications, First Edition.
Edited by Mark Wigglesworth and Terry Wood.

18.1.1
Economic and Organizational Advantage of Outsourcing

The main benefits of outsourcing are economic and organizational. Outsourcing makes economic sense, providing access to technology or capacity without the need for costly investment in equipment or personnel. Outsourcing also offers great flexibility, facilitating management of the inevitable peaks and troughs. In the demand for resources, strategic outsourcing can be a cost-effective alternative to in-house operations by providing access to the latest technology and/or sufficient operational capacity, by avoiding costly capital investment and technical obsolescence, and by decreasing depreciation [6, 7]. When contemplating outsourcing, a company will need to consider its own resources and how the service contractor can assist it in terms of financial resourcing (cost saving), physical resources (equipment/laboratory), as well as human resources (the people to do the job and their knowledge). By considering all of these aspects, the company can then evaluate to what extent using a contractor ties in with its needs and gives it the flexibility or speed that it cannot attain on its own [8].

18.1.2
Sourcing the Right Partner

The process of outsourcing starts with the identification of internal processes that are not part of the core business activity of the company, and this is followed by the identification of potential partners. The process of selection, negotiation of contract, management of the work, and adequate reporting is then proceeded with. Successful outsourcing operations should result in cost reduction, increased operational efficiency, and optimization of resource allocation. However, there are pitfalls to outsourcing, including poor partner selection and inadequate implementation.

Once the decision to outsource an activity is determined, the search for the right kind of partner to suit all needs and requirements can be protracted. The first aspect to consider is the 'scope' of the service contractor. Does it provide the services required? Some contractors offer a full range of services, while others may have a particular expertise. Potential partners need to have the capabilities required and to understand the quality implications. Does the service contractor offer the skills and capacity required? As well as looking for the basics, such as the company's business track record and scientific expertise, it is important to also consider questions of responsiveness and flexibility to changes in demand. Does the service contractor work to the right levels of quality and service? It is extremely important to conduct an audit of the service contractor's quality system and to check references from other companies with which it deals. The evaluation of the time required by in-house staff to manage the collaboration with the service contractor is important and will depend mainly on the service contractor. Choosing the right contractor, which should be suitably qualified and committed to team-work, will save time and money [6–9].

Most of the time, outsourcing is about building a long-term relationship with the service contractor, and the collaboration should be built on the key ingredients of quality, service, knowledge, culture, and cost. These ingredients can all impact on the overall cost of the collaboration.

Sourcing the right partner is the key to success. It takes time to develop the level of trust required for such an operation. Due diligence includes site visits and updated quotes. Facilities and IT investigations should be carefully conducted, and in-house experts should be engaged very early in the process, even if this generates some animated discussions!

The rest of this chapter will consider successively the different drivers, implications, and requirements regarding outsourcing compound and biosample management.

18.1.3 Compound Inventory – Cost of Ownership

Screening compound collections are very often compared to the 'crown jewels,' and in the past 5–10 years medium and large pharmaceutical companies have spent many millions of dollars investing in both compounds and the infrastructure to manage them. This infrastructure very often includes the construction of a brand-new facility and investment in both hardware and software, with the installation of new automated liquid compound storage linked to a powerful inventory database.

Compound storage, particularly automated liquid storage, the current standard for most high-throughput screening (HTS) operations, is expensive. On top of the massive capital investments in storage, laboratory space, and equipment, the company very often needs to reformat its current collection according to the specifications required by the new system selected [10, 11]. Quite often, the implementation of a new system causes significant down time and impacts other drug discovery activities.

Storage in the dry form (solid) affords a longer 'shelf life' than that achieved with storage in solution. However, the use of a solvent is generally a necessity for the automated dispensing of large numbers of very small samples, as required for HTS. Dimethyl sulfoxide (DMSO) is the most commonly used solvent due to its good solvating ability for compounds, relative chemical inertness, and relatively high freezing point. Samples are typically stored either at room temperature or in a frozen state (4 or −20 °C). Recent studies of liquid storage conditions and the stability of DMSO solutions have been reported in the literature and have helped in the development of better knowledge and understanding of the important factors causing compound decomposition. This permits significantly better sample integrity for compounds in DMSO solution [12–14]. In parallel, the development of efficient automated storage and retrieval systems for solid samples has contributed to a reduction in the number of compound management operations, although the equipment required to manage the solid collection is less sophisticated and very often not fully automated even for the biggest companies, making outsourcing to external operators an attractive alternative for the management of solid samples (see Section 18.1.4 below).

With an average of 1.8 million compounds per collection, the compound solution repository stores are focused on achieving and maintaining sample integrity. In addition to the storage format, time, temperature, and structural characteristics of the compound, many other factors may affect compound stability. In particular, these include humidity and repeated freezing and thawing.

Sample integrity is generally regarded as a measure of the quality of the material in the container and relates to (i) the chemical identity (structure) of the compound – Is it what it says on the label? (ii) the purity of the sample – Did any impurity appear as a consequence of compound decomposition during storage? – and (iii) the concentration of liquid sample (see additional information within Chapter 3).

A successful compound management system requires at least the following [15]:

- **Ensure compound accessibility.** Customers can access the compounds they need when they need them and in the required format. Sample delivery is accurate, precise, and reproducible.
- **Maintain compound integrity.** The quality of the repository compounds is very important to the drug discovery effort. Poor sample quality can cause many problems in biological assays [16].
- **Enable efficient compound usage.** Limit compound wastage during the compound management process.
- **Flexibility to adapt to new demand.** Be able to adapt to new strategies and technologies used for drug discovery.
- **Ability to maintain a dynamic collection.** Be able not only to add new samples but also to remove undesired compounds from the collection.
- **Availability and integrity of inventory data.** The lack of, or inaccuracy of, inventory data is a major source of inefficiency and frustration.

Today, compound management is recognized as a vital and integrated part of the drug discovery process. Nonetheless, despite recent efforts and major capital investments, most of the large pharmaceutical companies are still facing significant challenges with respect to disconnected inventories of research compounds located at various sites, particularly after a succession of mergers and acquisitions. The cost of maintaining several compound repository stores compared with having one unique compound repository is high. The multiplicity of screening collections and the consequent increase in file size often means that none of the facilities is large enough to accommodate the entire collection, so that the new company will have to go through an expansion phase, with a consequent increase in costs. There also remains the question of how best to make these collections available to the entire company. A compound only has value if it is usable! Compound collections will almost certainly need to be moved; management will have to decide where to, and consequently may have to make difficult decisions leading to internal conflict. One of the possibilities to solve these issues may be to take a neutral outsourcing path and to transfer the work to a third party such as a Contract Research Organization (CRO) [9, 17].

18.1.4
Areas of Outsourcing in Compound Management

Selected compound management process outsourcing is common in the marketplace; the tedious steps of weighing, reformatting, and plating of the client's existing compounds or newly purchased libraries can be handed over to the service providers. Most compound management groups are familiar with the use of service contractors. However, until recently outsourcing in compound management was mainly used only when internal capacity and flexibility were issues [18, 19]. For example, service contractors are very often involved during the implementation of new repositories or the movement of large numbers of samples, ensuring that day-to-day activities are maintained in an undisrupted manner, the tasks of physically moving, organizing, and preparing new samples then being most economically handled through outsourcing. Nevertheless, some customers may even prefer to incur extra fees to have these tasks performed in-house in order to avoid transfer of compounds and potential intellectual property and legal issues. Long-term insourced compound service is relatively new to most compound management service providers as well as to the industry players.

Service contractors specializing in compound management offer different types of services such as weighing, plating and reformatting, procurement, quality control, and storage. Although outsourcing offers the full range of expertise required in compound management, to date only two out of the top ten pharmaceutical companies have made the jump and decided to outsource their full inventory to service contractors, one of them having limited the contract to their solid sample library [20, 21]. Big pharmas tend to be conservative when 'one of their best assets,' the 'crown jewels,' are involved. Change may, however, be in the air! Most big companies have now become more familiar with outsourcing, in particular for chemistry-related functions and medicinal chemistry projects, to service contractors and collaborators. Some of these companies are even located in India or China, where the economic conditions create a comfortable environment for the customer. Pharmaceutical companies will inevitably learn and realize that compound collections can be treated in a similar manner to other functions that they have already successfully outsourced without any major issues.

18.1.5
Outsourcing to Exploit a Key Asset

Without any doubt, outsourcing compound inventory requires a change from traditional thinking. It is a strategic commitment involving difficult and risky management decisions. The number one risk is the externalization of management and storage of the compound libraries. Not physically having the compounds in-house may be perceived as loss of control; even if this is not true, it is an important strategic decision.

In addition to the 'loss of control' feeling there are two other aspects limiting the development of the outsourcing strategy: the need for these compounds in-house for screening operations and the maturity of the market place for such service suppliers. The industry has only a few vendors that can deliver a high-level, high-performance compound management service to a large enterprise. The number of contract organizations that conduct compound management services at the scale of large pharmaceutical companies with large storage and retrieval capacities is still limited, and quite often the contract will involve building new infrastructures such as warehousing, storage systems, and the development of robust data management systems.

For the two large pharmaceutical companies who have elected to outsource their screening collection management to external service providers, the drivers to help determine such a decision would have included:

- Need for consolidation after merger, or historical collection clean-up
- Fragmentation of compounds
- Distribution and format
- Fragmentation of compound request across sites
- Raising infrastructure cost and system cost maintenance
- Cutting the operational budget
- Significant investment for in-house compound management required
- Focus on R&D core competency.

The benefits are manifold: cost effectiveness with efficient speed and reliability, prevention of backlog on time-consuming processes, flexibility in resources allocation, overall lowering of operational budget, and one uniform inventory/database accessible by internal parties as well as the authorized external CRO.

The three main requirements when considering outsourcing the entire inventory are:

- **Physical security of the samples.** Screening collections are priceless assets which in many instances are irreplaceable. Not only should their storage be safe and protected from fire, extreme weather, or earthquakes (in some areas), but for some customers the storage area should also be separated from other business conducted at the contractor's site with limited access security and tracking of employees' hours and time working in the particular area.
- **Robust and reliable IT infrastructure.** In order to protect the intellectual property information, both IT security and interface are critical. The contractor should be able to demonstrate the presence of good firewall or 'Chinese wall' if separate database and server approach are not feasible for each customer. No structures should be stored in the contractor's system. All the operations should be completely bar code driven. The data exchange between internal and external systems should be in real time via a secure interface.
- **Back-up or disaster recovery plan.** A back-up copy of the screening collection stored at a different location may be an option to consider. Very often the

existence of both solid and the liquid samples of the same screening collection may be sufficient if they are physically separated. Each of these formats may be used as a back-up in case of loss of either one of them. In some cases, one of these formats (solid sample) is outsourced to a contractor while the second one (liquid sample) is managed internally.

Service levels should be superior to overall current store operations with better consistency, flexibility, and a real financial benefit. Regarding this latter point, companies should consider reducing cost while still maintaining the highest standards in compound management. Evaluating the total cost of ownership can raise some difficult issues. The company needs to be as objective as possible, but there will always be an element of subjectivity. Depending on the company, the saving of cost, mainly due to the reduction of both maintenance and full time equivalent (FTE), is estimated at one-third, particularly for the solid inventory, where basic warehouse and manual retrieval/weighing seem to be standard for service contractors. On the other hand, the financial benefit of outsourcing liquid sample management, which includes operations such as solubilization, formatting, storage, and retrieval, may be less obvious. The need to meet the high quality standards currently in place in the industry, together with the requests for sophisticated equipment, automation, specific consumables, and expensive storage and retrieval facilities, make this a more challenging proposition.

As the quality of the compounds within the screening collection is very important to drug discovery, pharmaceutical companies should expect the same level of quality of their samples as when they are handled in-house. One of the key considerations is to avoid jeopardizing the integrity of their screening collection by selecting a partner on the basis of price alone.

18.1.6 Future Developments

For major pharmaceutical companies, outsourcing the entire compound management activities to a third party is still very marginal, and it is difficult to foresee future developments. In some circumstances, such as post merger/acquisition re-organization, when compound collections need to be moved between sites or brought together into a single collection, the decision to use a 'neutral' outsourcing path rather than favor one of the legacy compound repository stores may help to prevent internal management conflicts and emotional discussions. This is an attractive option, but it would be extremely difficult to re-initiate an internal compound collection without significant new capital investment in the construction of a new storage area and the risk that internal expertise is lost as part of the process.

The decision to outsource such operations requires time and a clear business decision. The objectives of reducing internal fixed costs combined with adopting strategies for increased outsourcing of drug discovery activities may boost growth in the market for service contractors specializing in compound management. In this eventuality, big pharmas may be a good catalyst for this emerging market. Pharmaceutical companies are scrutinizing any proposed investment in capital and labor to see if it can be outsourced cost-effectively. However, small companies, start-ups, and academic screening institutes which are keen to minimize their capital investment may decide to outsource their compound management rather develop this activity in-house and may also contribute to this next focus for outsourcing contractors.

Alternatively, insourcing (or contracting in) activities is a practice which has started to gain in popularity. Initially, insourcing involves bringing in specialists to fill temporary needs or training existing personnel to perform tasks that would otherwise have been outsourced. Insourcing compound management activities may be more suitable for some organizations than outsourcing as they may be able to use their premises and equipment (both hardware and software), maintain in-house expertise, save on transportation costs, and exercise better control of the overall activities.

18.2 Outsourcing Biological Specimen Collections

Biosamples collected during the course of clinical trials are an important resource for evaluating future scientific information related to drug/disease response. There are examples of biosamples being used years after the close of a clinical trial to better describe subject response to therapy [22, 23]. Biosamples must be acquired and maintained under appropriate conditions to allow for future research. A biorepository must operate to assure the initial sample quality and continuing sample integrity during long-term storage.

Approximately 50% of pharmaceutical companies outsource some major component of their biological specimen collections [24]. The decision whether to create an internal or an external biorepository has been somewhat organic over the past 30 years. Most companies began collecting and storing preclinical biological specimens internally as part of discovery research programs. In many cases, as a drug development program advanced, any human clinical specimens collected and given consent for long term storage also became part of the specimen collections internally stored in discovery laboratories.

As companies began to use central laboratories more commonly, clinical specimen collections were often housed at the various central laboratories used in development programs (typically companies used multiple central laboratories for their various development needs). This created multiple repositories of specimens, often stored at a high premium externally. Additionally, by having multiple stored collections, the logistics of monitoring, maintaining adequate inventories, and cost containment became more difficult.

In order to solve the problem of expensive, large biosample collections, the industry either made the decision to internalize the biorepository, internalize a portion of the biorepository, or outsource the biorepository. While internalizing some or all of the biorepository has advantages from an oversight perspective, large biosample collections are difficult to maintain. Appropriate personnel, operating procedures,

IT infrastructure such as Laboratory Information Management System (LIMS), facilities, Occupational Safety and Health Administration (OSHA) precautions, and equipment must be maintained.

Approximately half of the industry is of the opinion that biospecimen maintenance is not a core business function for drug development companies and should be outsourced. Therefore, a model for maintaining samples in an external biorepository must be agreed upon and developed. Interestingly, there is no consistent model that is commonly employed. Some considerations for determining the model are discussed in more detail below.

A biorepository vendor is also required. Companies that now specialize fully in biostorage are one option (these companies vary in additional services required for use of biospecimens including aliquoting, extraction of analytes, and analysis options), central laboratories with long-term storage options, and academic collaborators with infrastructure for large sample collections are common options reviewed/employed. The internal biorepository model often helps dictate which option will work best in addition to requirements for specimen integrity, data protection/privacy, and ability to rapidly use high-quality specimens.

18.2.1 Outsourcing the Biorepository: Determining the Model

When deciding to outsource the biorepository, many models may be considered. The final model chosen will depend on types of trials the company routinely conducts (large global studies or primarily local studies), level of internal infrastructure the company wishes to invest in, and risk tolerance for sample quality and integrity.

- **Direct shipment of samples to the biorepository.** Samples may be collected from clinical trial sites and sent directly to the biorepository. In this model, the biorepository accepts responsibility for export and receipt of samples collected from global clinical trial sites. The company relies on the biorepository to provide inventory and tracking of clinical trial samples. The company may rely on the biorepository to provide sample collection instructions, training, and shipping materials to sites. This model provides maximum possibility of sample quality/integrity because it limits numbers of shipments and responsible parties for samples. It also requires less internal infrastructure to oversee, as the biorepository assumes oversight for ensuring correct inventory for the study. However, this model can be expensive if it requires shipment of samples intended for storage to a unique facility from global sites.
- **Routing samples from sites through central laboratory.** Samples may be shipped from clinical trial sites along with routine clinical trial samples (e.g., blood for complete blood counts) to a central laboratory. The central laboratory can then ship samples intended for long-term storage to the biorepository. The central laboratory assumes responsibility for export/receipt of biosamples from global clinical trial sites. This model decreases shipping costs, often by combining shipments of samples intended for the central laboratory. However, it introduces multiple shipments (impacting sample quality/integrity) and risk of loss of

chain of custody (due to multiple responsible parties) for samples if the central laboratory is not the final storage facility. This model requires some internal infrastructure to oversee the process, but much of the inventory work can be outsourced to the final biorepository vendor.

- **One biorepository versus multiple biorepositories.** Some companies may opt to centralize all biosample storage to a single facility. This model requires the least internal oversight and management. However, if a major disaster affects the single facility, there is a risk that all specimens could be lost. Additionally, some countries do not allow export of biosamples. Finally, a single biorepository will have more difficulty in negotiate pricing that is competitive with the external market. For many of these reasons, some companies will opt to use multiple biorepositories or biorepository locations to maintain biosample collections.

18.2.2
Key Competencies for Outsourced Biorepository Vendors

Once the model is selected, key competencies for outsourced biorepository vendors must be assessed. As with the compound inventory, key aspects of biorepository competencies include physical security of samples, robust and reliable IT infrastructure, and a robust backup or disaster recovery plan. In addition to these features, biological specimens require the following:

- **Privacy protection processes for samples.** As biosamples collected for future research or with permission for long-term storage are collected from individuals participating in clinical trials, robust measures for protection of subject privacy must be maintained. In order to ensure privacy protection, outsourced biorepositories often do not have access to clinical trial information and have only enough information associated with the sample for the company to identify and link data generated from the sample to a clinical trial profile.
- **Robust chain of custody procedures for samples.** In order to ensure that data derived from biosamples is associated with the corresponding clinical trial information, robust chain-of-custody procedures for samples that are accessioned into the biorepository must be maintained. In addition, any specimens derived from these samples (e.g., DNA from whole blood, peptides from plasma, etc.) must be correctly associated with the originating sample. Finally, a current inventory of samples, including accounting of any samples used during analyses, must be accurately maintained.
- **Maintain appropriate conditions for sample storage.** Maintenance of biological sample integrity to enable downstream assays is imperative for the biorepository [25, 26]. As companies typically collect multiple types of tissues for many applications, multiple storage temperatures and procedures are required at the biorepository.
- **Must maintain appropriate procedures to comply with global regulations/legislation.** Global regulations/legislation for collection and storage of

biosamples are undergoing considerable change due to increased concerns regarding usage of samples collected for future research. The chosen biorepository must be able to comply with global regulations/legislation in order to accept samples from multiple countries.
- **Sample coding.** Companies should consider whether they have requirements for sample coding. If samples are to have additional coding beyond the original accession number applied at the clinical trial site, consideration should be given to the question whether the biorepository itself will code the samples and whether the biorepository can track/maintain inventory of coded samples.

18.2.3 Internal Oversight for Outsourced Biorepository Vendors

Companies must consider an internal strategy for overseeing biorepository vendors. A management plan should be put in place that includes a communication plan, qualification requirements, and any specialty requirements necessary for storage of biospecimens. An internal group responsible for biorepository oversight must be formed. Frequency and format of reports from the biorepository, approval processes for new trial budgets, study-related activities, statements of work, and audits should be established.

Internal oversight must include the ability to accept requests for sample usage, ability to verify that subjects gave consent that would allow for the proposed analysis, and that all analyses proposed comply with any legal or regulatory requirements from the countries where the samples were originally collected. Internal groups must maintain up-to-date inventories of sample collections. A process to destroy samples at the end of an agreed storage time upon request by members of the ethics committee or regulatory authority or when they are no longer usable by clinical teams or researchers should be developed.

Finally, routine procedures to ensure quality of samples and appropriate inventory tracking at the biorepository should be put in place. Ongoing quality assurance should be in place to verify that sample collections are still available and usable by the company.

18.2.4 Lessons Learned

In determining the model for outsourced biospecimen storage, sophisticated models for specimen integrity, data privacy, and data transfer have been employed. Some of the scrutiny of biospecimen data protection methods may provide benefit to outsourced compound collections, either by improving the security of transfer or by facilitating the sharing of data. For example, by leveraging some of the more sophisticated internal oversight of external biospecimen collections, including integrated IT systems with the external vendor, more rapid access and tighter oversight of compound collections may be achieved.

18.3 Conclusions

The decision to outsource compounds and biosamples is complex but allows flexibility for the pharmaceutical industry. Careful planning and ongoing oversight must be implemented to ensure that appropriate conditions are being met. However, the outsourced model for compound and biosample stores has become an accepted, successful model for many companies within the industry.

Acknowledgments

We would like to thank Dr. John Mathew and Kimberly Matus for their input and helpful discussions.

References

1. Chaturvedi, S. Outsourcing in Pharmaceutical Industry. *http://www.bionity.com/articles/e/49803* (accessed 7 July 2010).
2. Clark, D.E. (2007) Outsourcing lead optimisation: constant change is here to stay. *Drug Discov. Today*, **12** (1/2), 62–70.
3. Clark, D.E. and Newton, C.G. (2004) Outsourcing lead optimisation – the quiet revolution. *Drug Discov. Today*, **9** (11), 492–500.
4. Crossley, R. (2004) The quiet revolution: outsourcing in pharma. *Drug Discov. Today*, **9** (16), 694.
5. Cavalla, D. (2003) The extended pharmaceutical enterprise. *Drug Discov. Today*, **8** (6), 267–274.
6. Mander, T. and Turner, R. (2000) To outsource in lead discovery: the devil or the deep blue sea? *J. Biomol. Screen.*, **5** (3), 113–117.
7. Klopack, T.G. (2000) Strategic outsourcing: balancing the risks and the benefits. *Drug Discov Today*, **5** (4), 157–160.
8. Modrate, E. (2008) The pitfalls of outsourcing analytical laboratory work. *Pharma*, **4** (5), 19–20.
9. Boucharens, S. (2010) Outsourcing compound inventory – trend or real benefits? *Pharm. Outsourcing*, 22–25.
10. Comley, J. (2005) Compound management on pursuit of sample integrity. *Drug Discov. World*, **6**, 59–78.
11. Fillers, W.S. (2004) Compound libraries – cost of ownership. *Drug Discov. World*, Summer, 86–90.
12. Kozikowski, B.A., Burt, T.M., Tirey, D.A., Williams, L.E., Kuzmak, B.R., Stanton, D.T., Morand, K.L., and Nelson, S.L. (2003) The effect of freeze/thaw cycles on the stability of compounds in DMSO. *J. Biomol. Screen.*, **8** (2), 210–215.
13. Cheng, X., Hochlowski, J., Tang, H., Hepp, D., Beckner, C., Kantor, S., and Schmitt, R. (2003) Studies on repository compound stability in DMSO under various conditions. *J. Biomol. Screen.*, **8** (3), 292–304.
14. Zitha-Bovens, E., Maas, P., Wife, D., Tijhuis, J., Hu, Q., Kleinoder, T., and Gasteiger, J. (2009) COMDECOM: predicting the lifetime of screening compounds in DMSO solution. *J. Biomol. Screen.*, **14** (5), 557–565.
15. Yates, I. (2003) Compound management comes of age. *Drug Discov. World*, Spring, 35–42.
16. Burr, I., Winchester, T., Keighley, W., and Sewing, A. (2009) Compound management beyond efficiency. *J. Biomol. Screen.*, **14** (5), 485–491.
17. Lease, T. and Stock, M. (2008) Compound management: the next focus for outsourcing? *Pharma*, **4** (5), 22–23.

18. Gutknecht, E. (2008) Working with outsourcing partners to achieve efficient and cost-effective compound management operations. IQPC – Pharma IQ's - Compound Management and Integrity, May 20–22, 2008, London.
19. Boucharens, S. (2010) Outsourcing compound inventory case study. IQPC – Pharma IQ's – Compound Management and Integrity, May 18–19, 2010, London.
20. Matus, K. (2008) Pfizer's decision to outsource the neat sample library. IQPC – Pharma IQ's – Compound Management and Integrity, May 20–22, 2008, London.
21. Meerpoel, L., Schroven, M., Goris, K., Demoen, K., and Marsden, S. (2006) Outsourcing to exploit a key asset. *Drug Discov. Today*, **11** (11/12), 556–560.
22. Ge, D., Fellay, J., Thompson, A.J. *et al.* (2009) Genetic variation in *IL28B* predicts hepatitis C treatment-induced viral clearance. *Nature*, **461**, 399–401.
23. Hughes, A.R., Mosteller, N., and Bansal, A.T. (2004) Association of genetic variation in HLA-B region with hypersensitivity to abacavir in some, but not all, populations. *Pharmacogenomics*, **5** (2), 203–211.
24. Franc, M.A., Warner, A.W., Cohen, N., Shaw, P.M., Groenen, P., and Snapir, A. on behalf of the Industry Pharmacogenomics Working Group (2010) Current practices for DNA sample collection and storage in the pharmaceutical industry and potential areas for harmonization: perspective of the industry pharmacogenomics working group. *Clin. Pharmacol. Ther.*, **89** (4), 537–545. (April 2011).
25. ISBER (2009) Definitions of Terms and Explanations of Best Practices, Guidelines, Standards, and Norms that Pertain to Repositories, *www.isber.org* (accessed 2010).
26. NCI (2007) National Cancer Institute Best Practices for Biospecimen Resources, *www.biospeciments.cancer.gov* (accessed 2010).

19
Sample Management Yesterday and Tomorrow

Terry Wood and Mark Wigglesworth

19.1
The Role of Sample Management

Most pharmaceutical and biotech companies that have an inventory of compounds for testing manage both solid and liquid storage facilities; similarly, most organizations have some requirements for storing biological samples with relatively few having significant biobanks. While solid compound repositories and biorepositories have to face an unusual set of challenges to maintain and operate such stores effectively, liquid repositories share many of these and also present an additional set of storage and operational requirements, particularly as the liquid store is invariably the supply source for any screening program that involves more than a few hundred compounds to test.

As a stand-alone science, sample management is a complex interaction of automation, chemistry, robotics, informatics, and, for stores of solubilized compounds, fluid dynamics. However, as a stand-alone science, sample management *per se* adds no value to any drug discovery program. No matter how neatly, nicely, and efficiently samples are stored, unless there is a strong understanding of the requirements of screening groups, and unless these groups are fully aware of what, when, and how the sample management team can deliver, the sample manager is simply running a warehouse, and no medicine was ever discovered by keeping samples in a warehouse.

The mantra of the compound manager is to deliver the right compound in the right format to the right place and at the right time, and ensuring that all these aspects are in place is what makes compound management an exciting and challenging key component of the drive to bring new medicines to market. The mantra seems simple, but if, like many major pharmas, the compound inventory contains millions of screenable entities, supplying the screening effort requires an extensive investment in automation plus a dedicated team of technicians, engineers, and informaticians supported by an infrastructure focused on delivery. For these reasons, sample management operations are being sourced externally to parent organizations, as Boucharens and Warner have illustrated in Chapter 18.

Management of Chemical and Biological Samples for Screening Applications, First Edition.
Edited by Mark Wigglesworth and Terry Wood.

The various ways in which compounds are screened, particularly in the field of high-throughput screening, are constantly evolving using the tools of the computational chemist and biologist to better define molecules with an increased chance of interacting with the biological target. While throwing every compound available at the target (full file screening) with the hope of discovering a lead compound or chemical series by serendipity is the only way of guaranteeing that an active compound (if it exists in the inventory) will be identified, it is a hugely expensive approach, both in time and materials. Two alternative processes can reduce the expenditure. The first is to screen the full file, but in a compressed format with multiple compounds per well, and the second is to screen a subset of the file, which may be selective for the target type or be representative of the file's diversity; both of these approaches are discussed in Chapters 2 and 15. It is the job of the compound manager to work with the screeners to identify which would be the more appropriate approach, and to set expectations on delivery.

19.2 Automation of Compound Management

The requirement to automate the retrieval of compounds from store, and to prepare them in a screenable format, has increased very much in line with the massive growth in the size of screening files held by most companies, also matching the advances in miniaturization, plate handling, and assay technology applied to the way compounds are screened. For example, Pfizer's compound file increased in size by a factor of 10 over the decade between 1999 and 2009, but in 2009 full file screens were completed in a quarter of the time taken 10 years prior to that.

In the high-throughput arena, provision of primary screening plates is very demanding, but the principal driver to automate compound stores came from the need to rapidly supply the resultant follow-up compounds required to confirm initial hits and to measure IC_{50} activity.

Toward the end of the millennium, both Glaxo and Pfizer were involved in programs to commission automated liquid stores in the UK. Pfizer in Connecticut had already installed an automated solid store at its research site in Groton, which provided a great source of prior knowledge to assist development of the functional design specification of the liquid store, built by (then) RTS Thurnall and installed in a purpose-designed building at the Pfizer site in Sandwich.

The RTS system consisted of 200 m of flex-link conveyor belt connecting a 600 m^3 freezer store with twin liquid handling units, via a defrost oven. Scientists were able to order compounds from their desktop, specify plate type, and sample concentration, and the system would automatically select and deliver the compounds and prepare the diluted screening plates with no further operator intervention. This presented a tremendous advantage to screening groups, who no longer needed to assign skilled experimental scientists to the preparation of their samples for testing. Automation made this a faster, more reliable, and consistent process, and above all, plate preparation could now boast an electronic audit trail.

The Automated Liquid Sample Bank (ALSB) as the RTS system was known, served the Pfizer business very well for a number of years, but like all good things, it eventually became obvious that without a huge investment in both hardware and software redesign the ALSB was coming to the end of its useful life. Principal reasons were that ALSB could only prepare plates in 96-well format, and most of Pfizer's microplate screens were being redeveloped to higher density formats. In addition, with a minimum aspirate volume of 10 μL, the system was not compound sparing, and samples were subjected to a freeze/thaw cycle each time they were ordered. Data was accruing suggesting that multiple freeze/thaw cycles were detrimental to compound integrity. Finally, even though the capacity of the system was 2.5 million samples, this was not enough to store the entire Pfizer screening file, and, moreover, emerging systems from other suppliers could deliver samples an order of magnitude faster. This system had become one of the 'Battleships' described by Gosnell in Chapter 7. As it was unable to adapt to modern requirements, the decision was taken to phase out ALSB.

What replaced ALSB at Sandwich was a major store supplied by REMP AG, being part of a global project undertaken by Pfizer which was still supplying its research teams 10 years post inception. Pfizer's huge screening file was multi-replicated into mini-tubes, over 30 copies of each compound, and distributed as single-shot samples in twin stores located in Sandwich UK and at Pfizer's major US site in Groton, Connecticut. These stores, each with a capacity of some 60 million tubes, were designed with identical operating systems and inventories, and were accessed via the same corporate ordering system. Single-use tubes eliminated any potential freeze/thaw cycle issues, and the sample volume and concentration contained in each tube was designed to support both directed (single concentration confirmatory) and end-point (IC_{50}) screens.

Delivery outputs supported both 96- and 384-well microplate formats, and with multiple replicates spread throughout the twin aisle store, with two picking robots on each aisle, pick rates of 10 000 samples per day could be attained, making selection of on-demand subsets of the file a less than challenging task. A key feature of the twin store project is that being identical they could back each other up if one should be out of commission for servicing or maintenance. This was very useful when a major software upgrade was installed, at first in the US, then in the UK REMP facilities, as by switching the ordering systems to access just the on-stream system there was no interruption or delay to the business, highlighting how important information technology is to the sample manager's chance of delivering success. It would benefit many aspects of sample management, including database design (Chapter 13), compound library selection (Chapters 2 and 15), automation choice (Chapter 10) as well as process delivery (Chapter 14) and the logistics of sample distribution (Chapter 8).

Major systems were being installed not only at Pfizer's UK research site but also on the UK campuses of major Pharmas such as GlaxoSmithKline, AstraZeneca, and Merck Sharpe & Dohme, as well as many other locations. Indeed the turn of the century saw an explosion in the provision of automated stores, both in Europe and the USA. These were commissioned to supply solids and liquids, the latter in a range of storage formats, and most utilized a wide variety of storage temperatures, concentrations, solvents, and containers, compared with other stores. Knowledge from the early compound stores was subsequently utilized in design and development of tissue sample stores such as the ones described by van Niekerk in Chapter 12. Interestingly, despite a wealth of publications, presentations, and lectures describing the relative advantages of each, the optimal operation has still not been identified that could be used to define an industry standard for liquid sample storage and for compound management best practice; this status quo is reflected in Chapters 3 and 4.

But what of the liquid store of tomorrow? Nano-technology, coupled with advances in screening methods, now permits assays to be run using nanoliters rather than microliters of compound solution. GlaxoSmithKline and many other pharmas have invested heavily in acoustic dispensing hardware, which allows noncontact cherry picking from individual wells. With sufficient storage pods and an efficient transport system, primary HTS plates, as well as subset collections, could be assembled on demand.

19.3 Compound Integrity

As stated in the previous section, successive freeze/thaw cycles can be detrimental to the integrity of a solubilized compound, and how to ensure samples are stored so as to minimize loss of integrity is a key consideration the compound manager must face. However, it is probably not the process of freezing and thawing *per se* that is the major factor leading to loss of integrity; it is more likely to be a combination of the water content of the solvent used (usually dimethyl sulfoxide (DMSO)), the environment in which the container is stored, and the process by which the liquid sample was initially prepared.

An analysis, commissioned by Pfizer, of existing (nominally) 30 mM DMSO stock solutions revealed some startling results. The concentration distribution of around 1000 compound solutions was broadly spread from 5 through to 30 mM, some concentrations being considerably higher. This observation was for solutions prepared from dry powder, as well as for samples prepared as liquids via high speed analoging. The majority of solutions presented analytical results less than 30 mM, with a mean around 20 mM. The variations were more than could be explained by the inherent variations in the analytical process, so what was going on? Compound integrity became a key phrase of compound managers.

As expected, there was not one simple, single answer to explain these results. Some anomalies in the process were identified – such as the chemistry laboratory weighing 1 mg of material into a 6 g vial, but the most likely cause was attributable to the hygroscopic nature of DMSO. Chris Lipinski made it absolutely clear that as the amount of water absorbed from the surrounding by DMSO increases, the freezing point of the mixture drops considerably [10], such that just a small percentage of water in the solvent could result in the DMSO solution remaining liquid even at

−20 °C. The 'wet' DMSO can and does result in precipitation of solute, and this can be partial, as the concentration drops to a level that can be supported by the amount of DMSO present – hence the lower than nominal concentrations. By hydrolysis, particularly if the atmosphere is not inert, compounds can degrade in the presence of water, either fully or partially to other components. Also, as the amount of water increases, even if the solute does not precipitate or degrade, there will be a resultant change in concentration as the volume of solvent increases.

Not surprisingly there was huge interest in these observation from both chemists and biologists, resulting in many modifications and improvements in the way compound solutions were prepared – particularly for IC_{50} measurements, a key metric in deciding which chemical series to investigate. Processes were modified to minimize DMSO water uptake: using liquid handling under an inert atmosphere, keeping DMSO in sealed, air-tight pressurized containment, and, for the all important IC_{50} work, preparing serial dilutions in 100% DMSO prior to transfer to the assay incubation medium. In parallel with these changes was an increase in the analytical support to confirm that samples were actually what they were expected to be, and quality and quantity control (QQC) became a second key phrase of the compound manager. The concepts and methods determining QQC are fully discussed in Chapters 3 and 4. In parallel to this, the quality of tissue samples is also a limiting factor to the progression of translational medicine as described by Miranda in Chapter 5.

Rubbish in–rubbish out is a good guideline – your assay results can only be as good as the material you are testing. Both GSK and Astra Zeneca rebuilt their screening files post analysis. Pfizer apply an automated purification process to all incoming samples, and run QQC on all samples reaching IC_{50} analysis, with purity data being recorded in the corporate compound management database. In the biological sample world, certification of reagents from cell line to tissue samples is required to ensure quality standards are met, but it seems that this area remains a limiting factor within biobanking as its complexity is not simply defined as for compound management. As yet the impact of variation in collection and preservation techniques is yet to be entirely understood. Whereas minimum standards should certainly be described and agreed more readily, the complexities of dealing with tissue types and experimental requirements means that the nuances and experimental variation they introduce will take a significant amount of collaborative resource to even blueprint.

However, not all was doom and gloom, as a further study at Pfizer monitoring the effect of multiple freeze/thaw samples on compound integrity in samples under various storage conditions showed that compound solutions, stored under the conditions of the sealed, individual samples in their new REMP store, were good for at least 20 freeze/thaw cycles. Also, as Novartis run a store operating at room temperature, with samples dissolved in 90% DMSO, with good sample integrity, it would suggest that the biggest detriment to compound integrity is multiple freeze/thaw cycles when samples are not protected from atmospheric moisture. As the field of biobanking expands, similar scientific experimentation is also sure to present similar 'eureka' moments of scientific and process insight.

19.4 Reduction of Redundancy

A screening file containing multi-millions of screenable entities will almost certainly have a significant degree of redundancy in that file, that is, compounds that are adding no value to the collection and will never contribute to a lead discovery program no matter what target they are screened against. This redundancy has many aspects. Duplication is very common in large collections, which can contain multiple batches of the same compound, a range of salt forms of the same compound, and often, due to variations in process, especially in companies that have compound management laboratories at a number of sites, replicates of exactly the same compound.

Another reduction opportunity is to explore molecular redundancy – many compounds that go through a registration process were never designed to be druggable molecules themselves, but may have been synthesized as process intermediates. These could contain reactive moieties or, because of their structural components, may be potential toxiphores. However, once registered into the system, they can become part of the 'corporate file' and hence enter the HTS stream. Such compounds can be promiscuously reactive and produce hits that the medicinal chemist will never follow up and will immediately discard from HTS hit lists in triage. However screening these compounds wastes a great deal of time and resources, which is why many pharmas have actively weeded out these 'ugly' chemical structures from their screening files.

At the other end of the reactivity spectrum are the dormant compounds – those structures that may have been through hundreds of primary screening campaigns but have never been flagged as a 'hit.' The first action with such compounds is to analyze the material to check if it is still what it purports to be – if it is not, then it can be safely discarded, but if it is, then a decision needs to be made on whether such a compound, with an extensive record of biological unreactivity, would demonstrate cost-benefit in adding it to even more screening campaigns?

If one accepts that the prime purpose of HTS is to identity novel chemical series for progression to therapeutic area projects, and not SAR generation around each hit, then a further portion of the primary file can be declared redundant and be removed from ongoing and future campaigns. Molecular redundancy within the screening file leads to biological activity being demonstrated by large numbers of structurally very similar compounds, as similar chemical structures are likely to share similar biological activities. Using computational methods, such as Tanimoto cut-offs, an optimal cluster size can be defined and redundancy calculated on a per compound basis such that those with many neighbors tend to be deselected, and those with fewer, kept.

A new approach to file enrichment is that compounds now need to earn a right to be part of the screening file by virtue of superior or under-represented chemical properties.

So, the primary file can be made as effective, more efficient, and with significant financial benefits by removal of duplicates, ugly structures, QQC failures, and over-represented similarity. Further opportunities exist to make a screening file even more likely to yield lead material by allocating compounds to various tiers. Pfizer took the opportunity to rebuild its primary, small-molecule HTS file into two tiers. Tier 1 contains compounds directed against standard targets to deliver medicines designed for oral delivery. Tier 2 contains the generally less attractive structures, such as macrocyclic compounds, high-end physiochemical failures (MW 600, clogP > 6), steroids, and peptides. These would be used against nonstandard targets such as protein–protein interactions and for nonstandard (e.g., inhaled) delivery. Moreover, the Tier 2 compounds are not intended to be used as a complete collection, but as subsets grouped into the classification category (examples above) into which the compound fell.

Finally the Tier 1 file could be tranched, or further subdivided, into groups that represent ranked hit ID potential, so that a screen could kick off with, say, the million compounds computationally designed to most likely identify a hit series, and proceed successively through other tranches until the screen leaders consider they have accrued sufficient lead material.

19.5 The Future of Sample Management?

19.5.1 Introduction

Currently within this book we have tried to bring together the disciplines of sample management, whereas physically, tissue biobanks, chemical libraries, and cell culture libraries are separate entities and are commonly dealt with in very segregated ways within large pharmaceutical companies and within Contract Research Organisations (CROs) across the globe. Yet similar automation, technology, and logistical skill sets are employed to manage these collections. Of course there are specialist areas of knowledge required for utilizing and managing these collections. However, it is certainly conceivable within an industry that is required to reduce cost base that these disciplines could become amalgamated either within individual companies or within cost-effective single contact CROs. Having said that, the trends outside of large pharmas and within the biobanking sector are ones of national biobanks that are government funded. As such, these institutions are in some ways protected from the pharmaceutical environmental pressures although not immune from global financial struggles. Hence it is entirely possible that two strong independent situations are generated whereby pharmas amalgamate these resources and government or grant-funded institutions establish national centers of biobanks and screening institutions as drug discovery and development tools. These models are in their infancy and in the next 10 years will almost certainly see large alterations in the way these Sample Management organizations impact drug discovery and indeed how drug discovery and development are conducted.

19.5.2
The Cost of Drug Discovery

It is commonly accepted that the cost of drug discovery has increased over recent years; estimates of the cost of a single new chemical entity (NCE) in 1987 placed it at ~$231 million [1, 2] whereas 20 years later the cost had increased around 10 times [3, 4]. It is also clear that the total spend on research and development is similarly increased over this time period, while the number of NCE launches has not significantly altered (FDA data). Hence, whatever the actual costs are, increased expenditure without increased return presents an unsustainable model of pharmaceuticals. Having already tried to grow its way out of this predicament with significant mergers and acquisitions during the 1990s, alternative approaches are now required. This changing environment has sparked the beginnings of significant alterations in pharma R&D, change which was predicted by Morgan Stanley in 2010 [5], although not even they, by their own admission, had predicted the rate of change we have seen [6]. In their analysis, Morgan Stanley claim that externalization of drug discovery is key to maintaining profitability of large pharmaceutical companies. By in-licensing compounds that reach Phase IIa they calculate a potential to triple the number of NCEs reaching the markets each year. Of course this is a complex business and there will be caveats to this, one perhaps being that the biotech industry can generate funding/collaborations to stimulate this level of innovation, although, as discussed below, there are also an increasing number of academic drug discovery institutes that may be capable of filling this gap in future. The pharmaceutical industry will also have to play its part by supporting innovation. It remains to be demonstrated that the industry is willing and able to commit to this [7]. Looking at the role of Sample Management Organizations is certainly not as simple as tracking the trends within pharmaceuticals. We are already a diverse collection of services and exist in many forms outside of this arena. Collectively this group of organizations has a role to play in shaping its own destiny as much as the large pharmaceutical companies do.

19.5.3
Independent Service Providers

There is certainly a market for independent service providers to manage operations for large companies, for example, the outsourcing of solid operations to Sigma by Pfizer, but also for smaller companies and academic groups. These groups may wish to gain access to the expert skill sets, quality of storage and automation, quality control of sample collections, dispense, and tracking of samples, as well as cost-effective solutions for short-term problems. The NIH screening program has also chosen to outsource its screening library management, this time to BioFocus. In this case, intellectual property is ultimately shared in a publication process, but even in this non-for profit organization the cost/skill base is driving strategic operational decisions. Such outsourced services are already prevalent within the biobanking arena. For small organizations the only way of accessing these capabilities while maintaining cost control is likely to be by cost sharing and outsourcing to a provider that manages multiple samples, that is, multiple in their owner origin and multiple in their sample diversity. There are certainly specialist providers in the market place currently, but these appear, for now at least, content with supporting a relatively limited area within the field, with no single overarching provider. As these companies grow and start to look at further possibilities themselves, we may well see expansion or amalgamations to create such companies.

19.5.4
Alternative Models of Drug Discovery

There are an increasing number of university-based screening institutes [8], for example, Scripps Institute (Florida, USA), Eskitis Institute (Brisbane, Australia), University of Washington (USA), Tel Aviv University (Israel), Dundee University (Scotland), as well as National screening centers such as the National Institute of Health (NIH) and the European screening center in Belgium. Expansion of these operations may be the way forward in terms of increasing medicine delivery. It has the potential to allow the pharmaceutical industry a further downsizing in its research activities while perhaps allowing it access to co-development deals for successful pipeline generation, the win-win situation here being that patients get the next generation of medicines, company share holders share in the financial reward they generate through a broad spectrum of medicines, and the average reimbursable cost would fall, allowing some benefit to be returned to the payer (in some cases the governments sponsoring the original science centers). This may also be achieved through the development of science park-based research where, for example, Information Technology companies have been able to externalize much of their product development with significant productivity and cost benefits.

Academic–Company partnerships are another way of maximizing benefit from sample assets. There are 68 academic institutes listed in the Society of Laboratory Automation and Screening academic screening institute directory [9], but only two are listed as having a library size of more than 500 K compounds. Hence, these academic screening institutes would almost certainly benefit from pharmaceutical library access and screening know-how. Additionally, if these organizations are to continue to build on their early success then they will need to learn from the mistakes pharmas have made rather than simply copy the model of small-molecule hit discovery. There seems no better way of gaining this experience than through mutually beneficial collaborations, where screening, library management, and other expertise can be provided to academic institutions in return for options to exploit successful hit discovery further down the drug discovery pipeline. There are already good examples of this type of collaboration in operation. For example, the

Medical Research Council-Technology (MRC-T)[1] group, has recently confirmed a deal with AstraZeneca whereby they gain access to a 100 000 compound set for screening at targets within the MRC-T. In return there will undoubtedly be options to re-internalize the development of hits and or co-develop these compounds enabling AstraZeneca to exploit their asset, develop collaboration with academic groups, and potentially reduce competition. By allowing academic groups access to these compounds in this way, enabling them to progress drug discovery with rich tools from a diverse and expertly created compound set, the authors wonder what effect this will have on the future of externally created compound sets, where profitability from selling them is an underlying argument for their creation. It is certainly suggested in the literature that this type of compound sharing through carefully controlled collaborations will expand as pharmaceutical companies seek to further reduce cost and exploit the assets they have, yet there is a balance to be struck between carefully guarding the content of these sets and allowing them to be utilized externally. The answers to the question 'will the market for CRO and Small Medium Enterprise (SME) screening demand creation of open-access screening collections?' probably lies in the hands of the pharmaceutical giants that currently hold such collections. They are beginning to indicate that they are prepared to share their treasure trove, but do they know how to do this or that they could control what happens next? It will take some brave or desperate decisions to rapidly progress this within today's research culture. However, we certainly believe that this will happen given time, and there appears no better time for such groundbreaking decisions to be made.

19.6 Concluding Remarks

It is certainly not foreseeable that the discipline of sample management will quickly or easily disappear, given that they are well-established successful methods of biological research and innovative and safe medicine delivery. However, the type of technological frame shift that could spark this would be on the scale of enabling whole-tissue HTS, where biobanks could expand at the expanse of cell collections or where small-molecule research is eroded by advances in safety profiles for biological agents making them the medicine of choice. Hence, sample management is destined to evolve, but it is certainly required as an integral part of medicine development for the foreseeable future. We do not envisage that technology within sample management will generate the next frame shift, but the interaction with screening/clinical customers and the cost pressures on these services may do this. In many western countries we may also need to address cultural, ethical, and legal perspectives on our human-selves before medical research can embrace the use of

1) MRC-T has previously developed IP rights and successfully spun out companies and profitable antibody therapies such as Tysabri and Actemra.

tissue bio-banking and sample management and enable the twenty-first century to truly improve life and the history of medicine. For now at least we are a long way from this as a reality, but we should not lose sight of the fact that we within sample management are all service organizations that enable the expansion of scientific knowledge, ultimately with the end of improving the quality and span of human life.

References

1. Dimasi, J.A. (1992) Rising research and development costs for new drugs in a cost containment environment. *Pharmacoeconomics*, **1**, 13–20.
2. Dimasi, J.A., Hansen, R.W., Grabowski, H.G., and Lasagna, L. (1991) Cost of innovation in the pharmaceutical industry. *J. Health Econ.*, **10**, 107–142.
3. Adams, C.P. and Brantner, V.V. (2010) Spending on new drug development. *Health Econ.*, **19**, 130–141.
4. Adams, C.P. and Brantner, V.V. (2006) Estimating the cost of new drug development: is it really 802 million dollars? *Health Aff. (Millwood)*, **25**, 420–428.
5. Baum, A., Verdult, P., Chugbo, C.C., Abraham, L., Mather, S., Bradshaw, K. and Nieland, N. (2010) Morgan Stanley Research Report: Exit Research and Create Value.
6. Baum, A., Verdult, P., Chugbo, C.C., Mather, S. and Nieland, N. (2010) Morgan Stanley Research Report: Research Shrinkage. Even Faster Than We Envisaged.
7. Hunter, J. (2010) Is the pharmaceutical industry open for innovation? *Drug Discov. World*, Fall, p. 9.
8. Frearson, J.A. and Collie, I.T. (2009) HTS and hit finding in academia – from chemical genomics to drug discovery. *Drug Discovery Today*, **14**, 1150–1158.
9. *https://www.slas.org/screeningFacilities/facilityList.cfm* (accessed 2011).
10. Rasmussen, D.H. and Mackenzie, A.P. (1968) Phase diagram for the system water-dimethylsulphoxide. *Nature*, **220**, 1315–1317.

Index

Management of Chemical and Biological Samples for Screening Applications, First Edition.
Edited by Mark Wigglesworth and Terry Wood.